SYSTEMS ARCHITECTURE

Third Edition

Stephen D. Burd
University of New Mexico

**COURSE
TECHNOLOGY**
™
THOMSON LEARNING

Australia • Canada • Mexico • Singapore • Spain • United Kingdom • United States

COURSE TECHNOLOGY
™
THOMSON LEARNING

Systems Architecture
Stephen D. Burd

Publisher
Kristen Duerr
Managing Editor
Jennifer Locke
Product Manager
Margarita Donovan

Development Editor
Deb Kaufmann
Production Editor
Daphne Barbas
Marketing Manager
Toby Shelton

Editorial Assistant
Janet Aras
Text Designer
GEX Publishing Services
Cover Designer
Efrat Reis

Disclaimer
Course Technology reserves the right to revise this publication and make changes from time to time in its content without notice.

ISBN 0-619-03418-1

DEDICATION

To Bernard A. Wallace, Jr., thank you for a lifetime of friendship.

CONTENTS

CHAPTER 3
Data Representation   61

CHAPTER 5
Data Storage Technology   **155**

CHAPTER 7
Input/Output Technology   **239**

CHAPTER 8
Data and Network Communication Technology 275

CHAPTER 9
Computer Networks   **325**

CHAPTER 11
Operating Systems

CHAPTER 12
File Management Systems

CHAPTER 13
Internet and Distributed Application Services **489**

PREFACE

INTENDED AUDIENCE

This text is intended for undergraduate students majoring or concentrating in information systems (IS) and as a reference for IS professionals. The text provides a technical foundation for systems design, systems implementation, hardware and software procurement, and computing resource management. Computer hardware and system software topics that are most useful to IS students and professionals are described at an appropriate level of detail. For some topics, the reader will gain sufficient knowledge to solve technical problems. For other topics, the reader will gain sufficient knowledge to enable effective communication with technical specialists.

Computer science students are exposed to computer hardware and system software technology in many undergraduate computer science courses. There are many texts designed for such computer science courses that focus specifically on a subset of the topics in this book. However, coverage of computer hardware and system software in an IS curriculum is usually limited. Most IS curricula emphasize applications development, systems analysis and design, programming, and database management. A brief overview of computer hardware and system software may be provided in an introductory IS course and some specific technical topics may be covered in other courses, but there is usually only one course exclusively devoted to computer hardware and system software.

The topical coverage in this text closely matches the requirements for IS'97.4—Information Technology Hardware and Software in the most recent recommended IS curriculum.[1] This text is specifically designed for IS'97.4 but it can also supplement other texts in system design and computer resource management courses. With respect to system design, the text covers many technical topics that must be addressed when selecting and configuring computer hardware and system software. With respect to computer resource management, the text provides the broad technical foundation needed to effectively manage computer resources.

READER BACKGROUND KNOWLEDGE

Because IS'97.4 is placed early in the recommended IS curriculum, this text makes few assumptions about reader background knowledge. Unlike many computer science texts, the reader is not assumed to have an extensive background in

---

[1]*IS'97 Model Curriculum and Guidelines for Undergraduate Degree Programs in Information Systems*, published by the Association for Computing Machinery, Association for Information Systems, and Association of Information Technology Professionals, 1997, http://www.is-97.org. As of this writing, the next revision of IS'97 is scheduled for publication in 2002.

mathematics, physics, or engineering. Where necessary, background information in these areas is presented in appropriate depth.

The reader is not assumed to know any particular programming language. However, classroom or practical experience with at least one programming language is helpful to fully comprehend the coverage of CPU instruction sets, operating systems, and application development software. Programming examples are provided in several programming languages and in pseudocode.

The reader is not assumed to have detailed knowledge of any particular operating system. However, as with programming experience, practical experience with at least one operating system is helpful. Lengthy examples from specific operating systems are purposely avoided but there are some short examples from MS-DOS, UNIX, and recent Windows versions.

The reader is not assumed to have any knowledge of low-level machine instructions or assembly language programming. Assembly instructions are described in several chapters, but a generic assembly language is used and no detailed coverage of assembly language program organization is provided.

CHANGES IN THE THIRD EDITION

The goals of the third edition are to modernize the topical coverage and to improve organization and readability. Achieving these goals required many changes from the second edition including:

► Revising much of the second edition material to reflect current technology

► Rewriting every chapter to improve organization, clarity, and readability

► Adding additional system software coverage for networking, distributed resources, and distributed applications

► Eliminating topics seldom covered in courses using this text including operating system file I/O, user interface service calls, and job control language

RESOURCES FOR INSTRUCTORS

Systems Architecture, Third Edition includes the following resources to support professors in the classroom. These tools can be found in the Instructor's Resource Kit, which is available from the Course Technology Web site (www.course.com) and on CD-ROM.

► **Instructor's Manual** The Instructor's Manual provides materials to help professors make their classes informative and interesting. It includes teaching tips, discussion topics, and solutions to end-of-chapter materials.

► **Classroom Presentations** Microsoft PowerPoint presentations are available for each chapter of this book to assist instructors in classroom lectures or to make available to students

▶ **ExamView®** ExamView® is a powerful testing software package that allows instructors to create and administer printed, computer (LAN-based), and Internet exams. ExamView® includes hundreds of questions that correspond to the topics covered in this text, enabling students to generate detailed study guides that include page references for further review. The computer-based and Internet testing components allow students to take exams at their computers, and also save the instructor time by grading each exam automatically.

▶ **MyCourse.com** MyCourse.com is an online syllabus builder and course-enhancement tool. Hosted by Course Technology, MyCourse.com adds value to courses by providing additional content that reinforces what students are learning.

Most importantly, MyCourse.com is flexible, allowing instructors to choose how to organize the material—by date, by class session, or by using the default organization, which organizes content by chapter. MyCourse.com allows instructors to add their own materials, including hyperlinks, school logos, assignments, announcements, and other course content. Instructors using more than one textbook can build a course that includes all of their Course Technology texts in one easy-to-use site. Instructors can start building their own courses today at www.mycourse.com/instructor.

WORLD WIDE WEB SITE

A support site for this book is located at http://averia.mgt.unm.edu. The site provides:

▶ The Instructor's Manual
▶ Interactive end of chapter questions and answers
▶ Web resource links for most text topics and research problems
▶ Text updates and errata

TEXT ORGANIZATION

This text is organized into four chapter groups. The first group contains two chapters that provide overviews of computer hardware, software, and networks and describe sources of technology information. The second group contains five chapters covering hardware technology. The third group contains two chapters that cover data communications and computer networks. The fourth group contains five chapters covering software technology and system administration.

The chapters are intended for sequential coverage though other orderings are possible. The prerequisites for each chapter are described below. Alternate chapter orders can be constructed based upon these prerequisite relationships. Chapter 2 should always be covered before other chapters. Some chapters and sections may be skipped without loss of continuity.

There should be time to fully cover between 9 and 12 chapters in a three credit hour undergraduate course. The text contains 14 chapters to provide flexibility in course outline and topics. Topics in some chapters may be covered in other courses in a specific curriculum. For example, Chapters 8 and 9 are often covered in a separate networking course and Chapter 14 is often covered in a separate systems administration course. The instructor can choose specific chapters to best match the overall curriculum design and teaching preferences.

CHAPTER DESCRIPTIONS

Chapter 1: Computer Technology: Your Need to Know. This chapter briefly describes how knowledge of computer technology is used in the systems development life cycle. The chapter also describes sources for computer hardware and system software information and provides a list of recommended periodicals and Web sites. The chapter can be skipped entirely or assigned only as background reading.

Chapter 2: Introduction to Systems Architecture. This chapter provides brief overviews of computer hardware, system and application software, and computer networks. The chapter describes primary classes of hardware components and computer systems and describes the differences between application and system software. The chapter introduces many key terms and concepts that are used throughout the text.

Chapter 3: Data Representation. This chapter describes primitive CPU data types and common coding methods for each type. Binary, octal, and hexadecimal numbering systems and a limited set of data structures are also described. Chapter 2 is a recommended prerequisite.

Chapter 4: Processor Technology and Architecture. This chapter describes CPU architecture and operation including instruction sets and assembly language programming. The chapter covers traditional von Neumann architectural features including instruction and execution cycles, word size, clock rate, registers, and instruction formats. Semiconductor and microprocessor fabrication technology are also described. Chapters 2 and 3 are necessary prerequisites.

Chapter 5: Data Storage Technology. This chapter describes primary and secondary storage implementation with semiconductor, magnetic, and optical technologies. Principles of each storage technology are described first, followed by details of storage hardware based on the technology. Chapters 3 and 4 are necessary prerequisites. Chapter 2 is a recommended prerequisite.

Chapter 6: System Integration and Performance. This chapter describes communication among computer system components and various performance enhancement methods. The chapter begins with a discussion of the system bus and bus protocols followed by a description of device controllers, mainframe channels, and interrupt processing. Performance enhancement method descriptions include buffering, caching, and compression. Chapters 4 and 5 are required prerequisites and Chapters 2 and 3 are recommended prerequisites.

Chapter 7: Input/Output Technology. This chapter describes I/O devices including keyboards, pointing devices, printers and plotters, video display terminals, graphic displays, optical scanners, and audio I/O devices. The chapter also covers fonts, image representation, color representation, and image description languages. Chapter 3 is a necessary prerequisite. Chapters 2, 5, and 6 are recommended prerequisites.

Chapter 8: Data and Network Communication Technology. This chapter describes data communication technology beginning with analog and digital signals, transmission media, and bit encoding methods. Communication and protocol topics include serial and parallel transmission, synchronous and asynchronous transmission, channel sharing methods, and error detection and correction. Chapters 2 and 3 are recommended prerequisites.

Chapter 9: Computer Networks. This chapter describes computer network architecture and hardware. The chapter begins with network topology, addressing, and routing. The chapter then describes media access control and network hardware devices such as routers and bridges. The chapter concludes with a discussion of IEEE and OSI networking standards and an in-depth look at TCP/IP. Chapters 3, 4, and 8 are necessary prerequisites. Chapter 2 is a recommended prerequisite.

Chapter 10: Application Development. This chapter begins with a brief overview of the software development process and software development tools. The chapter then describes programming languages, compilation, link editing, interpretation, and symbolic debugging. The final section describes advanced application development tools including CASE tools, code generators, and integrated development environments. Chapters 2, 3, and 4 are necessary prerequisites.

Chapter 11: Operating Systems. This chapter begins with an overview of operating system architecture and resource allocation. The chapter then describes how the operating system manages the CPU, processes, threads, and memory. Chapters 2, 4, and 5 are necessary prerequisites. Chapter 10 is a recommended prerequisite.

Chapter 12: File Management Systems. This chapter begins with an overview of file management functions including the differences between logical and physical secondary storage access. The chapter then describes file content and structure and directories. Next, the chapter describes storage allocation, file manipulation, and access controls. The chapter concludes with file migration, backup, and recovery. Chapters 5 and 11 are necessary prerequisites. Chapter 10 is a recommended prerequisite.

Chapter 13: Internet and Distributed Application Services. This chapter begins by discussing distributed computing and network resource access. The chapter then describes networking software starting with the lowest OSI layers. The chapter also describes services for tracking distributed resources and applications. Chapters 2, 8, and 9 are necessary prerequisites. Chapters 4, 11, and 12 are recommended prerequisites.

Chapter 14: System Administration. This chapter begins with an overview of system administration and the strategic role of hardware and software resources in an organization. The chapter then describes the hardware and software acquisition process. Next, the chapter describes requirements determination and performance modeling methods. The last section discusses physical aspects of computer installation and operation. Chapters 2, 4, 8, 11, and 12 are recommended prerequisites.

Appendix: Units of Measurement. The appendix briefly describes common measurement units, abbreviations, and usage conventions for data storage capacity and data transfer.

ACKNOWLEDGMENTS

The first edition of this text was a revision of another text entitled *Systems Architecture: Software and Hardware Concepts*, by Leigh and Ali. Some of their original work has endured through all three editions of this text. I am indebted to Leigh, Ali, and Course Technology for providing the starting point for all editions of this text.

I thank everyone who contributed to this edition and helped to make previous editions a success. Kristen Duerr and Jennifer Locke shepherded this text through the second and third editions and I thank them for their hard work and continued faith in me. I thank Deb Kaufmann for helping me to improve the content and writing quality in this edition. I thank Daphne Barbas and all development and production team members for all their excellent work on this project. Thanks also to the peer reviewers of the third edition: Susan E. Kruck, James Madison University; Ruth Schwartz, Virginia State University; Tate Redding, University of Maryland; and Carol Thomas, Dartmouth College.

I thank students in the undergraduate MIS concentration at the Anderson Schools of Management, University of New Mexico, who have used manuscripts and editions of the text over the last 12 years. Student comments have contributed significantly to improving the text. I also thank my faculty colleagues, Ranjit Bose, William I. Bullers, Jr., Laurie Schatzberg, and Alex Seazzu, and the Anderson Schools Dean, Howard Smith, for their continued support and encouragement of my textbook writing activities.

Finally, I'd like to thank Dee, my wife, and Alex and Amelia, my children. Developing this text has been a long and time-consuming process that has often impinged on family time and activities. My success as an author would not be possible without my family's love and support.

Chapter **1**

Computer Technology: Your Need to Know

Chapter Goals

► Describe the activities of information systems professionals

► Describe the technical knowledge of computer hardware and systems software needed to develop and manage information systems

► Identify additional sources of information for continuing education in computer hardware and systems software

The words you are reading now result, in part, from computer-based information systems: the author, editors, production team, and distribution team all relied on computer systems to help organize, produce, and then convey this information. Although many different kinds of computer information systems were involved, they all share similar technology: each consists of computer hardware, system and application software, data, and communications capabilities. In this chapter you learn why you need to study hardware and software technology if you work, or plan to work, in information systems. You also learn about additional sources of information that can help to expand your knowledge of hardware and software and keep it current.

TECHNOLOGY AND KNOWLEDGE

The world is filled with complex technical devices that ordinary people use every day, such as automobiles and televisions. Fortunately, we do not need detailed understanding of how these devices work in order to use them. Imagine what it would be like if three months of training were required to use a refrigerator, or if detailed understanding of mechanics and electronics were required to drive an automobile. It took years of training to use the earliest computers in the 1940s, but today, even though computers have become increasingly complex and powerful, they also have become easier to use. As a result, computers have proliferated beyond their original scientific applications into business, classrooms, and homes. If computers have become so easy to use, then why do you need to know anything about their inner technology?

ACQUIRING AND CONFIGURING TECHNOLOGICAL DEVICES

The knowledge required to purchase and configure technically complex devices is far greater than the knowledge required to use them effectively. Many people can use complex devices such as automobiles, home theatre systems, and computers, but few people feel comfortable purchasing or configuring them. Why is this so?

When you walk into a store or visit a Web site to purchase a computer, you are immediately confronted with a wide range of choices. Examples include processor type and speed, hard disk speed and capacity, amount of memory, and choice of operating system. To make an informed choice, you must know your preferences and requirements (for example, the application software that you plan to use and whether the computer will be upgraded or discarded in a year or two). You also must have sufficient knowledge of the available alternatives to determine their compatibility with your preferences and requirements. To gain sufficient knowledge of the alternatives, you must understand related technical terms (for example, megahertz, gigabyte, IDE, and PCI), technical documentation, product and technology reviews, and the advice of friends, experts, and salespeople.

An information systems professional will face computer acquisition, upgrade, and configuration choices that are significantly more complex. Larger computer systems employ more complex technology than smaller ones. There are many more components and, therefore, more complex configuration, compatibility, and administrative issues. And, of course, the stakes are higher. Employers and users will rely on the expertise of the IS professional and invest substantial sums of money based on recommendations. Are you (or will you be) able to meet the challenge?

INFORMATION SYSTEM DEVELOPMENT

The steps necessary to develop an information system are called the **systems development life cycle (SDLC)**. Figure 1-1 shows the SDLC steps commonly used to develop large and complex information systems: survey, analysis, design, implementation, and evaluation. The following paragraphs explore what knowledge of computer hardware and software are required at various stages of the SDLC.

Figure 1-1 ▶

Steps in the systems development life cycle

Systems Survey

The **systems survey** examines user information needs, ways in which those needs are being addressed (if any), and the feasibility of developing or acquiring an information system to address those needs. The survey normally is conducted within a relatively short time frame of either days or weeks. If unmet or poorly met needs are found, the costs and benefits of alternative systems that may address those needs are determined. Finally, it is decided whether the estimated net benefit of any of the alternatives justifies continuing to the remaining SDLC steps.

Technical knowledge of computer hardware and systems software is required to assess the degree to which user needs currently are being met, and to estimate the resources required to address unmet needs. For example, an analyst surveying a point of sale system in a retail store may pose questions such as:

- How much time is required to process a sale?
- Is the existing system easy for a salesperson to use?
- Is sufficient information being gathered (for example, for marketing purposes)?
- Can the existing hardware handle peak sales volumes (for example, holidays)?
- Can the existing system be expanded easily?
- What are the current hardware operating costs?
- Are there cheaper hardware alternatives?

All of these questions directly or indirectly require technical knowledge of hardware and systems software.

Processing time and ease of use depend on the capabilities of the hardware and software and the structure of the user interface. Determining whether a system can respond to peak demand requires detailed knowledge of hardware processing and storage capabilities, operating systems, and application software. Expansion possibilities depend on the unused capacity in the hardware and the limitations of both hardware and software. Determining whether cheaper alternatives exist requires technical knowledge of a wide range of hardware and software options.

The systems survey requires a broad range of technical knowledge. The time limit inherent in the survey, however, usually precludes detailed use of this knowledge. For example, it is rare to see exacting calculations of processor performance or storage capacity made during this phase. Similarly, cost estimates are rarely made to the dollar.

Systems Analysis

Systems analysis is a detailed examination of user needs and the extent to which they are being met. This step of the SDLC also can be called **needs analysis** or **requirements analysis**. Various information-processing requirements, including data contents, data storage, calculations, and system performance, are described in detail. Systems analysis usually takes weeks or months and produces a **requirements document** as output.

Many of the questions posed during the systems survey are reexamined and answered in greater detail during systems analysis. For example, general requirements for ease of use and response time are the starting point for developing detailed descriptions of user interfaces and data access speed. Estimates of current and future transaction volumes help determine detailed data-processing and storage requirements.

Systems Design

During **systems design** the exact configuration of all hardware and software components is determined (see Figure 1-2). These selections affect all other design decisions, and provide a blueprint for later systems implementation. Specific systems design tasks include selecting:

► Computer hardware (processing, storage, input/output, and network components)
► Systems software (an operating system and/or network communication software)
► Application program development tools (programming languages and database management systems)

Other design decisions are of a more limited scope and depend on earlier decisions. Examples include:

► Files or database design (such as grouping data elements into records and files, indexing, and sorting)
► Program design
► User interface design (input screen formats, report formats, and command dialogues)
► Design of system backup and recovery mechanisms

Figure 1-2 ▶

Detailed systems
design tasks in
the systems
development
life cycle

Technical knowledge of computer hardware and systems software is most important during the early stages of systems design. Selecting hardware components requires a detailed knowledge of their capabilities and limitations. When multiple hardware components are integrated in a single system, the compatibility among elements must be evaluated. The hardware and overall performance requirements of the system affect the choice of operating system. Both hardware and systems software also must be considered in terms of their compatibility with other information systems and computer technology in the organization.

Selecting appropriate program development tools requires knowledge of information system requirements and the capabilities of the hardware and operating system. Development tools (and the software components built with them) vary widely in their efficiency, power, and compatibility. Tool selection also will affect future system development projects.

Systems Implementation

During **systems implementation**, a system is constructed based on the "blueprints" created during systems design. Hardware and systems software are acquired, installed, configured, and tested. Application programs and files are developed, installed, and tested. Then the entire system is tested to ensure that all of its components work together and that the system correctly meets the users' needs as defined during the analysis phase.

Technical knowledge of computer hardware and systems software is needed to complete many implementation tasks. Installing and configuring hardware and systems software is a specialized task that requires a great deal of understanding of the components being installed and the purposes for which they will be used. Both hardware and software usually must be configured to a particular operating environment. This requires that adjustments be made to the components and that the system then be tested to determine if the adjustments are sufficient. This time-consuming process can require many iterations. Technical expertise also is needed to complete tasks such as formatting storage devices, setting up system security, installing application programs, and establishing accounting and auditing controls.

Systems Evaluation and Maintenance

As the system is used over time, various problems can and will arise. Errors that escaped detection during installation and testing may appear. The system may become overloaded due to inadequate estimates of processing volume. Additional needs for information can become apparent, necessitating additional data collection, storage, and/or processing.

Relatively minor system changes (such as correcting program errors or minor processing changes) normally are handled as maintenance changes. Maintenance changes may require extensive technical knowledge, and some technical knowledge may be required to classify a proposed change properly as major, or minor. Will new processing requirements be the "straw that breaks the camel's back" in terms of hardware or software capacity? Do proposed changes require programming tools that are incompatible with the current systems design or configuration? The answers to these questions determine whether the existing system will be modified or replaced by an entirely new system.

If the existing system is to be modified, the programs and/or files to be changed are identified, modified, and tested. The specific technical knowledge requirements depend heavily on the specific hardware and software components affected by the change. In the event that an entirely new system is required, a new SDLC will be initiated, requiring the technical knowledge at each step, as described earlier.

MANAGEMENT OF COMPUTER RESOURCES

Thus far, the need for technological knowledge has been discussed in the context of developing a single information system. But consider the complexities and knowledge required to manage the computer resources within a large organization.

Existing information systems in a single organization may number in the hundreds or thousands. Many new development projects or major upgrades of existing systems can be in progress at any one time.

This environment requires increased attention to two very important technological issues—compatibility and future trends. Both have become increasingly important issues as the number of computer systems and applications has increased. Most information systems are integrated tightly with other systems through both software and hardware. For example, accounts payable and accounts receivable programs may share a common hardware platform and operating system. Data from both systems can be input to a financial reporting system on an entirely different computer. Also, data from many sources within the organization can be stored in a common database and accessed via a communications network.

The manager of a collection of information systems must contend with a great deal of technical complexity. He or she must ensure that each new system not only operates correctly by itself, but also operates smoothly with all of the other systems in the organization. The manager also must ensure that hardware and software acquisitions provide a good foundation for both current and future systems.

Given the rapid pace of change in computer technology, a manager must have a broad understanding of current technology, as well as future technology trends. Will the computer purchased today be compatible with the hardware available three years from now? Will the communications network purchased today have the capacity to meet future needs? Should the organization invest only in "tried and true" technologies or should it risk the purchase of cutting edge technologies in hopes of greater performance or competitive advantage?

The answers to these questions require a great depth of technical knowledge —far more knowledge than any one person possesses. Typically, a manager confronted by these questions will rely on the advice of experts and other sources of information. Even so, a manager responsible for such decisions must have a sufficient base of technical knowledge to understand the information and advice provided by others.

ROLES AND JOB TITLES

A large number of people can be classified loosely as computer professionals, and an even larger number use computers in the workplace, school, or home. There is a bewildering array of job titles, specializations, and professional certifications to accompany this wide range of roles. The following sections attempt to classify computer professionals into groups, explain some of their common characteristics, and describe the computer hardware and systems software knowledge needed by members of each group.

Application Development Personnel

Many computer professionals create application software for specific processing needs. These professionals have many different job titles, including programmer, systems analyst, and systems designer. Each role contributes to a different part of the systems development life cycle. A **systems analyst** primarily is responsible for conducting surveys, determining feasibility, and defining and documenting user requirements. A **systems designer** primarily is responsible for procuring hardware and procuring or designing applications software. An **application programmer** primarily is responsible for software implementation and testing.

Many professionals engage in activities that do not neatly fit into these job descriptions. For example, it is very common for a systems analyst to be responsible for survey, analysis, design, and management of a development project. Programmers often perform limited analysis and design tasks in addition to writing and testing programs.

Further confusion about job titles and descriptions arises from differences in the types of software applications that are developed and the training required for each type. Applications can be classified loosely into two types—information processing and scientific/technical. Information processing applications typically process business transactions or provide information to managers. Developers of these applications usually have college or technical degrees in management or business with a specialization in information processing.[1]

Scientific applications meet the data processing and numerical modeling needs of researchers in areas such as astronomy, meteorology, and physics. Technical applications typically are concerned with the control of hardware devices. Examples of these include applications for the control of robots, flight navigation, and scientific instrumentation. Developers of scientific and technical applications typically have degrees in **computer science** or some branch of engineering.

Application developers need technical knowledge of computer hardware and systems software as described earlier in the Information Systems Development section. Developers of technical applications typically need in-depth hardware knowledge due to the hardware control characteristics of many technical applications. Scientific application developers also require in-depth hardware knowledge if the applications they develop push the boundaries of hardware capabilities, for example, simulating weather patterns on a global scale over long periods of time.

---

[1] Other names for the field include management information systems, data processing, and business computer systems.

Systems Software Personnel

Another large class of computer professionals, called **systems programmers**, is responsible for the development of systems software such as operating systems, compilers, database management systems, and network security monitors. Systems programmers typically have degrees in computer science or computer engineering. Organizations with large amounts of computer equipment and software frequently employ systems programmers as hardware/software consultants and to perform hardware and software installation and configuration. Larger numbers of systems programmers are employed by organizations that develop and market systems software and/or application tools, such as Microsoft, ORACLE, and Corel.

Systems programmers require in-depth knowledge of system software. They also require in-depth knowledge of computer hardware because many types of system software, such as operating systems, directly control computer hardware. For this reason, computer science programs usually cover the subjects within this textbook in several different courses.

Hardware Personnel

Personnel responsible for hardware design, installation, and maintenance are employed by vendors of computer hardware. Lower-level personnel usually have a technical degree and/or vendor-specific training, while higher-level personnel often have a degree in computer science or computer engineering. Personnel who design computer systems often have a degree in computer science or a related branch of engineering.

Hardware personnel require extensive knowledge of computer hardware including processing, data storage, input/output, and networking devices. Hardware designers require the greatest depth of knowledge, far exceeding the contents of this text. Personnel who install and maintain hardware must know the hardware subjects covered in this text and, in addition, require product or vendor-specific knowledge.

Systems Management Personnel

The proliferation of computer hardware, networks, and software applications over the last five decades has created a need for large numbers of managers and administrators. The job descriptions for such positions vary widely from one organization to another due to differences in organizational structure, and in the amount and types of computing and information resources employed by the organization. Some of the more common job titles in this category are computer operations manager, network administrator, database administrator, and chief information officer.

A **computer operations manager** is responsible for the operation of a large information-processing facility. Such facilities typically have one or more large computer systems and all related peripheral equipment housed in a central location. Large databases, thousands of application programs, dozens to hundreds of employees, and significant amounts of batch processing are typical of such installations. Examples of organizations with this type of computing facility include large banks, credit reporting bureaus, the Social Security Administration, and the Internal Revenue Service. The management of day-to-day operations is extremely complex. Scheduling, staffing, security, system backups, maintenance, and upgrades are some of the more important responsibilities of a computer operations manager.

A computer operations manager can have many technical specialists on staff. Staff members usually have considerable technical knowledge within narrow specialties such as data storage hardware, data communications, mainframe operating systems, and performance tuning. A computer operations manager requires a broad base of technical knowledge to understand and manage all of the hardware and software resources he or she controls. The manager also must be capable of understanding the advice provided by staff members.

The title of **network administrator** typically implies one of two possible roles. The first is responsibility for the network infrastructure of a large distributed organization. A multinational corporation, with dozens or hundreds of worldwide locations, typically will implement a network to tie together its databases, application software, and computer hardware. The design, operation, and maintenance of a large network is a complex task requiring substantial technical expertise in computer hardware, telecommunications, and systems software. The role of network administrator in this environment is an important and high-level position. Technical knowledge requirements are similar to those of a computer operations manager, though the emphasis is on network and data communication technology.

In the second role, the title of network administrator usually implies the management of a local area network in smaller organizations. Such networks connect anywhere from a half dozen to a few hundred computers (mostly microcomputers) and provide access to one or more shared databases. In smaller organizations, the network administrator can be responsible for many tasks other than operating and maintaining the network itself. These tasks might include the installation and maintenance of end user software, hardware installation and configuration, training users, and assisting management in the selection and acquisition of software and hardware resources. This position is one of the fastest growing areas of employment for information systems professionals. It also is one of the most demanding positions in terms of breadth and depth of required skills and technical knowledge.

Since the early 1980s, many organizations have realized the economic and strategic importance of the data stored in their computer systems. This data can be exploited for many purposes, including coordinating organizational subunits and developing effective marketing strategies. The technology by which large collections of data, called databases, are managed and accessed is specialized and highly complex. This complexity, combined with managerial recognition of the importance of data resources, has resulted in the creation of many positions with the title of **database administrator**. The title implies more than just technical expertise; it also implies that he or she has the responsibility for helping the organization realize the potential benefits of its investment in stored data.

The role of **chief information officer** (CIO) is a high-level management position. It is a position that typically exists only in large organizations with substantial investments in computer and software technology. Many of the previously defined positions (database administrator, network administrator, and computer operations manager) report to the CIO. The responsibilities of this position include all computers, networks, and software owned and operated by an organization. As with a database administrator, this position implies a responsibility for strategic planning and the effective use of information and computing technology in the broadest sense of those terms.

A CIO cannot possibly be an expert in every aspect of computer technology relevant to his or her organization. But the CIO must possess a broad enough base of technical knowledge to interact effectively with all the technical specialists employed by the organization. The CIO also must possess knowledge that transcends mere technical capability. He or she must be able to see technology as an integrated whole that supports the mission and objectives of the organization. The CIO also must have a vision of how technology is changing and how best to respond to and take advantage of those changes.

COMPUTER TECHNOLOGY INFORMATION SOURCES

This book will provide a foundation of technical knowledge for a career in information system development or management. Unfortunately, that foundation will erode quickly because of the rapid pace of change in computer and information technology. You constantly will need to update your knowledge simply to stay current.

There are many resources you can use to keep your knowledge current. Newspapers and periodicals can provide a wealth of information regarding current trends and technologies. Training courses offered by hardware and software vendors will teach you the specifics of current products. Additional coursework and self-study can keep you up to date on technologies and trends less specifically geared toward particular products. By far, the most important of these activities is reading periodical literature.

Periodical Literature

The volume of literature on computer topics is huge. It often is difficult to determine which sources are most important and appropriate to information systems professionals. Because of differences in training and focus among information systems professionals, computer scientists, and computer engineers, sources of information that are oriented toward one specialty are difficult reading for the others. For example, a detailed discussion of user interaction and validation of requirement models aimed at an information systems audience may be beyond the training of a computer scientist or engineer. Similarly, detailed descriptions of optical telecommunications theory may be beyond the training of an information systems professional.

These differences pose a problem for information systems professionals who need current information on hardware and systems software technology. Much of the available literature is oriented toward computer scientists and/or computer engineers. The content and detail of this material often is beyond the training of most information systems professionals.

The periodicals listed below are good information sources for information systems professionals. They avoid excessive technical depth, while still providing enough detail to be useful.

► *ACM Computing Surveys* — an excellent source of information on the latest research trends in computer software and hardware. The articles are written as in-depth summaries of technologies or trends geared toward a readership with a moderate level of familiarity with computer hardware and software.

► *Computerworld* — a weekly magazine primarily geared toward computer news items. Provides coverage of product releases, trade shows, and occasional coverage of technologies and trends.

► *Communications of the ACM* — a source of information about research topics in computer science. Many of the articles are highly technical and specialized to research, but some are geared toward a less research-oriented audience.

► *Computer* — a publication of the Institute of Electrical and Electronics Engineers (IEEE) devoted to coverage of computer hardware and software. Many of the articles are research-oriented but provides occasional coverage of technologies and trends for a less technically oriented audience.

The periodicals listed are only a small sample of the available literature. A complete printed list of today's recommended periodicals quickly would become out of date due to rapid change in computer technology and the computer-related publishing industry. A current list of recommended periodicals can be found on the textbook Web site (http://averia.unm.edu).

Most periodical publishers have a Web site that provides content and services beyond those provided in printed periodicals, including:

► Content from back issues of the printed periodical

► Additional current content that doesn't appear in the printed periodical

► Search engine for back issues and other content

Web-based periodicals usually provide Web links to article references and other related material. For example, an article that refers to another article in a back issue frequently will contain a hypertext link to the back issue. Reference lists and bibliographies also can be hyperlinked. The task of accessing background and reference material is reduced to the click of a mouse. Similarly, reviews of specific software or hardware typically contain links to product information and specifications maintained on vendor or manufacturer Web sites.

Figure 1-3 shows the Web site for *Computerworld* magazine.

Figure 1-3 ►

Computerworld home page

Technology-Oriented Web Sites

Computer technology is a big business and there are millions of computer professionals worldwide. There are many Web sites that are devoted to serving the information needs of computer professionals. Table 1-1 lists some of the most widely used sites. Check the textbook Web site for updates to this table.

Table 1-1 ►

Technology-
related Internet
sites

Site name	Internet name	Comments
c/net	www.cnet.com	This site primarily is oriented to consumers though it does have substantial content for IS professionals.
Earth Web	www.earthweb.com	This site contains a broad range of information for IS professionals.
Gartner Group	www.gartnergroup.com	The Gartner Group is a consulting and research company specializing in services for CIOs and other executive-level decision makers.
internet.com	internet.com	The site specializes in Internet technology and Internet-related business. It primarily is a news channel, though it does have other resources for IS professionals.
ITworld	itworld.com	This site contains a broad range of information for IS professionals.

Consolidation among periodical publishers has created large corporate families of technology-related publications. Many technology-oriented Web sites are owned by, or closely associated with, periodical publishers. For example, TechWeb is owned by CMP Media, which publishes *Data Communications*, *Internet Week*, *UNIXWorld*, and many other periodicals. Technology-oriented Web sites provide a common interface to these publication families. They also enable a publisher to provide content and services that transcend single paper-based publications, such as cross-referencing publications, online discussion groups, employment services, and Web-based interfaces to hardware, software, and technology service vendors.

Technology Web sites can make money in several ways. A few companies charge customers directly for Web-based information and services (for example, Gartner Group). But most companies earn revenue in other ways, including:

► Advertising (note the ads in Figure 1-3)
► Direct sales of goods and services
► Commissions on goods and services sold by advertisers and partners

Any site that generates revenue from advertising, referrals, commissions, or preferred partner arrangements may have biased content. Bias can appear in many forms, including:

► Ordering of content, links, or search results that favor organizations that have paid a fee to the Web site owner
► Omission of information from organizations that have not paid a fee to the search provider
► Omission of information that is against the interests of organizations that have paid a fee to the search provider

Note that these same biases may be present in general Internet search engines. It is not always obvious to the reader which, if any, of these biases are present in any particular Web site.

Unbiased information is present on the Web though it is not always easy to find. The old saying that "you get what you pay for" applies well to the Web. High-quality, unbiased information is the product of intensive research. Some computer professional societies and governmental organizations do this as a public service, but much of the information you will find is produced for a profit. Expect to pay directly for completely unbiased information. When dealing with publicly accessible information sources, be sure to use information from multiple unrelated sources in order to balance the biases of each individual source.

Vendor and Manufacturer Web Sites

Most vendors and manufacturers of computer hardware and software have rushed to create a Web presence (see Figure 1-4) because many purchasers of computer hardware and software already are online. Vendor Web pages are oriented to sales, but they usually contain detailed information on products either directly or as links to manufacturer Web sites. Manufacturer Web sites contain detailed information on their products. Technical support and other customer support services also are provided via these Web sites.

Figure 1-4 ►

A typical computer hardware manufacturer's Web page

Computer Technology: Your Need to Know

It is important to remember that these sites exist primarily as marketing and customer support tools. The positive aspect of manufacturers' Web sites is that they provide a wealth of technical information for specific products. This information is far more extensive and current than was ever available through paper-based product brochures and technical documents. The breadth and depth of this information potentially allows information system professionals to make faster, better, and more informed choices. The down side of manufacturers' Web sites is that the information provided often is biased to the vendors' products and the technologies on which they are based. As with any source of information, a reader must consider the motives and objectivity of the information provider. Hardware and software manufacturers typically are not in the business of providing unbiased information. They are in the business of selling what they produce. A visitor to these Web sites should expect the content of a manufacturer or vendor Web site to be biased toward the products it produces or sells.

In the best case, biased content may manifest itself as marketing hype that a reader must wade through or filter out to get to the real information. In the worst case, the information may be biased purposely by content, omission, or both. Many sites contain technology overviews, often called white papers, and similar information that easily is confused with independent research. Others contain links to a biased subset of supporting reviews or research. It is the readers' responsibility to balance potentially biased vendor and manufacturer information with information from unbiased sources.

Professional Societies

There are several professional societies that are excellent sources of information about computer technology. These include:

- ▶ *AITP*—The membership of the **Association for Information Technology Professionals** (**AITP**) consists primarily of IS managers and application developers. AITP has local chapters throughout the country and publishes several periodicals including *Information Executive*. Their Web site is www.aitp.org.

- ▶ *ACM*—The **Association for Computing Machinery** (**ACM**) is a well-established organization with a primary emphasis on computer science. ACM has dozens of special interest groups. ACM sponsors hundreds of research conferences each year and publishes many periodicals and technical books. The membership represents a broad cross section of the computer community, including hardware and software manufacturers, educators, researchers, and students. Their Web site is www.acm.org.

► *IEEE Computer Society*—The **Institute for Electrical and Electronics Engineers (IEEE) Computer Society** is a subgroup of the IEEE that specializes in computer and data communication technology. The membership largely is composed of engineers with an interest in computer hardware, but there are many members with software interests and many with academic and other backgrounds. The IEEE Computer Society sponsors many conferences (many jointly with the ACM). Its publications include several periodicals, such as *IEEE Computer*, and a large collection of technical books and standards. Their Web site is www.computer.org.

SUMMARY

► Technical knowledge of computer hardware and systems software is required to develop information systems. The type and depth of required knowledge differs among phases of the systems development life cycle (SDLC). A broad base of knowledge is required in the survey phase to evaluate feasibility and related issues. During the design phase, more in-depth knowledge about hardware and systems software is required to derive detailed specifications.

► Technical knowledge also is required to manage the computer resources of an organization. Particular attention must be given to compatibility and future trends. Compatibility is important because computer hardware and systems software typically are shared among organizational units and subsystems. Future trends must be considered in acquisitions due to the long-term nature of hardware and software investments.

► Technical knowledge constantly must be updated due to changes in hardware and software technology. Information systems professionals must engage in continuing education and study to keep pace with these changes. Training can be obtained through vendors, educational organizations, and self-study. Self-study relies heavily on periodical literature and Web resources. Care must be exercised in the selection of periodical literature due to the differences in intended audience and required background and training.

► Information about computer hardware and software is readily available on the World Wide Web. Sites of interest include those maintained by publishers, vendors, manufacturers, and professional organizations. The amount of available information is substantial, but care must be exercised because the available information may be biased and/or incomplete.

► In this chapter you learned why you need to understand computer technology and how you can keep that knowledge current. In the next chapter you will take a broad look at hardware, software, and networking technology. We will examine concepts and terms that will be explored in detail in the rest of the text. Your journey through the inner workings of modern information systems is about to begin.

Key Terms

application programmer
Association for Computing
   Machinery (ACM)
Association for Information
   Technology Professionals
   (AITP)
chief information officer (CIO)
computer operations manager
computer science

database administrator
Institute for Electrical and
   Electronics Engineers
   (IEEE) Computer Society
needs analysis
network administrator
requirements analysis
requirements document
systems analysis

systems analyst
systems design
systems designer
systems development life cycle
   (SDLC)
systems implementation
systems programmer
systems survey

Vocabulary Exercises

1. Students of _____ generally focus on system software, while students of information systems generally focus on _____ .

2. Configuration of hardware and system software occurs during the _____ phase of the SDLC.

3. Professional organizations with which an information systems student or professional should be familiar include _____ , _____ , and _____ .

4. Selection of hardware and system software occurs in the _____ phase of the SDLC.

5. A(n) _____ typically is responsible for a large computer center and all of the software executed therein.

6. Computer specialties most concerned with hardware and the hardware/software interface are _____ and _____ .

7. An initial estimate of hardware and system software requirements is made during the _____ phase of the structures SDLC.

8. The job titles of persons directly responsible for developing application software include _____ , _____ , and _____ .

Review Questions

1. In what way(s) is the knowledge needed to operate complex devices different from the knowledge needed to acquire and configure them?

2. What knowledge of computer hardware and systems software is necessary to complete the survey phase of the systems development life cycle successfully?

3. What knowledge of computer hardware and systems software is necessary to complete the design phase of the systems development life cycle successfully?

4. What additional technical issues must be addressed when managing a computer center or local area network, as compared to developing a single information systems application?

Research Problems

1. The Bureau of Labor Statistics (BLS) compiles employment statistics for a variety of job categories and industries. BLS also produces predictions of employment trends by job category and industry. Most of the information produced by the BLS is available via the World Wide Web. Access the BLS Web site (http://stats.bls.gov) and investigate the current and expected employment prospects for information systems professionals.

2. Assume that you are an IS professional and you have been asked by your boss to prepare a briefing for senior staff on the comparative advantages and disadvantages of three competing tape drive technologies—digital linear tape (DLT), Advanced Intelligent Tape (AIT), and Linear Tape Open (LTO). Search each of the technology Web sites in Table 1-1 for source material that may aid you in preparing the briefing. Which site(s) provide the most useful information? Which site(s) allow you to find useful information with relative ease?

3. Read several dozen job listings at www.itcareers.com or a similar site. Which jobs (and how many) require a broad knowledge of computer technology? At what level are those jobs? Try searching the listings based on a few specific keywords, such as programmer or network. Look closely at the hiring companies and where their job postings show up in the search results. Can you draw any conclusions about how the listings are ordered?

Chapter 2

Introduction to Systems Architecture

Chapter Goals

- ► Discuss the development of automated computing

- ► Describe the general capabilities of a computer

- ► Describe computer system components and their functions

- ► List computer system classes and their distinguishing characteristics

- ► Define the role and function of application and system software

- ► Describe the economic role of system and application development software

- ► Describe the components and functions of computer networks

Computer systems are a complex combination of hardware, software, and network components. The term **systems architecture** describes a computer system's components and the interaction between them. The term architecture is a misnomer, because it implies a concern with only static structural characteristics. In this text you also learn about the dynamic behavior of a computer system, that is, how its components interact as the computer operates.

This chapter lays the foundation for the text with a brief discussion of the major components and functions of hardware, software, and networks. Each component is described in detail in later chapters. Avoid the temptation to rush through this chapter, because it is as important to know how all the components of a computer system relate as it is to know their internal workings.

AUTOMATED COMPUTATION

A simple definition of a computer is any device that can:

► Accept numeric inputs
► Perform computational functions, such as addition and subtraction
► Communicate results

This definition captures the basic functions of a computer, but it also can apply to people and simple devices, such as calculators. The basic functions of an automated computer can be implemented by many methods and/or devices. For example, a modern computer may implement computation electronically (using transistors within a microprocessor), implement storage optically (using a laser and the reflective coating on an optical disc), and implement communication using a combination of electronics and mechanics, for example, the mechanical and electrical components of a printer. Some simple experimental computers have even used quantum physics to implement data storage and computation.

Mechanical Implementation

Early mechanical computation devices were designed to perform repetitive mathematical calculations. The most famous of these machines was called the difference engine, built by Charles Babbage in 1821 (see Figure 2-1). The difference engine computed logarithms through the movement of gears and other mechanical components. Many other mechanical computation machines were developed well into the twentieth century and were in use as recently as the 1960s.

Figure 2-1 ▶

Charles
Babbage's
difference engine

The common element in all such computation devices is a mechanical representation of a mathematical calculation. A simple example is a clock driven by a spring and pendulum. Each swing of the pendulum allows a gear to move one step under pressure from the spring. As the pendulum swings, the gears advance the hands of the clock. A mechanical clock actually is a simple adding machine: each swing of the pendulum adds one unit to the current time.

Computational functions beyond addition also can be implemented mechanically. Multiplication of whole numbers, for example, can be mechanically implemented as repeated addition. A machine capable of addition can perform multiplication by executing the addition function multiple times (6 times 3 can be calculated by adding 6 plus 6, temporarily storing the result, and then adding 6 a third time). Appropriate combinations of moving parts also can be used to implement complex functions such as logarithms and trigonometric functions.

There are inherent limitations and shortcomings of mechanical computation including:

▶ Complex design and construction
▶ Wear, breakdown, and maintenance of mechanical parts
▶ Limits on operating speed

Automated computation with gears and other mechanical parts requires a complex set of components that must be designed, manufactured, and assembled to exacting specifications. As the complexity of the computational function increases, the complexity of the mechanical device that performs it also increases, exacerbating problems of design, construction, wear, and maintenance.

Electrical Implementation

Much as the era of mechanical clocks gave way to the era of electrical clocks, the era of mechanical computation eventually gave way to electrical computers. The biggest impetus to change from mechanical to electrical computing devices came during World War II. The military faced many problems, such as navigation and breaking codes, and large numbers of calculations were required to solve those problems. The mechanical devices of the time were simply not up to the task, forcing the design and construction of electrical computation devices. The ENIAC, one of the first general-purpose electrical computers, was built in 1946 and filled an entire wall.

In an electrical computing device, the movement of electrons performs essentially the same functions as gears and wheels in mechanical computers. Storage of numerical values is accomplished by the storage of magnetic charges, rather than by the position of gears and wheels. When necessary, electro-mechanical devices are used to convert physical movement into electrical signals, or vice versa. For example, a keyboard converts the mechanical motion of keystrokes into electrical signals. The ink pumps in an ink-jet printer convert electrical signals into the mechanical motion that forces ink out of a nozzle and onto paper.

Electrical computers addressed most of the shortcomings of mechanical computation. They were faster due to the relatively high speed at which electrons move. With improvements in the design and construction of electrical devices, electrical computers became more reliable and easier to build than their mechanical counterparts. Electrical computers made it possible to perform complex calculations at speeds previously thought impossible. Larger and more complex problems could be addressed and simple problems could be solved much faster.

Optical Implementation

Light also can be used as a basis for computation. A particle of light, called a photon, moves at a high rate of speed. As with electrons, the energy of a moving photon can be harnessed to perform computational work. Light can be transmitted over appropriate conductors, such as laser light through a fiber-optic cable. Data can be represented as pulses of light and stored either directly (such as an image stored as a hologram) or indirectly by materials that reflect or don't reflect light, such as the surface of a CD or DVD).

Optical devices are at the leading edge of computer hardware technology. Optical data communication now is common in computer networks that cover relatively large distances. Optical devices are currently in use for storing and retrieving large amounts of data. Some input/output devices (laser printers and optical scanners) are based on optical technologies and devices. Optical and hybrid electro-optical devices have been developed for communication among computer system components and for computer processing. At this time, these devices are largely experimental. However, it is expected that optical and hybrid electro-optical technologies will find much wider application in the computer hardware of the twenty-first century.

Technology
Focus

Quantum Computing

Current computer technology is based on principles of classical physics including electronics, magnetism, and optics developed during the seventeenth to twentieth centuries. These principles are based on mathematical rules that describe the behavior of matter at the level of atoms, molecules, and larger units. Mechanical, electrical, and optical processors use these mathematical rules to implement automated computation. Data is represented by clearly defined binary physical states, such as magnetic polarity, called bits. Each bit has only two possible states that are represented by the values zero and one.

Quantum physics describes the behavior of matter at the subatomic level. Subatomic physical behavior also is described by mathematical rules. A computer that manipulates subatomic particles can implement corresponding mathematical functions. Automated computation can be based on either classical or quantum physics.

Quantum particles, however, exhibit behaviors that are not seen in larger units of matter. For example, a quantum particle can be in two physical states at the same time, therefore representing two numeric values simultaneously. An atom or molecule can only represent one value. The data stored in a single quantum particle is called a quantum bit or **qubit**.

When a quantum computing device operates on a qubit it operates on both values simultaneously, producing two results. A computation with a single qubit input produces twice as many results as a computation with a single bit input. The doubling of computational results scales up exponentially with the number of qubits. For example, a conventional computation on four bits produces a single result, but a quantum computation on four qubits produces 16 (2^4) results because it actually is operating on 16 different inputs at the same time.

A quantum computing device is built from subatomic parts. This creates obvious problems of design and fabrication. Additional problems arise from other quantum behaviors, including the **quantum measurement problem**. The multiple states of a qubit are disturbed by any attempt to measure their value. When measured, a qubit reverts to only one of its states, losing the data held in the other state. To perform repeated calculations, a quantum computer must either avoid measurement until the last computation or it must reconstruct intermediate results after each quantum instruction is executed.

Scientists and engineers have been experimenting with quantum computing devices during the last decade. One such device is shown in Figure 2-2. It is an ion trap used to implement qubits with calcium ions. Lasers are used to manipulate the qubits and perform quantum calculations. Researchers have experimented with different ways of implementing qubits and quantum computations. These experiments are still in their infancy and a commercially viable quantum computer is estimated to be more than a decade away.

Figure 2-2 ▶

An experimental ion trap used to store qubits

COMPUTER CAPABILITIES

All computers are automated computing devices, but not all automated computing devices are computers. The primary characteristics that distinguish a computer from other automated computation devices include:

▶ General-purpose processor capable of performing computation, data movement, comparison, and branching functions

▶ Storage capacity sufficient to hold large numbers of program instructions and data

▶ Flexible communication capability through the use of multiple media and devices

Each of these characteristics is considered in the following sections.

Processor

A **processor** is a device that performs data manipulation and/or transformation functions including:

▶ Computation (addition, subtraction, multiplication, and division)

▶ Comparison (less than, greater than, equal to, and not equal to)

▶ Data movement among memory, mass storage, and input/output devices

An **instruction** is a signal or command to a processor to perform one of its functions. When a processor performs a function in response to an instruction, it is said to be **executing** that instruction.

Each instruction directs the processor to perform one simple task (for example, add two numbers). Complex functions are performed by executing a sequence of instructions. To add a list of 10 numbers, a processor must be instructed to add the first number to the second and to temporarily store the result. It then must be instructed to add the stored result to the third number and to store that result temporarily. Instructions must continue to be issued and executed until all 10 numbers have been added. Most useful computational tasks, such as recalculating a spreadsheet, are accomplished by executing a long sequence of instructions called a **program**. A program is a stored set of instructions for implementing a specific task, such as calculating payroll and then generating paychecks or electronic fund transfers. Programs can be stored and reused over and over again.

A processor can be classified as either general-purpose or special-purpose. A **general-purpose processor** is capable of executing many different instructions in many different sequences or combinations. The task performed by a general-purpose processor can be altered by changing the program that directs its actions. A general-purpose processor can be instructed to do many different tasks such as payroll calculation, text processing, or scientific calculation simply by supplying it with an appropriate program.

A **special-purpose processor** is designed to perform only one specific task. In essence, it is a processor with a single internal program. Many commonly used devices such as microwave ovens, compact disc players, and computer printers contain special-purpose processors. Although such processors and/or the devices that contain them can be called computers, the term computer usually refers to a device containing a general-purpose processor that can execute a variety of programs.

Formulas and Algorithms Some processing tasks require little more than a processor's computation instructions. For example, consider the following calculation:

$$Gross\_Profit = Quantity\_Sold \times Selling\_Price - Selling\_Expenses$$

The only processor functions needed to compute Gross_Profit are multiplication, subtraction, and storing and accessing intermediate results. In statement

30 (see Figure 2-3), Quantity_Sold and Selling_Price are multiplied and the result is stored temporarily as Intermediate_Result. In statement 40, Selling_Expenses is subtracted from the Intermediate_Result. The Gross_Profit calculation is a **formula**. A processor executes a sequence of computation and data movement instructions to solve a formula.

Figure 2-3 ▶

A BASIC
program to
compute gross
profit

```
10      INPUT QUANTITY_SOLD
20      INPUT SELLING_PRICE
30      INTERMEDIATE_RESULT = QUANTITY_SOLD * SELLING_PRICE
40      GROSS_PROFIT = INTERMEDIATE_RESULT - SELLING_EXPENSES
50      OUTPUT GROSS_PROFIT
60      END
```

Computer processors also can perform a more complex class of processing tasks called algorithms. An **algorithm** is a program in which different sets of instructions are applied to different data input values. The program must make one or more decisions to determine what instructions to execute in order to produce correct data output values. Depending on the data input values, entirely different subsets of instructions may be executed. In contrast, all of the instructions that implement a formula always are executed in the same order regardless of the data input.

The procedure for computing United States income tax is an example of an algorithm. Figure 2-4 shows a set of income tax computation formulas. Note that different income values require different formulas in order to calculate the correct amount of tax. A program that computes tax based on this table must execute only the set of instructions that implements the appropriate formula for a particular income value. The program examines the income value and decides which set of instructions to execute to compute the tax.

Figure 2-4 ▶

A tax table

Single - Schedule X

If gross pay is:		The tax is:	
Over-	**But not over-**		**of the amount over-**
$0	$26,250	15%	$0
26,250	63,550	$3,937.50 + 28%	26,250
63,550	132,600	14,381.50 + 31%	63,550
132,600	288,350	35,787.00 + 36%	132,600
288,350	--------	91,857.00 + 39.6%	288,350

Introduction to Systems Architecture

Comparisons and Branching Decisions within a processing task are based on numerical comparisons. Each numerical comparison is called a **condition**, and the result of evaluating a condition is either true or false. In the tax example, the income value is compared to the valid income range for each formula. Each formula is implemented as a separate instruction or set of instructions within the program. When a comparison condition is true, the program **jumps** or **branches** to the first instruction that implements the corresponding formula.

Figure 2-5 illustrates a BASIC program that uses comparison and branching instructions to calculate income taxes. In statements 20, 50, 80, and 100, INCOME is compared to a constant. The comparison result (either true or false) determines which instruction is executed next. If the comparison condition is false, then the next program instruction is executed. If the comparison condition is true, then the program jumps to a different point in the program by executing a GOTO, or branch, instruction.

Figure 2-5 ▶

A BASIC program to calculate income taxes

```
10      INPUT INCOME
20      IF INCOME > 26250 THEN GOTO 50
30      TAX = INCOME * 0.15
40      GOTO 150
50      IF INCOME > 63550 THEN GOTO 80
60      TAX = 3937.50 + (INCOME - 26250) * 0.28
70      GOTO 150
80      IF INCOME > 132600 THEN GOTO 110
90      TAX = 14381.50 + (INCOME - 63550) * 0.31
100     GOTO 150
110     IF INCOME > 288350 THEN GOTO 140
120     TAX = 35787.00 + (INCOME - 132600) * 0.36
130     GOTO 150
140     TAX = 91857.00 + (INCOME - 288350) * 0.396
150     OUTPUT TAX
160     END
```

Comparison and branching are the fundamental bases for managing the processing activity beyond simple calculations. Comparison instructions are part of a group of **logic instructions**, implying a relationship to intelligent, decision-making behavior. A general-purpose processor within a computer is more restricted in its comparative abilities than a human being. A person can compare complex objects and phenomena and also handle uncertainty in the resulting conclusions, whereas a computer only can perform simple comparisons (equality, less than, and greater than) with numeric data where the results either are completely true or completely false. Despite these limitations, comparison and branching instructions are the building blocks of all computer intelligence. They also are the capabilities that distinguish a computer processor from the processors of simpler, automated computation devices such as calculators.

Storage Capacity

A computer stores a variety of things, including:

► Intermediate processing results
► Data
► Programs

Because computers break complex processing tasks into small parts, a computer needs to store intermediate results (for example, the variable Intermediate_Result on line 30 in Figure 2-3). Programs that solve more complex, real-world problems may generate and access hundreds, thousands, or millions of intermediate results during program execution.

Larger units of data such as customer records, transactions, and student transcripts also must be stored for present or future use. A user may need to store and access thousands or millions of data items. A large organization may need to storage and access trillions of data items. This data may be used by currently executing programs, held for future processing needs, or held as a historical record.

Programs also must be stored for present and future use. A simple program can contain thousands of instructions. Complex programs can contain millions or billions of instructions. A small computer may store thousands of programs, and a large computer may store millions of programs.

Storage requirements vary for intermediate processing results, data, and programs, depending on whether they are currently in use or being held for future use. Program execution speed is increased when intermediate results, currently needed data, and instructions are stored in the processor or in devices that can be accessed rapidly by the processor. Programs and data currently not needed can be stored on slower or more distant storage devices. Storage needs also vary depending on whether it is necessary to transport them physically from place to place, such as on a CD-ROM or floppy disk.

Input/Output Capability

Processing and storage capabilities are of little use if a computer cannot communicate with users or other computers. A computer's input/output devices must be flexible and encompass a variety of communication modes—sound, text, and graphics for humans and electrical or optical communication for other computers. A typical small computer has up to a dozen input/output devices, including a video display, keyboard, mouse, printer, and modem or network interface. A large computer can have many more input/output devices (for example, multiple printers and network interfaces) and may use devices of substantially more power and complexity than a smaller computer.

COMPUTER HARDWARE

Not surprisingly, computer hardware components parallel the processing, storage, and communication capabilities just outlined (see Figure 2-6). Computer hardware has four major functions, including:

▶ Processing—execution of computation, comparison, and other instructions to transform data inputs into data outputs

▶ Storage—storage of program instructions and data for temporary, short-term, and long-term use

▶ External communication—communication with entities outside the computer system, including users, system administrators, and other computer systems

▶ Internal communication—communication among internal and peripheral hardware components such as processors, disk drives, video displays, and printers

Figure 2-6 ▶

The primary functions of computer hardware

All computer systems include hardware to implement each function. However, each function is not necessarily implemented within a single device. For example, a computer processor implements both processing and storage functions, and external communication is implemented by many different hardware devices, such as keyboard, video display, modem, network interface device, sound generating device, and speakers.

Figure 2-7 shows hardware components of a computer system. The number, implementation, complexity, and power of these components can vary substantially from one computer system to another, but the functions performed by the components are similar.

Figure 2-7 ▶

Components of a
computer system

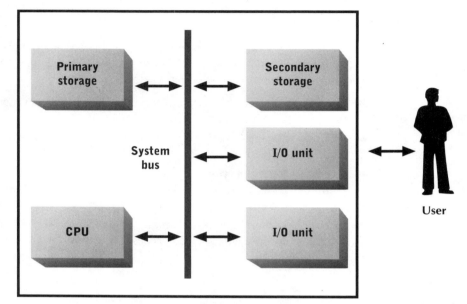

The **central processing unit** (**CPU**) is a general-purpose processor that executes all instructions and controls all data movement within the computer system. Instructions and data for currently executing programs flow to and from **primary storage.** **Secondary storage** holds programs that are not currently being executed, as well as groups of data items that are too large to fit in primary storage. Secondary storage can be composed of several different devices (for example, multiple hard disk drives, Zip drives, or tape drives), although only one device is shown in Figure 2-7.

Various **input output** (**I/O**) **units** implement external communication functions. Two I/O units are shown in Figure 2-7 though a typical computer has many more. Each hardware component is connected via the **system bus**. The system bus is the internal communication channel that connects all other hardware devices.

Central Processing Unit

Figure 2-8 illustrates the components of the central processing unit, including the:

► Arithmetic logic unit
► Registers
► Control unit

Figure 2-8 ►

Components of the central processing unit (CPU)

The **arithmetic logic unit** (**ALU**) contains electrical circuits that implement each instruction. Executing a math or logic instruction causes data (electrical signals) to flow through the appropriate circuitry. A CPU can implement dozens or hundreds of different instructions. Simple arithmetic instructions include addition, subtraction, multiplication, and division. More advanced computation instructions such as exponentiation and logarithms also can be implemented. Logic instructions include various forms of comparison (equal, greater than, less than) and other instructions that will be discussed in Chapter 4. The CPU contains a small number of internal storage locations called **registers**, which each can hold a single instruction or data item. Registers store data or instructions that are needed immediately and/or frequently. For example, two numbers that are about to be added together are each stored in a register. The ALU reads these numbers from the registers and stores the result of the addition in another register. Because they are located within the CPU, the contents of registers can be accessed quickly by the other CPU components.

The **control unit** has two primary functions:

► Control the movement of data to and from CPU registers and other hardware components

► Access program instructions and issue appropriate commands to the ALU

As program instructions and data are needed, they are moved from primary storage to registers by the control unit. The control unit examines incoming instructions to determine how they should be processed. Computation and logic instructions are routed to the ALU for processing. Data movement instructions to/from primary storage, secondary storage, and I/O devices are processed by the control unit itself.

System Bus

The system bus is the primary pathway for data transmission among hardware components. The capacity of this channel is a critical factor in the overall performance of the computer system. A powerful CPU needs a high-capacity system bus to keep it supplied with instructions and data from primary storage. Bus capacity also is critical to secondary storage and I/O device performance.

Primary Storage

Primary storage contains millions or billions of storage locations that hold program instructions that currently are being executed, as well as data currently being processed by those instructions. Primary storage also is referred to as **main memory,** or simply as **memory**. Program execution requires continual movement of both instructions and data between main memory and the CPU. Because the CPU is a relatively fast device, it is desirable to implement main memory using devices that are capable of rapid access.

In current computer hardware, main memory is implemented with silicon-based semiconductor devices commonly referred to as **random access memory (RAM)**. RAM provides the access speed required by the CPU and allows the CPU to read or write to specific memory locations. Unfortunately, RAM is relatively costly and this limits the amount of main memory that one can afford to include in a computer system.

Another problem with RAM is that it does not provide permanent storage. When the power is turned off, the contents of RAM are lost. This characteristic of RAM is called **volatility**. Any type of storage device that cannot retain data values indefinitely is said to be volatile. In contrast, storage devices that permanently retain data values are said to be nonvolatile. Due to both the volatility and limited capacity of primary storage, a computer system requires auxiliary means of storing data and programs over long periods of time.

Secondary Storage

Secondary storage refers to nonvolatile storage devices with large capacities. Secondary storage devices hold:

► Programs not currently being executed

► Data not needed by currently executing programs

► Data needed by currently executing programs that does not fit within available primary storage

Within a typical information system, the number of programs and amount of data is quite large. A typical computer system must have a large amount of secondary storage in comparison to its primary storage. For example, a microcomputer may contain 500 million primary storage locations and 50 billion secondary storage locations. Differences in the content and implementation of the various storage devices are summarized in Table 2-1.

Table 2-1 ►

Comparison of storage types

Storage type	Implementation	Contents
CPU registers	Very-high-speed electrical devices within the CPU	Currently executing instruction; a few dozen items of related data
Primary storage	High-speed electrical devices (RAM) outside of the CPU	Programs currently being executed; hundreds to millions of data items
Secondary storage	Low-speed electromagnetic or optical devices (for example, magnetic and optical disc)	Programs not currently being executed; billions to trillions of data items

Relatively slow devices are used to implement secondary storage. Using slow devices keeps the total cost of secondary storage within acceptable limits. The most common devices used to implement secondary storage are magnetic discs, optical discs, and magnetic tape. Magnetic disk storage provides relatively fast access, compared to optical disc and magnetic tape storage. Optical disc provides greater storage capacity than magnetic disk at reduced cost. Magnetic tape provides the slowest but cheapest method of storage.

Input/Output Devices

There is a great variety of input/output devices. The variety is the result of the many different forms of human-to-computer and computer-to-computer communication. From an architectural perspective, each I/O device is a separate hardware component attached to the system bus. The methods of communication between different I/O devices and other computer hardware devices are similar, but their internal implementation varies depending on the exact form of communication they support.

I/O devices can be classified broadly into human-oriented and computer-oriented communication devices. Examples of input devices for human use include keyboards, pointing devices such as a mouse or trackball, and voice recognition devices. The purpose of such devices is to accept input from a human (voice, touch, or physical movement) and convert that input into something the computer can understand (electrical signals). Output devices for human use include video displays, printers, plotters, and devices for speech and sound output. All of these devices convert electrical signals into a format that a person can understand, such as pictures, words, or sound. Computer-oriented input/output devices include modems and network interface units. These devices implement communication among computer systems, or between a computer and a distant I/O device.

COMPUTER SYSTEM CLASSES

Computer systems are available in many configurations that vary in CPU power, storage capacity, I/O capacity, number of simultaneous users, and intended application software. Computer systems can be loosely classified into the following categories:

► Microcomputer

► Minicomputer

► Mainframe

► Supercomputer

A **microcomputer** is a computer system designed to meet the information-processing needs of a single user. A microcomputer also can be called a **personal computer** (**PC**) or **workstation**. Examples of processing performed with microcomputers include word processing, computer games, and small-sized to medium-sized application programs, such as programs to compute an individual's income tax, perform budgeting for a home or small business, and compute payroll for a small business.

To some people, the terms personal computer and workstation are interchangeable. To others, the term workstation implies something more powerful than a typical microcomputer. This is especially true of microcomputers in scientific and engineering organizations. Processing tasks in such organizations require more computer hardware power than do typical business and home processing.

Examples of hardware-intensive tasks include complex mathematical computation, computer-aided design (CAD), and manipulation of high-resolution video images. Particularly with respect to processor speed and storage capabilities, the power of a workstation often is similar to that of a minicomputer. The overall design of a workstation is targeted toward a single-user operating environment, as opposed to the multiple-user orientation of minicomputers.

A new type of microcomputer emerged in the late 1990s. A **network computer** is a microcomputer with minimal secondary storage capacity and little or no installed software. An ordinary microcomputer loads its operating system and application software from secondary storage devices directly attached to the computer. A network computer establishes a connection to a server when it is powered on. All operating system and application software is supplied by the server via a computer network. Network computers are slightly less expensive than ordinary microcomputers. Server-based software installation and configuration makes it easier to reconfigure network computers than ordinary microcomputers.

A **minicomputer** is designed to provide information processing for multiple users and to execute many application programs simultaneously. Minicomputers are designed so that up to 100 users can interact with the computer system at the same time. Supporting multiple users and programs requires fairly extensive capabilities for processor speed, storage, and input/output. It also requires more sophisticated system software than is typically installed on microcomputers.

There are many ways of supporting multiple users. Multiple users may be connected to a computer system with relatively simple I/O devices such as video display terminals. In this case, all processing, data storage, and network communication is handled by the shared computer system. Multiple users also can be supported by sharing one or more resources (for example, printers, databases, and Web pages). A minicomputer can support several dozen people using video display terminals or a few hundred simultaneous accesses to shared resources.

A **mainframe** computer system is designed to handle the information processing needs of a large number of users and applications. A mainframe can support hundreds of people using video display terminals, as well as executing hundreds or thousands of programs at one time. A mainframe also can support thousands of simultaneous accesses to shared resources. A typical use may involve 100 users entering customer orders, several programs generating periodic reports, dozens of users querying the contents of a large corporate database, and an operator making backup copies of disk files—all at the same time.

The feature that distinguishes mainframes from other computer classes is the ability to move large quantities of data quickly from one place to another. Data can be moved quickly among up to a dozen CPUs, several dozen secondary storage devices, and hundreds or thousands of users connected via a network. Fast

CPUs and large amounts of primary and secondary storage also are required, but a mainframe primarily is optimized for data movement.

Business
Focus
Bigger Isn't Always Better

In 1952, computer scientist H. A. Grosch asserted that computing power, as measured by millions of instructions per second (MIPS) is proportional to the square of the cost of hardware. According to Grosch, large and powerful computers always will be more cost efficient than smaller ones. The mathematical formula that describes this relationship is called **Grosch's Law**. For many years, computer system managers pointed to this law as justification for investments in ever larger computers.

Today, the emergence of distinct classes of computers, expanded abilities to change the configuration and capabilities of a given computer, an increase in software costs relative to hardware costs, a vast expansion in the size of computer databases, the widespread adoption of graphical user interfaces, and the widespread use of computer networks have invalidated Grosch's Law.

In Grosch's time there was basically only one class of computer—the mainframe. Purchasing larger machines within a single class offers more CPU power for the money and thus Grosch's law appears to hold. But if all classes of computers are considered as a group, Grosch's law no longer holds. *The cost of CPU power actually increases on a per-unit basis as computer class increases.*

Also, CPU power is an incomplete measure of real computer system power. For example, minicomputers and mainframes differ little in raw CPU power, but they differ substantially in their ability to store, retrieve, and move large amounts of data.

Cost-efficiency comparisons are even more difficult when networked computers are considered. In Grosch's time, the hardware and software to support computer networking did not exist. But it now is possible to network multiple machines and operate them as if they were a single larger machine. Many government and business organizations have constructed networked systems of microcomputers and minicomputers, and avoided smaller numbers of mainframes due to the inherent cost efficiencies.

Within a given class or model of computer system, a deliberate choice can be made to cost effectively tune the performance of the system to a particular application or class of applications. For example, in a transaction-processing application, the system may be configured with fast secondary storage and I/O subsystems while sacrificing CPU performance. The same system may be cost effectively tailored to a simulation application by providing a large amount of high-speed main memory, a CPU with a high computational capacity, and a high-capacity communication channel between memory and the CPU.

Secondary storage and I/O performance would be far less important, and less costly alternatives could be used.

Another difficulty in applying Grosch's Law today is the reality of costly software. Developing software requires a substantial commitment of labor and many other resources. Coupled with the decreasing cost of computer hardware, this has made the cost of computer hardware a less significant factor in the total cost of an information system. Although hardware costs should be managed effectively, a user or manager is far more likely to maximize the cost effectiveness of the system by concentrating on its software components.

A **supercomputer** is designed for one purpose—to quickly perform many mathematical computations (billions, trillions, or more). Supercomputers are used for computationally intensive applications such as simulation, 3-D modeling, weather prediction, computer animation, and real-time analysis of large databases. Such tasks require multiple CPUs with the highest possible computational speed. Storage and communications requirements also are extremely high, but are secondary to computational speed. Supercomputers employ the very latest (and expensive) computer technology.

The term **server** can describe computers as small as microcomputers and as large as supercomputers. The term does not imply any minimum set of hardware capabilities, rather, it implies a specific mode of use. Specifically, it implies a computer system that manages one or more shared resources such as file systems, databases, Web sites, printers, and high-speed CPUs, and allows users to access those resources over a local or wide area network.

Server hardware capabilities depend on the resources being shared and the number of simultaneous users. For example, an ordinary microcomputer may be more than sufficient for sharing a file system and two printers among a dozen users on a local area network. More demanding resources, like large databases accessed by thousands of users, may require a mainframe server. Supercomputers sometimes are used as servers to augment the computational capabilities of workstations used for computer aided design and animation.

Table 2-2 summarizes the configuration and capabilities of each class of computer system. Each class is represented by a typical model available in 2000. Specifications and performance are in a constant state of flux. Rapid advances in computer technology lead to rapid improvements in performance, as well as a redefinition of both the classes themselves and the specifications of computers within each class. There also is a technology pecking order. That is, the newest and most expensive hardware and software technology usually appears first in supercomputers. As experience is gained and costs decrease, those technologies move downward to the less powerful and less costly computer classes.

Table 2-2 ▶

Representative
products in
various
computer
classes (2000)

Class	Typical Product	Typical Specifications	Approximate Cost	Processor(s) Speed
Micro	IBM PC 300PL	128 million main memory cells 20.4 billion disk storage cells Single CPU Single user	$2,000	300 million instructions per second
Workstation	IBM RS/6000 43P Model 150D	256 million main memory cells 18 billion disk storage cells Single CPU High-performance video I/O Single user	$13,000	750 million instructions per second
Mini	IBM pSeries 640	1024 million main memory cells 54 billion disk storage cells 2 CPUs Tape drive	$25,000	1.5 billion instructions per second
Mainframe	IBM S/390 X67	32 billion main memory cells 100 billion disk storage cells 6 CPUs 32 high-capacity I/O channels Up to 1000 interactive users	$750,000	2.5 billion instructions per second
Super	IBM RS/6000 SP	64 billion main memory cells 1 trillion disk storage cells 64 CPUs No interactive users	$2,000,000	50 billion floating point operations per second

Another factor that blurs the lines among various classes of computers (particularly for microcomputers, workstations, and low-end minicomputers, and for high-end minicomputers and low-end mainframes), is the modern capability to configure subsystems such as processors, primary and secondary storage, and input/output of varying power and cost. Although microcomputers have been defined as single-user machines, high powered versions of these machines often can be used for multiuser applications simply by upgrading the capacity of a few of the subsystems and utilizing more powerful system software. Similar possibilities exist for upgrading the performance of a minicomputer to that of a mainframe.

Technology
Focus

IBM POWER-Based Computer Systems

International Business Machines (IBM) first introduced the POWER CPU architecture in 1990. With this CPU, IBM broke away from its previous designs and laid the foundation for the next decade of computer system development. Early versions of the PowerPC were used in many Apple Macintosh systems though no POWER CPUs are used in current microcomputers. IBM has incorporated POWER CPUs in current systems ranging from workstations to supercomputers.

The POWER architecture has been through several generations and variations since 1990. The PowerPC RS64 IV and the POWER3-II are the latest POWER processors. Both processors incorporate IBM's latest microprocessor technologies, including:

► 0.18 micron fabrication technology

► Copper chip wiring

► Large on-chip and off-chip memory caches

► Simultaneous execution of multiple instructions

Sample computer systems based on POWER CPUs are summarized in Table 2-3. The RS/6000 44P Model 170 is a high-end graphical workstation intended for 3-D modeling and visualization. Typical applications include computer-aided design, computer-aided animation, and data visualization. The system incorporates sophisticated graphical I/O hardware to speed image display and refresh.

Table 2-3 ►

IBM system configurations (year 2000) for POWER CPUs

Model	System Type	Number of CPUs	CPU Speed	Memory Capacity	Disk Capacity
RS/6000 44P Model 170E	Workstation	1	400 MHz	2 billion bytes (GB)	36 GB
RS/6000 Model M80	Minicomputer	4	500 MHz	16 GB	Variable
pSeries 680	Mainframe	24	600 MHz	96 GB	875 GB
ASCI White	Supercomputer	8192	400 MHz	6 trillion bytes (TB)	160 TB

The RS/6000 M80 is a minicomputer based on two-to-eight PowerPC RS64 III processors. Internal disk capacity is limited to 36 GB but the system can be expanded well beyond those limits with external storage units. The M80 is primarily intended as a server for computationally intensive applications such as factory floor control and data analysis.

The pSeries 680 is a mainframe with 6-to-24 PowerPC RS64 IV processors. Typical applications include large-scale Web and e-commerce servers, large database servers, and data mining and analysis. The system primarily is designed for applications that require large amounts of data movement and computation.

ASCI White is an RS/6000 SP supercomputer designed and built for the U.S. Department of Energy. The system was installed at Lawrence Livermore National Laboratories in 2000 and was the fastest computer in the world at that time. ASCI White can perform 12.3 trillion numerical computations per second. It is used for simulation applications in nuclear weapon design and testing.

Business

Focus

Do We Need a Mainframe?

Bauer Industries (BI) is a manufacturer and wholesale distributor of jewelry-making machinery and components. The company has 200 employees and approximately 2500 products. The current inventory of automated systems is small for a company of its size. Applications include order entry, inventory control, and general accounting functions like accounts payable, accounts receivable, payroll, and financial reporting. BI's applications currently run on an older Compaq minicomputer (VAX 6000). Dedicated character-based video display terminals are used to interact with application programs.

BI has grown rapidly over the last decade. Increased transaction volume has strained existing computer hardware capability. Typical order-entry response time is now 30 to 60 seconds compared with 2 to 5 seconds a few years ago. The order-entry system fully utilizes memory and CPU capacity during daytime hours. All non-essential processing has been moved to nighttime hours to minimize order-entry response time. The VAX 6000 currently is configured with the maximum possible disk space and memory.

Bauer Industries management realizes the seriousness of the situation, but they are uncertain how to address it. There are additional factors complicating the decision, including:

► An expected four-fold increase in business over the next 10 years

► Relocation to new facilities within the next year

► Plans to place the product catalog on the Web and move to Web-based ordering

➤ A desire on the part of some members of management to substantially increase and modernize automated support of operating and management functions

Management has identified three viable options to address the current problem:

1. Purchase a used VAX 6000 (they are no longer manufactured), and partition the application software between the two computer systems.

2. Accept an offer from Compaq to upgrade to an AlphaServer 4000 series computer system.

3. Develop a modern client-server-based architecture consisting of workstations or network computers, one or more medium-sized servers, and a high-speed local area network.

Option 1 is least disruptive to the status quo but has two obvious disadvantages. The first is the use of outdated hardware. Parts for VAX 6000s only are available from abandoned machines. The second problem is data overlap among application programs. The current operating and application software will not support access to data stored on one of the machines by applications executing on the other machine. Thus, data will have to be stored on both machines and periodically synchronized to reflect updates made by different application programs. Estimated two-year costs are $75,000 for hardware and $100,000 for labor.

Option 2 theoretically preserves the existing application software and allows some hardware, such as video display terminals, to be reused. It also provides sufficient capacity to operate current software at many times the current transaction volume. However, the new hardware requires a new version of the VMS operating system. The new operating system version was designed to support applications executing under older versions on VAX hardware. But management has heard of many problems in migrating application software. Estimated two-year costs are $350,000 for hardware and $150,000 for labor.

Option 3 is the most expensive and the biggest departure from current operating methods. Migration problems are similar to option 2 if one or more AlphaServers are acquired. Existing video display terminals can be used initially and gradually replaced with workstations or network computers. Option 3 appears to hold the most promise for implementing a new generation of application software. But the company has little experience or expertise in building or maintaining client-server application software. Estimated (guesstimated?) two-year costs are $150,000 for hardware and $400,000 for labor (including hiring or contracting three new IS staff memebers for two years).

Questions

1. What problems would you anticipate within two-to-five years if option 1 is implemented? Option 2? Option 3?

2. Management historically has opted for lowest-cost solutions. How can you justify the higher costs of options 2 and 3 to them?

3. Which alternative would you recommend to management? Why?

THE ROLE OF SOFTWARE

The primary role of software is to translate user needs and requests into computer instructions that, when executed, will produce a result that satisfies the need or request. People usually ask computers to perform relatively complex tasks such as "generate and print my company's income statement," "spell check my term paper," or "give me a report on weekly sales" (see Figure 2-9).

Figure 2-9 ▶

The role of software as a translator between user requests and CPU instructions

The need or idea that motivates a request for computer processing is stated at a relatively high (abstract) level. The individual tasks that are performed (instructions that are executed) to satisfy a request are very specific and low level. Because CPU instructions are very small units of work, the number of instructions that must be executed to satisfy a processing request is large.

Software is complex because it performs a complex translation process that bridges two gaps:

▶ Human language to machine language

▶ High-level abstraction to low-level detail

People think, speak, and write in their native language, which has specific vocabulary, grammar, and semantics. Computers communicate in binary languages that consist of zeros and ones, which are grouped together by specific grammatical rules to form meaningful statements or instructions.

Software Types

A program is a set of detailed instructions that directs a computer to perform a complex task. Much like a recipe, a program can be written down, stored, and retrieved when needed. Computer programs usually are stored in files on secondary storage devices. When needed, they are retrieved by name, copied into primary storage, and then executed by the CPU.

An **application program,** or **application software,** is a stored set of instructions for responding to a very specific request, much as you might access a specific recipe to prepare a particular dish. Examples of application programs in a system devoted to payroll processing include programs to print checks, enter new employee information, and produce annual tax reports.

Other programs can be created and stored for more general purposes. These **utility programs** contain instructions for performing relatively basic tasks that are a necessary component of many application-oriented tasks, such as printing a text file, opening a new window on a video display, and copying a file from one disk to another. Although these tasks are relatively basic, they can still require thousands of CPU instructions to complete.

System software is a collection of programs that are used to:

► Implement utility functions needed by many application programs

► Allocate computer resources to application programs

► Manage computer resources

The most significant distinction between application and system software is specificity of purpose and use. Application software is targeted to specific information-processing tasks like generating customer credit card bills. System software is targeted to general-purpose tasks that span many application programs and users.

Most application software is intended for end user interaction. In contrast, most system software does not interact with end users. Most utility programs used by application software operate invisibly in the background. Programs that allocate computer resources among application programs also operate in the background. System administrators may set policies that govern allocation rules and methods, but the allocation programs do not interact directly with users or system administrators during execution.

Utility programs used to manage computer resources can be used directly by either end users or system administrators. Examples of user-oriented management programs include those that copy files and directories, install software on personal computers, and create and manage personal computer connections to Internet service providers. Examples of management programs intended for system administrators include those that create and modify user accounts, install hardware devices on servers, and implement computer security policies.

Figure 2-10 illustrates the interaction between the user, application software, utility programs within system software, and computer hardware. User input and output is limited to communication directly with the application program. Application programs, in turn, communicate with system software to request basic services (for example, opening a file or creating a new window). System software translates a service request into a sequence of machine instructions, passes those instructions to the hardware for execution, receives the results, and passes the results back to the application software.

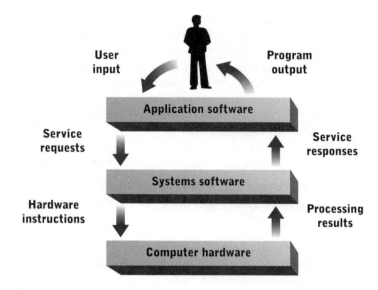

Figure 2-10 also illustrates what commonly is called a layered approach to software construction and operation. Application software is layered above system software, which is in turn layered above the hardware. A key advantage of the layered approach to software is that users and applications programmers do not need to know the technical details of computer hardware. Instead, they interact with hardware through a set of standardized service requests. Those requests then are translated into actual hardware instructions that are needed to satisfy the requests. Knowledge of the machine's physical details is embedded into the system software and hidden from the user and application programmers. This advantage is commonly referred to as **machine independence** or **hardware independence**.

System Software Layers

Figure 2-11 shows a more detailed view of software layers and their relationship to hardware functions. System software functions have been divided into four layers:

► System management—utility programs used by end users and system administrators to directly manage and control computer resources

► System services—utility programs used by system management and application programs to perform common functions

► Resource allocation—utility programs that allocate hardware and other resources among multiple users and programs

► Hardware interface—utility programs that control and interact with individual hardware devices

Figure 2-11 ►

Software layers
and their
relationship to
hardware
functions

The principle of software layers applies not only to the relationship between application and system software but also to the relationship among various components of system software. Machine independence is achieved by placing all hardware interface functions within a single system software layer. In theory, system software can be modified to operate on new computer hardware by modifying only the hardware interface layer. In practice, machine independence is not always that simple, but a more detailed discussion of that topic is included in a later chapter.

Resource allocation is an important, but relatively invisible, function of system software. A modern computer system can support multiple users and hundreds or thousands of programs executing simultaneously. Under such conditions, resources such as files and I/O devices cannot always be immediately provided to a program or user. System software balances the competing demands of users and processes for access to resources. It does this in such a way that users and programs do not interfere with one another, while also ensuring that all users and programs receive the resources they need.

Operating Systems

An **operating system** is a collection of programs that provides utility functions needed by users and application programs, allocates computer resources among multiple users and application programs, and controls access to computer hardware.

Note that this definition is very similar to the earlier definition of system software. Also, note that the definition spans a very large range of functions. Examples of operating systems include Windows, UNIX, Linux, and OpenVMS. Operating systems are the most important and pervasive type of system software, but they are not the only type of system software.

An operating system provides many general-purpose functions needed by most users, administrators, and application programs. But some system software functions are not needed as often as others. For example, most application programs need to access files, so all operating systems include file management and access utilities. A small number of application programs need to access large databases. Because this need is not pervasive, few operating systems include utilities to manage and access large databases. Instead, a separate type of system software called a database management system is used for that purpose.

Operating systems include utility programs that meet the needs of most or all users, administrators, and application programs. Functions in most operating systems include:

▶ Program storage, loading, and execution
▶ File manipulation and access
▶ Secondary storage management
▶ Network and interactive user interfaces

Some operating systems include additional functions. For example, the OS/400 operating system used with IBM AS/400 computer systems includes database management functions. The Windows 2000 Server operating system includes Web server and FTP server software.

Application Development Software

Since their inception, computers have been used to automate tasks normally performed by people. Early programmers developed application programs without assistance from the computer itself. They wrote programs that consisted entirely of binary CPU instructions. There were no intermediaries such as compilers, interpreters, programming languages, or operating system services. As computers became more powerful, it became possible to write much larger application programs; but such programs were too complex to develop without automated assistance. **Application development software** describes programs used to develop other programs. Though the term implies otherwise, most application development programs also can be used to develop system software or other application development programs.

Earlier in the chapter, software complexity was said to result from the need to bridge the gaps between human-oriented and machine-oriented language, and between abstract, high-level statements of need and the low-level machine instructions that satisfy those needs. The same challenge of translating human intelligence into computer instructions applies to the process of developing programs.

A **program translator** is a program that translates instructions in a **programming language** into CPU instructions. Examples of programming languages include FORTRAN, Java, C++, and BASIC. A program translator doesn't just translate one language into another, it also translates high-level program instructions into detailed groups of CPU instructions. Modern programming languages allow programmers to express complex processing tasks in a single statement or instruction. For example, consider the BASIC program shown earlier in Figure 2-3. Hundreds or thousands of CPU instructions must be executed to perform that task. Programming languages free programmers from the need to individually specify every CPU instruction.

Program translators are not the only type of application development software. Other types include:

▶ **Program editors**—writing tools, similar to word processors, but customized for writing programs instead of documents

▶ **Debugging tools**—tools that simulate program execution and allow programmers to trace errors

▶ **System development tools**—tools that allow systems analysts and designers to develop models of entire information systems; models then are used as the starting point for developing application programs

Modern application development software provides a toolkit used by developers of application and system software. Like any tool, application development software increases the productivity of programmers and the quality of what they produce. The complexity of modern software demands application development tools of at least equal complexity.

ECONOMICS OF SYSTEM AND APPLICATION DEVELOPMENT SOFTWARE

The system and application development software we use today bears little resemblance to software used in the past. Programming languages, program development tools, and operating systems were not developed until the late 1950s, and were primitive in comparison to their modern equivalents.

Why is modern software so advanced? The answer derives from three economic facts of computer hardware and software:

▶ System software requires hardware resources

▶ The cost per unit of computing power has decreased at nearly an exponential rate

▶ Software is more cost effective when it is reused many times

These facts underlie a shifting tradeoff between the cost of developing and supporting application programs and the cost of computer hardware resources.

In the early years of computers, application development was mostly a manual process. Programmers developed programs using binary CPU instructions written with pencil and paper. They desk-checked their program and, when it looked correct, they used a plugboard, paper tape punch, or paper card punch machine to convert the written zeros and ones into something the computer could read. Then they tested the program to see if it worked as they expected. If too many rounds of testing and correction were required, they were criticized for using too much valuable computer time. Computer hardware simply was too expensive to have CPU time wasted debugging software.

As the cost of computer hardware decreased, the economic balance between programming labor and computer hardware resources shifted. Tedious programming using binary CPU instructions and extensive manual checks of code became more expensive than writing and executing programs with automated assistance. The trends toward greater automated support for application development and computer operation, as well as software reuse, had begun.

When a programmer uses a programming language, application development software, and a computer to develop a new program, the programmer is substituting software and hardware resources for the labor required to write binary CPU instructions. As programming languages and development tools become more powerful, application development software becomes more complex and consumes ever greater quantities of computer hardware resources. Those hardware resources rapidly become cheaper than the extra labor resources that would be needed if the application development tools were not employed. As you can see in Figure 2-12, the balance of hardware and software cost has shifted over time because of the declining cost of hardware and the emergence of system software. Hardware typically is the cheapest component of current information systems, and system and application software are nearly equal components in the total cost.

Figure 2-12 ▶

The change over time in the relative cost of hardware, application software, and system software in a typical information system

Both system and application development software also increase economic utility by promoting software reuse. Each system service is a utility program, and application software reuses those programs each time it makes a service request. Each time a program translator translates a generic instruction, such as Print Customer_Name into the equivalent set of CPU instructions, it is reusing the small utility program that implements a PRINT statement. Modern application development tools provide more overt examples of software reuse by providing libraries of utility programs and generating program skeletons from templates.

Technology
Focus

Intel CPUs and Microsoft Operating Systems

IBM introduced the first mass-market microcomputer in 1981. The computer contained an Intel 8088 CPU and ran the Microsoft MS-DOS 1.0 operating system. By the late 1980s, Intel and Microsoft had become the standard-setters for microcomputers, and they have continued in that role.

Microsoft has taken an increasingly active role in defining the scope of micro-computer software. Microsoft also pushed microcomputer operating system functionality into new areas, such as graphical user interfaces, distributed soft-ware, and Internet browsing. Each new version of Windows has become the de facto standard for microcomputer operating systems.

Early versions of the MS-DOS operating system did little to support the user interface of application programs. Application program developers implemented their own interface styles, resulting in a wide variation among application pro-gram interfaces. Microsoft brought some consistency to the look and feel of application programs with the introduction of the Windows operating system. Windows provides application program services such as windows, menus, font management, and interaction with high-resolution video displays. Over the years Microsoft has tightened user interface standards. Currently, application programs must pass a rigorous set of tests performed by Microsoft to receive certification as Windows compatible.

Figure 2-13 shows a time line for Intel CPUs and Microsoft operating system versions. Notice the relationship between operating system releases and new CPU releases. Major advances in operating system function typically have lagged behind the introduction of supporting CPU technology by approximately three years. This is because software development is a long and labor-intensive process. Also, operating system development depends on hardware other than the CPU, such as video displays, network communication, primary storage, and secondary storage, that may be slower to develop.

Figure 2-13 ▶

The timing of
new CPU
introductions by
Intel and new
operating system
introductions by
Microsoft

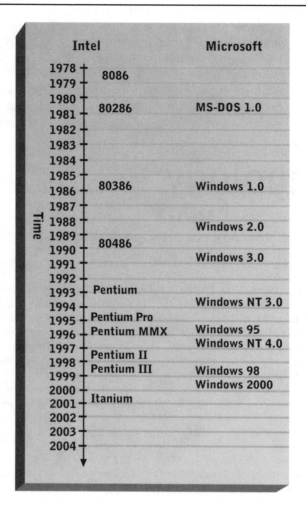

The development of Windows is a good example of how software development depends on hardware technology. As Intel CPUs became more powerful, the Windows operating system became more complex.

▶ 80286—supplied the needed processor power for early versions of Windows, which was too complex an operating system to run on the original IBM PC platform (the Intel 8086). Several enhancements to the PC platform, such as a wider/faster system bus and the EGA video display standard, provided the rest of the required hardware capabilities.

- 80386—Although the capabilities of the 80286 were sufficient to run early versions of Windows, they were not necessarily optimal; application programs frequently ran slower than on the 80286. Many users switched to computers based on the 80386 and newer supporting technologies, such as VGA displays and 32-bit communication channels. The 80386 replaced earlier approaches to memory addressing that effectively had limited application programs to no more than 640 kilobytes of memory, included hardware support for executing multiple programs simultaneously, simplified the partitioning of primary storage among programs, and provided mechanisms to protect programs from interfering with one another. It also simplified the implementation of operating system functions such as task switching and virtual memory management. Windows 3.0 was the first operating system to take advantage of the new capabilities of the 80386.

- 80486—The 80486 soon followed with increased processing power, and application programs became larger and more complex. Windows 95 was developed to more fully take advantage of the chip's capabilities. Although Windows 95 did not require any specific capabilities of CPUs beyond the 80386, it did benefit greatly from increases in raw processing power, and fueled a rapid move by users of 80386-based and 80486-based computers to Pentium-based computers.

- Pentium—Windows NT implemented much more sophisticated versions of the memory and hardware management capabilities found in Windows 3.x and Windows 95. It also was designed to run on hardware platforms other than IBM-compatible PCs. Portability across hardware platforms required increased CPU speed and faster memory access. The 80486 barely was adequate, but the Pentium was more powerful. Later generations of the Pentium and other hardware improvements, such as higher-speed system buses and RAM, were used by Microsoft to add many new features to Windows NT, resulting in Windows 2000.

In the push-pull relationship of hardware power and software capability, hardware is the driving force. To a great extent CPU makers can safely assume that their product improvements will quickly prove useful even if they are not immediately needed by most users. In fact, software makers often develop software that pushes the limits of current hardware, confident that hardware manufacturers will provide more power. More demanding software in turn fuels users' drive to acquire new and more powerful machines, and the cycle continues. Competition among software developers guarantees continual and fairly rapid assimilation of new hardware technology.

COMPUTER NETWORKS

Computers once were processing islands; people physically carried data and programs to them. The widespread linking of computers through networks and the pervasive impact of the Internet have forever changed the landscape of computing. Today, a computer isn't considered useful if it doesn't have the ability to communicate with virtually every other computer on the planet.

A **computer network** is a set of hardware and software components that allows information, software, and hardware resources to be shared among multiple users and computer systems. A computer network also supports many types of personal communication, such as e-mail, newsgroups, or collaboration on projects. The number and complexity of computer network functions and components has grown as network technology has matured and become more pervasive.

Network functions simply are extensions to the hardware and software functions discussed earlier in the chapter (see Figure 2-14) Those extensions allow a computer system to interface with a physical network and external resources managed by other computer systems.

Figure 2-14 ▶

Computer network functions and their relationship to computer hardware and software

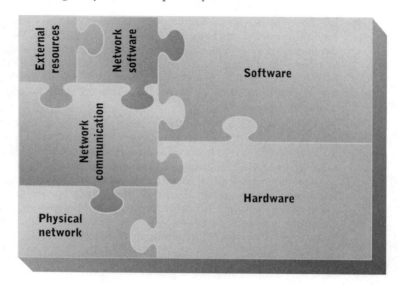

External Resources

A key to understanding computer networks is to understand their primary purpose —allowing one computer to use resources on another. Each computer on a network has access not only to its own local resources, but to the resources of other computers on the network, called distributed resources. Much of the complexity of modern networks arises from the huge quantity of distributed resources and the difficulties inherent in finding, accessing, and managing those resources.

In the early days of networks, the only significant distributed resource type was raw data. As networks have matured and expanded, so have the types of resources that can be accessed. Data now can be accessed in many different forms, including text files, sound and video, databases, and Web pages. Programs can be accessed and executed on remote machines or retrieved via the network and executed locally. Hardware devices of all types also can be shared across a network.

Network Software

Recall that a key function of system software is allocating resources among multiple users and programs. Allocation is simple when a single user and program access local resources. Allocation is more complex when multiple users or programs compete for hardware and other resources on a single computer system. System software complexity, however, is minimized by the fact that all of the resources are managed locally.

Allocating and accessing resources is very complex when a user or program can request resources that are not present on the local computer system and also are not managed by locally installed system software. In that case, system software must:

► Find the requested resources on the network

► Negotiate resource access with distant resource allocation software

► Receive and deliver the resources to the requesting user or program

If a computer system makes its local resources available to other computers, then system software must implement an additional set of functions to:

► Listen for resource requests

► Validate resource requests

► Deliver resources via the network

In essence, system software plays two roles in each network resource access—request and response. Many computer systems fill both roles and must implement both system software functions.

System software implements the intelligence needed to make and respond to external resource requests. Most operating systems include support for both functions. Specialized response functions often are provided as optional operating system components or as stand-alone system software packages. Examples include Web server software, e-mail distribution software, and Internet security software.

Network Communication and the Physical Network

A computer system requires at least one hardware device to implement its connection to a network. Network communication devices were once exotic and expensive but they now are commonplace. Network communication devices differ from I/O devices used to communicate with people in two important ways. First, they usually are simpler because they do not need to convert data represented electronically into another form. Second, they should support communication at relatively high speeds so access to external resources is not significantly slower than access to local resources.

Each computer system's network communication hardware is attached to a physical network. The physical network is a complex combination of communication protocols, methods of data transmission (cables and/or wireless), and network hardware devices (network interface cards in each connected computer or device, hubs, routers, and other devices to make the physical data connections, as well as networked printers or other resources). The physical network can be implemented with a dizzying array of technologies and architectural approaches, which will be discussed in more detail later in the book.

SUMMARY

► A computer is an automated device for performing computational tasks. It accepts input data from the external world, performs one or more computations on the data, and then returns results to the external world. Early computers had extremely limited capabilities and were mechanical devices. Modern computers have more extensive computational capabilities and are implemented with a combination of electrical and optical devices. The advantages of electrical and optical implementation include speed, accuracy, and reliability.

► Computer capabilities include processing, storage, and communication. A computer system contains a general-purpose processor that can perform computations, compare numeric data, and move data among storage locations and I/O devices. A command to the processor to perform a function is called an instruction. Complex tasks are implemented as a sequence of instructions called a program. Computer system behavior can be changed by using a different program.

► A computer system consists of a central processing unit (CPU), primary storage, secondary storage, and input/output (I/O) devices. The CPU performs all computation and comparison functions and directs all data movement. Primary storage holds programs and data that are currently in use by the CPU. Primary storage generally is implemented by electrical devices called random access memory (RAM). Secondary storage consists of one or more devices with high capacity and relatively low cost and access speed. Secondary storage holds programs and data that are not currently in use by the CPU. I/O devices allow the CPU to communicate with the external world. They may allow communication with humans or with other computers.

► A computer system can be classified as a microcomputer, minicomputer, mainframe, or supercomputer. Microcomputers are designed for use by a single user. Minicomputers and mainframes are designed to support many programs and users simultaneously. Mainframe computers are designed for large amounts of data access and movement. Supercomputers are designed to perform large amounts of numeric computation very quickly.

► The role of software is to translate user processing requests into machine instructions. The two primary types of software are application software and system software. Application software consists of programs that satisfy specific user processing needs such as payroll calculation, accounts payable, and report generation. System software consists of utility programs that are designed to be general in nature and can be used many times by many different application programs and/or users.

▶ An operating system is the most important type of system software on most computer systems. An operating system provides administrative utilities, utility services to application programs, resource allocation functions, and direct control over hardware. System and application development software reduce the cost of developing and deploying application programs, and allow organizations to substitute relatively cheap computer hardware for relatively expensive labor.

▶ A computer network is a set of hardware and software components that permits information, software, and hardware resources to be shared among multiple users and computer systems. Computer networks contain software and hardware components. Networking software allows a computer system to find and retrieve resources on other computer systems or to respond to requests from other computer systems for access to local resources. Network hardware implements direct communication among computer systems.

▶ In this chapter you've learned about the basic elements of computer system architecture—hardware, software, and networks. We've introduced many new terms and concepts but we haven't discussed them in detail. You'll get a chance to digest and expand your knowledge of these topics in later chapters.

▶ Figure 2-15 shows key computer system functions that have been covered so far, and identifies the chapter(s) that explore each function in depth.

Figure 2-15 ▶

Topics covered in the rest of the textbook

Key Terms

algorithm
application software
application development
  software
application program
arithmetic logic unit (ALU)
branches
central processing unit (CPU)
computer network
condition
control unit
debugging tools
executing
formula
general-purpose processor
Grosch's Law
hardware independence

input output (I/O) units
instruction
jumps
logic instructions
machine independence
main memory
mainframe
memory
microcomputer
minicomputer
network computer
operating system
personal computer (PC)
primary storage
processor
program
program translator

program editors
programming language
quantum measurement problem
qubit
random access memory (RAM)
registers
secondary storage
server
special-purpose processor
supercomputer
system development tools
system bus
system software
systems architecture
utility program
volatility
workstation

Vocabulary Exercises

1. A(n) _____ generally supports more simultaneous users than a(n) _____. Both are designed to support more than one user.

2. A(n) _____ is a storage location implemented within the CPU.

3. A problem-solving procedure that requires the execution of one or more comparison and branch instructions is called a(n) _____.

4. The term _____ describes any storage device that holds its data content indefinitely.

5. A(n) _____ is a computer that manages shared resources and allows other computers to access them through a network.

6. A program that solves a(n) _____ requires no branching instructions.

7. The primary components of a CPU are the _____, _____, and a set of _____.

8. Primary storage also can be called _____ and generally is implemented using _____.

9. A set of instructions that is executed to solve a specific problem is called a(n) _____.

10. A(n) _____ typically is implemented with the latest and most expensive technology.

11. The _____ is the plumbing that connects all computer system components.

12. _____ or _____ allows computer system hardware to change without changing application or service layer programs.

13. Resource allocation and direct hardware control is the responsibility of a(n) _____ .

14. _____ software is general purpose. _____ software is specialized to a specific user need.

15. A(n) _____ is a set of hardware and software components that enables information, software, and hardware resources to be shared among multiple users and computer systems.

Review Questions

1. What similarities exist among mechanical, electrical, and optical methods of computation?

2. What shortcomings of mechanical computation were addressed by the introduction of electrical computing devices?

3. What shortcomings of electrical computation will be addressed by the introduction of optical computing devices?

4. What is a CPU? What are its primary components?

5. What are registers? What is/are their function(s)?

6. What is main memory? In what way(s) does it differ from registers?

7. What are the differences between primary and secondary storage?

8. How does a workstation differ from a microcomputer?

9. How does a supercomputer differ from a mainframe computer?

10. What class(es) of computer system(s) normally are used to implement a server?

11. What is Grosch's Law? Does it hold today? Why or why not?

12. How can a computer system be tuned to a particular application?

13. What characteristics differentiate application software from system software?

14. In what ways does system software make the development of application software easier?

15. In what ways does system software make application software more portable?

16. What are the primary components of an operating system?

17. Why has the development of system software paralleled the development of computer hardware?

18. List at least five types of resources that can be shared among computers on a local area or wide area network.

19. Describe the relationship between the resource allocation and management functions of system software and external resources accessible via a network. What system software functions must be provided to access external resources?

Research Problems

1. Obtain a catalog or access the Web site of a major distributor of microcomputer equipment such Computer Discount Warehouse (www.cdw.com). Assume that you have a budget of $2,500 to purchase workstation hardware, exclusive of peripheral devices such as a scanner or printer. Select or configure a system that provides optimal performance for the following types of users:

   A home user who uses word-processing software such as Microsoft Word or Corel WordPerfect, a home accounting package such as Quicken and/or TurboTax, and children's games.

   An accountant who uses office software such as Microsoft Office, statistical software such as SYSTAT, and who regularly downloads large amounts of financial data from a corporate server for statistical and financial modeling.

   An architect who uses typical office software and computer-aided design software.

   Pay particular attention to the adequacy of CPU power, disk space, and I/O capabilities.

2. Assume that you have been asked to recommend a computer system to use as a server for a 250-person office. The server must provide file access of approximately 10 GB, access to several shared printers, access to the Internet (news, e-mail, and Web site services), and database services (database size is approximately 50 GB) to a variety of application programs. Assume further that Microsoft products such as Windows 2000 Server and Microsoft SQL Server will be the primary software used on the server.

   Obtain product information from IBM (www.ibm.com), Compaq (www.compaq.com), and Dell Computer Corporation (www.dell.com). Determine which of their products best meet the server requirements.

3. Table 2-2 probably will be out of date before this text is in print. Access the manufacturer's Web site (www.ibm.com) or contact a sales representative and update Table 2-2 with current representative models of each computer class. Access the student support site for this textbook (http://www.averia.unm.edu) and download copies of Table 2-2 from earlier textbook editions. What is the rate of change in CPU speed, memory and disk capacity, and cost?

Chapter 3

Data Representation

Chapter Goals

▶ Describe numbering systems and their use in data representation

▶ Compare and contrast various data representation methods

▶ Describe how nonnumeric data is represented

▶ Describe common data structures and their uses

Computers manipulate and store a variety of data such as numbers, text, sound, and pictures. This chapter describes how data is represented and stored within computer hardware. It also describes how simple data types are used as building blocks to create more complex data structures such as arrays and records. Understanding data representation is key to understanding hardware and software technology.

DATA REPRESENTATION AND PROCESSING

People can understand and manipulate data represented in a variety of forms. For example, we can understand numbers represented symbolically as Arabic numerals (for example, 8714), Roman numerals (for example, XVII), and simple lines or tick marks on paper (for example, ||| to represent a value of 3). We can understand words and concepts represented using pictorial characters (电 脑), alphabetic characters ("computer" and ЛПНРШАФЕТ), or directly as sound waves (a word spoken in a particular language). People also extract data content from visual images (photographs and motion pictures) and from the senses of taste, smell, and touch. Much of the brain's processing power and flexibility are due to the rich variety of data representations it can recognize and understand.

To be manipulated or processed by the brain, external data representations such as printed text must be converted to an appropriate internal format and transported to the brain's "processing circuitry." Our sensory organs convert inputs such as sight, smell, taste, sound, and skin sensation into electrical impulses that are transported through the nervous system to the brain. Processing within the brain occurs as networks of neurons exchange data electrically. Any data and information processor, whether organic, mechanical, electrical, or optical, must be capable of:

► Recognizing external data and converting it to an appropriate internal format
► Storing and retrieving data internally
► Transporting data among internal storage and processing components

Note that these capabilities correspond roughly to computer system components described in Chapter 2—I/O units, primary and secondary storage, and the system bus.

AUTOMATED DATA PROCESSING

Computer system data processing currently is implemented with electrical switches. Electrical switches are combined to form processing circuits that are in turn combined to form processing subsystems and entire CPUs. This can be seen as an equation:

$$A + B = C$$

where data inputs A and B, represented as electrical current, are transported through processing circuits (see Figure 3-1). The electrical current that emerges from the circuit represents a data output, C. Automated data processing thus combines physics and mathematics.

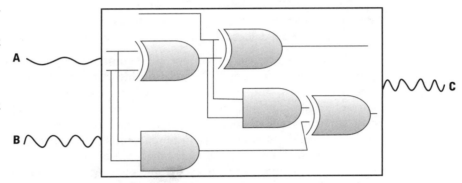

The physical laws of electricity, optics, and quantum mechanics are described by mathematical formulas. If the behavior of a device is based on well-defined, mathematically described laws of physics, then, in theory, that device can be used to implement a processor to perform the equivalent mathematical function. This relationship of mathematics and physics underlies of all automated computation devices from mechanical clocks (using the mathematical ratios of various gears) to electrical microprocessors (using the mathematics of electrical voltage and resistance). As you saw in the previous chapter, in the case of quantum mechanics, we understand the mathematical laws but don't yet know how to build reliable and cost-effective computing devices based on those laws.

Basing computer processing on mathematics and physics has limits. Processing operations must be based on mathematical functions such as addition and equality comparison, use numerical data inputs, and generate numerical outputs. From a human perspective, computers are limited because they can only represent and process numeric data. Concepts such as "mother," "friend," "love," and "hate" have no obvious numerical representation, so there is no direct way to process them using computer circuitry.

Although many researchers are studying ways to represent nonnumeric data numerically so it can be processed by computers, as well as ways to implement nonnumeric processing devices, the vast majority of computers currently in use are designed to perform only numerical computing. Even simple nonnumeric data like characters or tiny components of pictures can be processed only by first converting them to numeric form.

BINARY DATA REPRESENTATION

Computers represent data using binary numbers for two primary reasons:

► Binary numbers represented as electrical signals can be reliably transported among computer system components.

► Binary numbers represented as electrical signals can be processed by two-state electrical devices that are relatively easy to design and fabricate.

Reliable data transport is important if computer applications are expected to produce accurate outputs. The relationship between binary numbers, binary electrical signals, and reliable transport is fully described in Chapter 8. Given current technology, binary signals and processing devices represent the most cost-efficient tradeoff among capacity, accuracy, reliability, and cost. Future technological developments could fundamentally alter this tradeoff.

Binary numbers also are well-suited to computer processing because they correspond directly with values in **boolean logic**. This form of logic is named for nineteenth-century mathematician George Boole, who developed methods of reasoning and logical proof using sequences of statements that could be evaluated only as true or false. Similarly, a computer can perform logical comparisons of two binary data values to determine whether it is true or false that one data value is greater than, equal to, less than, less than or equal to, not equal to, or greater than or equal to another value. As discussed in the previous chapter, this primitive logical capability is the means by which a computer can implement intelligent behavior.

Using binary numbers does not mean that the only two numbers a computer can represent and process are zero and one. Both computers and humans can combine multiple digits to form a single data value to represent and manipulate larger numbers. Decimal and binary notation are alternate forms of a positional numbering system. In a positional numbering system, numeric values are represented as strings of digits. The symbol used to represent a digit and the digit position within a string determine its value. The value of the entire string is the sum of the values of all digits within the string.

For example, in the decimal number system, the number 5,689 is interpreted as follows:

$$(5 \times 1000) + (6 \times 100) + (8 \times 10) + 9 =$$
$$5,000 + 600 + 80 + 9 = 5,689$$

The same series of operations can be represented in columnar form, or with positions of the same value aligned in columns:

$$
\begin{array}{r}
5,000 \\
600 \\
80 \\
+\quad 9 \\
\hline
5,689
\end{array}
$$

For whole numbers, values are accumulated from right to left. In the above example, the digit 9 is in the first position, 8 is in the second position, 6 is in the third, and 5 is in the fourth.

The maximum value, or weight, of each position is a multiple of the weight of the position to its right. In the decimal number system, the first (rightmost) position is that of the ones (10^0), and the second position is 10 times that of the first position (10^1). The third position is 10 times the second position (10^2 or 100). The fourth is 10 times the third position (10^3 or 1,000), and so on. In the binary numbering system, each position is two times the previous position. Thus position values for whole numbers are 1, 2, 4, 8, and so forth. The multiplier that describes the difference between one position and the next is the **base** of the numbering system. Another term for base is **radix**. The base, or radix, of the decimal number system is 10. The base, or radix, of the binary numbering system is 2.

The fractional part of a numeric value is separated from the whole part by a period, though in some countries a comma is used instead of a period. In the decimal numbering system, the period or comma is called a **decimal point**. In numbering systems other than decimal, the term **radix point** is used for the period or comma that divides the fractional part of a value from the whole part. A decimal point is one type of radix point. An example of a decimal value with a radix point is:

$$5,689.368$$

The fractional portion of this real number is .368 and its value is interpreted as follows:

$$
(3 \times .1) + (6 \times .01) + (8 \times .001) =
$$
$$
0.3 + 0.06 + 0.008 = 0.368
$$

Proceeding toward the right from the radix point, the weight of each position is a fraction of the position to its left. In the decimal (base 10) numbering system, the first position to the right of the decimal point represents tenths (10^{-1}), the second position represents hundredths (10^{-2}), the third represents thousandths (10^{-3}), and so forth.

In the binary numbering system the first position to the right of the radix point represents halves (2^{-1}), the second position represents quarters (2^{-2}), the third represents eighths (2^{-3}), and so forth. As with whole numbers, each position has a weight that is 10 (or 2) times that of the position to its right. Table 3-1 shows a comparison between decimal and binary representations of the numbers 0 through 10.

Table 3-1 ▶

Binary and decimal notations for the values 0 through 10

	Binary system (Base 2)					Decimal system (Base 10)			
Place	2^3	2^2	2^1	2^0		10^3	10^2	10^1	10^0
Values	8	4	2	1		1000	100	10	1
	0	0	0	0	=	0	0	0	0
	0	0	0	1	=	0	0	0	1
	0	0	1	0	=	0	0	0	2
	0	0	1	1	=	0	0	0	3
	0	1	0	0	=	0	0	0	4
	0	1	0	1	=	0	0	0	5
	0	1	1	0	=	0	0	0	6
	0	1	1	1	=	0	0	0	7
	1	0	0	0	=	0	0	0	8
	1	0	0	1	=	0	0	0	9
	1	0	1	0	=	0	0	1	0

The number of digits required to represent a particular value depends on the base of the numbering system. The number of digits required to represent a value increases as the base of the numbering system decreases. Therefore, values that can be represented in a relatively compact format in decimal notation may require lengthy sequences of binary digits. For example, the decimal value 99 requires two decimal digits but seven binary digits. Table 3-2 summarizes the number of binary digits required to represent various decimal values.

You can convert a binary number to its decimal equivalent by multiplying the value of each position by the decimal weight of that position, then summing the results. Figure 3-2 shows how the binary number 101101.101 is converted to its decimal equivalent, 45.625.

Table 3-2 ▶

Binary data representations for up to 16 positions

Number of bits (n)	Number of values (2^n)	Numeric range (decimal)
1	2	0...1
2	4	0...3
3	8	0...7
4	16	0...15
5	32	0...31
6	64	0...63
7	128	0...127
8	256	0...255
9	512	0...511
10	1024	0...1023
11	2048	0...2047
12	4096	0...4095
13	8192	0...8191
14	16384	0...16383
15	32768	0...32767
16	65536	0...65535

Figure 3-2 ▶

Computing the decimal equivalent of a binary number

In computer terminology, each digit of a binary number is called a **bit**. A group of bits that describe a single data value is called a **bit string**. The leftmost digit, which has the greatest weight, is called the **most significant digit**, or **high order bit**. Conversely, the rightmost digit is the **least significant digit**, or **low order bit**. A string of eight bits is called a **byte**. A byte generally is the smallest unit of data that can be read from or written to a storage device.

The following mathematical rules define addition of positive binary digits:

$$0 + 0 = 0$$
$$1 + 0 = 0$$
$$0 + 1 = 1$$
$$1 + 1 = 10$$

To add two positive binary bit strings you first must align their radix points as follows:

$$101101.101$$
$$\underline{+10100.0010}$$

The values in each column are added separately, starting with the least significant, or rightmost, digit. If a column result exceeds one, then the excess value must be carried to the next column and added to the values in that column.

The result of adding the two numbers above is:

```
 111      1
 101101.101
+ 10100.0010
 1000001.1100
```

The result is the same as that obtained by adding the values in base 10 notation:

Binary	Real fractions	Real decimal
101101.101 =	45⅝ =	45.625
+ 10100.0010 =	20⅛ =	20.125
1000001.1100 =	65¾ =	65.750

Binary numbers usually contain many digits and are difficult for people to remember and manipulate without error. Higher-level programming languages such as C and Java provide program translators that automatically convert decimal numbers into binary numbers when generating CPU instructions and data values. However, programmers sometimes must deal directly with binary numbers, such as when they program directly in machine language or for some operating system utilities. To minimize errors, and to make dealing with binary numbers easier, numbering systems that are even multiples of two are sometimes used. These number systems include:

► Hexadecimal
► Octal

Hexadecimal Notation

Hexadecimal numbering uses 16 as its base or radix (hex = 6, and decimal = 10). There are not enough numeric symbols (Arabic numerals) to represent 16 different values, however. The digits of the hexadecimal numbering system include several English letters to represent the larger values (see Table 3-3).

Table 3-3 ►

Hexadecimal and decimal values

Base 16 digit	Decimal value	Base 16 digit	Decimal value
0	0	8	8
1	1	9	9
2	2	A	10
3	3	B	11
4	4	C	12
5	5	D	13
6	6	E	14
7	7	F	15

The primary advantage of **hexadecimal notation**, as compared to binary notation, is its compactness. Numeric values expressed in binary notation require four times as many digits as those expressed in hexadecimal notation. For example, the data content of a byte requires eight binary digits (such as 11110000) but only two hexadecimal digits (such as F0). This compact representation greatly reduces programmer error.

Hexadecimal numbers often are used to designate memory addresses. For example, a 64 kilobyte (KB) memory region contains 65,536 bytes (64 × 1,024 bytes/KB). Each byte is identified by a sequential numeric address. The first byte always is address 0. Therefore, the range of possible memory addresses is 0 to 65,535 in decimal numbers, 0 to 1111111111111111 in binary numbers, and 0 to FFFF in hexadecimal numbers. As you can see from this example, hexadecimal addresses are more compact than decimal or binary addresses due to the higher radix of the number system.

When reading a numeric value in written text, the base of the number may not be obvious. For example, when reading an operating system error message or a hardware installation manual, should the number 1000 be interpreted in base 2, 10, 16, or something else? In mathematical expressions the base of a number usually is specified with a subscript. For example,

$$1001_2$$

indicates to the reader that 1001 should be interpreted as a binary number and

$$6044_{16}$$

indicates that 6044 should be interpreted as a hexadecimal number.

When writing about computers, the base of a number sometimes is stated by affixing a letter to the end of the number. For example,

$$1001B$$

indicates a binary number and

$$6044H$$

indicates a hexadecimal number. No letter normally is used to indicate a decimal (base 10) number. Some programming languages also use the prefix 0x to indicate a binary number (0x1001 is equivalent to 1001_2).

Unfortunately, these conventions are not observed consistently. Often it will be left to the reader to guess the correct base by the content of the number or the context in which it appears. Any numeral other than a 0 or 1 indicates that the number cannot be binary. Similarly, the use of letters A through F indicates that the contents are expressed in hexadecimal. Bit strings usually are expressed in binary, and memory addresses usually are expressed in hexadecimal.

Octal Notation

Some operating systems and machine programming languages, such as those used with some IBM mainframe computers, use **octal notation**. Octal notation uses the base 8 numbering system and has a range of digits from 0 to 7. Octal numbers always are one-third the length of corresponding binary numbers and double the length of corresponding hexadecimal numbers.

GOALS OF COMPUTER DATA REPRESENTATION

Although all modern computers internally represent data with binary digits, they don't necessarily represent larger numeric values with positional bit strings. Positional numbering systems are convenient for people to interpret and manipulate, because the sequential processing of digits in a string parallels the way the human brain functions and because people traditionally are taught to perform computation in a linear fashion. For example, positional numbering systems are well-suited to addition and subtraction of numbers in columns using pencil and paper. Computer processors, however, operate differently from a human brain. Data representation tailored to human capabilities and limitations may not be best suited to computers.

Any representation format for numeric data represents a balance among several factors, including:

► Compactness
► Accuracy
► Range
► Ease of manipulation
► Standardization

As with many computer design decisions, alternatives that perform well in one dimension or category often perform poorly in others. For example, a data format with a high degree of accuracy and a large range of representable values usually is difficult and expensive to manipulate. It also is not very compact.

Compactness

In general, the more values you can represent, the more digits (larger bit strings) are required, that is, compactness and range are inversely related. A large number of digits also requires relatively large amounts of storage (disk and RAM) and processing capability (CPU). Greater storage and processing requirements directly translate into increased costs. Larger data formats require more expensive processing and storage devices.

Accuracy

Although concise coding minimizes the complexity and cost of hardware, it can do so at the expense of accurate data representation. The accuracy, or precision, of representation increases with the number of data bits that are used. As shown earlier in Table 3-2, the number of data bits also limits the largest and smallest number that can be directly represented within the machine.

It is possible for routine calculations to generate quantities that either are too large or too small to be contained within the finite circuitry of a machine (that is, within a fixed number of bits). For example, the fraction 1/3 cannot be represented within a fixed number of bits because it is a nonterminating fractional quantity (0.333333333 with an infinite number of 3s). In such cases, the quantities must be manipulated and stored as approximations. Each approximation introduces a degree of error. If approximate results are used as inputs for other computations, errors can be compounded. In some cases, the extent of error can be significant. Therefore, it is possible for a computer program to be without apparent logical flaws and yet produce inaccurate results.

If all data types were represented in the most concise form possible, then approximations required in some instances would introduce unacceptable margins of error. If a large number of bits were allocated to each data value instead, machine efficiency and performance would be sacrificed and hardware cost would be increased substantially. The most favorable tradeoffs can be achieved by using an optimum coding method for each type of data and/or for each type of operation that must be performed. This is the main reason for the variety of data representation formats used in most modern CPUs.

Ease of Manipulation

When discussing computer processing, *manipulation* refers to executing processor instructions such as addition, subtraction, and equality comparison. *Ease* refers to machine efficiency. The efficiency of a processor depends on its complexity (that is, the number of its primitive components and the complexity of the wiring that binds them together). Efficient processor circuits perform their function quickly due to a relatively small number of components and the short distance that electricity must travel. More complex devices require longer times to perform their function.

Data representation formats vary in their ability to efficiently implement processor circuits. As an example, consider the difficulty most people have performing computations on fractions, compared to working with decimal numbers. The decimal format is more efficiently processed by people than the fractional format.

Unfortunately, there is no one best representation format for all types of computation operations. For example, representing numbers as logarithms simplifies multiplication and division for people and computers because $\log A + \log B = A \times B$ and $\log A - \log B = A \div B$. Logarithms complicate the other operations such as addition and subtraction, and they can substantially increase a number's length (for example, $\log 99 \approx 1.9956351945975499153402557777533$).

Standardization

Data must be communicated among devices in a single computer system and among other computer systems via local and wide area networks. To ensure correct and efficient data transmission, data formats must be suitable for a wide variety of devices and computer systems. This especially is true for alphanumeric data. For this reason, various organizations have created standard data encoding methods for communication among computer systems and their components. Adhering to such standards provides computer users with the flexibility to combine hardware from different vendors with minimal data communication problems.

CPU DATA TYPES

The CPUs of most modern computers can represent and process five basic, or primitive, data types:

► Integer

► Real number

► Character

► Boolean

► Memory address

Some CPUs include variants of these five types for specialized application software. The representation format for each data type is designed to balance compactness, accuracy, ease of manipulation, and standardization.

Integers

An **integer** is a whole number, or value, that does not have a fractional part. For example, the values 2, 3, 9, and 129 are integers while the value 12.34 is not. Integer data formats can be signed or unsigned. The value of an unsigned integer always is assumed to be positive. A signed integer uses one bit to represent whether the value is positive or negative. The choice of which bit value (zero or one) represents which sign (positive or negative) is arbitrary. The sign bit normally is the high order bit in a numeric data format. In most data formats, the sign bit is one for a negative number and zero for a nonnegative number (note that zero is a nonnegative number).

The sign bit occupies a bit position that would otherwise be available to store part of the data value. Using a sign bit reduces the largest positive value that can be stored in any fixed number of bit positions. For example, the largest positive value that can be stored in an 8-bit unsigned binary format is 255, or 2^8-1. If a bit is used for the sign, then the largest positive value that can be stored is 127 or 2^7-1.

With unsigned numbers, the lowest value that can be represented always is zero. With signed numbers, the lowest value that can be represented is the negative of the highest value that can be stored (for example, -127 for 8-bit signed binary). The choice between signed and unsigned numeric data formats determines the largest and smallest values that can be represented.

Signed integers can't be stored as ordinary binary numbers because some means of representing the sign is required. Unsigned integers can be stored as ordinary binary numbers, because no bits are required to represent a radix point or a positive or negative sign. Many CPUs don't provide separate storage formats for signed and unsigned integers because extra data types add to CPU complexity, increasing the CPU's cost and reducing its efficiency.

Excess Notation   One format that can be used to represent signed integers is **excess notation**. Excess notation always uses a fixed number of bits with the leftmost bit representing the sign. The lowest value with a one as its leftmost digit represents zero. As shown in Table 3-4, all non-negative values have a one as the high-order bit, and negative values have a zero in this position.

Table 3-4 ►

Excess notation using four bits

Bit string	Decimal value	
1111	7	
1110	6	
1101	5	
1100	4	
1011	3	Nonnegative values
1010	2	
1001	1	
1000	0	
0111	-1	
0110	-2	
0101	-3	
0100	-4	
0011	-5	Negative values
0010	-6	
0001	-7	
0000	-8	

Two's Complement Notation   In the binary number system, the complement of 0 is 1 and the complement of 1 is 0. The complement of a bit string is formed by substituting 0 for all values of 1 and 1 for all values of 0 (the complement of 1010 is 0101). This transformation is the basis of **two's complement** notation. In this notation, nonnegative integers are rendered as ordinary binary numbers. For example, a two's complement representation of 7_{10} using four bits is: 0111.

Negative integer values are represented using the following transformation:

Complement of positive value + 1 = negative representation

Parentheses are a common mathematical notation for showing the complement of a value; if A is a numeric value, (A) represents its complement. In 4-bit two's complement representation, -7_{10} is represented as:

$$(0111) + 0001 =$$
$$1000 + 0001 =$$
$$1001 = -7_{10}$$

In this example, 0111 is converted to its complement, 1000. Then, a binary value of 1 is added to the least significant (or rightmost) digit. The result, 1001, is the two's complement representation of -7_{10}.

Two's complement notation is awkward for most people but it is highly compatible with digital electronic circuitry for the following reasons:

► The leftmost bit represents the sign.

► A fixed number of bit positions are used.

► Only two logic circuits are required to perform addition on single-bit values.

► Subtraction can be performed as addition of a negative value.

In the latter two reasons, processing circuits can be implemented with fewer components than for other integer data formats, resulting in cheaper processors and faster integer computations. For these reasons, virtually all CPUs in use today represent and manipulate signed integers using two's complement format.

Range and Overflow   Most modern CPUs use either 32 or 64 bits to represent a two's complement value (we'll assume 32 bits for the remainder of this text unless otherwise specified). A small positive value like one occupies 32 bits even though, in theory, only one bit is required. The unneeded bits are padded with leading zeros. Processor and data communication circuitry are more efficiently implemented for fixed-width formats. Although people easily can deal with numeric values of varying lengths, computer circuitry is not nearly as flexible. The additional complexity required to handle variable-length data formats results in unacceptably slow performance in electronic processing and storage devices.

The **numeric range** of a two's complement value is $-(2^{n-1})$ to $(2^{n-1}-1)$, where n is the number of bits used to store the value. The exponent is n-1 because one bit is used for the sign. For 32-bit two's complement format, the numeric range is $-2,147,483,648_{10}$ to $2,147,483,647_{10}$. With any fixed-width data storage format, it is possible that the result of executing a computation will be too large to fit within the format. For example, executing an addition instruction with $2^{31}-1$ and any other positive number as data inputs will generate a result that is too big to be stored in 32 bits. This condition is referred to as **overflow** and is treated as an error by the CPU. Executing a subtraction instruction also can result in overflow, for example, $-(2^{31}) - 1$. Overflow occurs when the absolute value of a computational result contains too many bits to fit into a fixed-width data format.

As with most other aspects of CPU design, data format length is one design factor that needs to be balanced with others. Large formats reduce the chance of overflow by increasing the maximum absolute value that can be represented, but many of these bits are wasted when smaller values are stored. If bits were free then there would be no tradeoff. But extra bits increase processor complexity and storage requirements, which, in turn, increases computer system cost. A CPU designer chooses a data format width by balancing numeric range; the chance of overflow during program execution; and the complexity, cost, and speed of processing and storage devices.

To avoid overflow and to increase accuracy, some computers and programming languages define additional numeric data types called **double precision** data formats. A double precision data format combines two adjacent fixed-length data items to hold a single value. Double precision integers sometimes are called **long integers**.

Overflow also can be avoided by careful programming. If the programmer anticipates that overflow is possible, the units of measure for program variables can be made larger. For example, calculations on centimeters could be converted to meters or kilometers, as appropriate. Converting to different units of measure may not be practical if data values are restricted to whole numbers. If numbers with fractional parts are required, then a non-integer data format is required.

Real Numbers

A **real number** can contain both whole and fractional components. The fractional portion is indicated by digits to the right of the radix point. For example, the following computation uses real number data inputs and generates a real number output:

$$18.0 \div 4.0 = 4.5$$

The equivalent computation in binary notation is:

$$10010 \div 100 = 100.1$$

Representing a real number requires some means to separate the whole and fractional components of the value (that is, the equivalent of a written radix point). A simple way to accomplish this is to define a storage format in which a portion of the bit string holds the whole portion and the remainder of the bit string holds the fractional portion. Figure 3-3 illustrates such a format with a sign bit and fixed radix point.

Figure 3-3 ▶

32-bit storage
format for real
numbers using a
fixed radix point

The format illustrated in Figure 3-3 is structurally quite simple due to the fixed location of the radix point. The advantage of this simplicity is realized when performing computations. Unfortunately there is a tradeoff between processing efficiency and numeric range for the storage format in Figure 3-3. Although the sample format uses 32 bits, its numeric range is substantially less than 32-bit two's complement. Only 16 bits are allocated to the whole portion of the value. Thus the largest possible whole value is 2^{16}-1, or 65,535. The remaining bits are used to store the fractional portion of the value, which can never be greater than or equal to one.

We could increase the numeric range of the format by allocating a larger number of bits to the whole portion. If the total size of the format is fixed at 32 bits, however, then the reallocation would reduce the number of bits used to store the fractional portion of the value. Reallocating bits from the fractional portion to the whole portion reduces the precision of fractional quantities. Reduced precision, in turn, reduces computational accuracy when those values are used as data inputs.

The nature of fixed-width data formats combined with a fixed-position radix point forces a tradeoff between numeric range and fractional precision. For complex application programs, 128 or more bit positions are used in order to get satisfactory range and precision, at the expense of large amounts of data storage and substantially increased CPU complexity.

Floating Point Notation One way of dealing with the tradeoff between range and precision is to abandon the concept of a fixed radix point. To represent extremely small (precise) values, move the radix point far to the left. For example, the value:

0.0000000013526473

has only a single digit to the left of the radix point. Similarly, very large values can be represented by moving the radix point far to the right as in:

1352647300000000.0

Note that both examples have the same number of digits. The first example trades range of the whole portion for increased fractional precision, and the second example trades fractional precision for increased whole range by having the radix point float left or right. Values either can be very large or very small (precise) but not both at the same time.

People tend to commit large numbers of representational and computation errors when manipulating long strings of digits. To minimize errors, people often write large numbers in a more compact format called scientific notation. In scientific notation the two numbers on the previous page are represented as $13,526,473 \times 10^{-16}$ and $13,526,473 \times 10^{8}$. Note that the base of the number system (10) is used as part of the multiplier. The exponent attached to the base can be interpreted as the number and direction of positional moves of the radix point, as illustrated in Figure 3-4. Negative exponents indicate movement to the left, and positive exponents indicate movement to the right.

Figure 3-4 ▶
Conversion of scientific notation to decimal notation

Real numbers are represented in computers using **floating point notation**. Floating point notation is similar to scientific notation except that 2 (rather than 10) is the base. A real value is derived from a floating point bit string according to the following formula:

$$\text{value} = \text{mantissa} \times 2^{\text{exponent}}$$

The mantissa holds the bits that are interpreted to derive the digits of the real number. By convention, the mantissa is assumed to be preceded by a radix point. The exponent field indicates the position of the radix point.

Many CPU-specific implementations of floating point notation are possible. Differences among these implementations may include the length and coding formats of the mantissa and exponent, and the location of the radix point within the mantissa. Although two's complement can be used to code either or both fields, other coding formats may represent better design tradeoffs. Prior to the 1980s there was little compatibility in floating point format among different CPUs. Incompatible formats resulted in an inability to transport floating point data among different computers.

The Institute of Electrical and Electronics Engineers (IEEE) addressed this problem by defining standard formats for representing floating point data. IEEE Standards 754 and 854 define 32-bit and 64-bit floating point coding formats, respectively. Both formats have been adopted by microprocessor manufacturers and are implemented in most modern computers. For the remainder of this chapter all references to floating point coding formats will refer to the IEEE 32-bit format unless otherwise specified.

The IEEE 32-bit floating point format is illustrated in Figure 3-5. The leading sign bit applies to the mantissa, not the exponent. The 8-bit exponent is coded in excess notation (this implies that its first bit is a sign bit). The 23-bit mantissa is coded as an ordinary binary number. It is assumed to be preceded by a binary 1 and the radix point. This extends the precision of the mantissa to 24 bits, although only 23 actually are stored.

Figure 3-5 ▶

IEEE 32-bit floating point format

Range, Overflow, and Underflow The number of bits in a floating point string and the formats of the mantissa and exponent fields impose limits on the range of values that can be represented. The number of digits in the mantissa determines the number of significant (that is, nonzero) digits in the largest and smallest values that can be represented. The number of digits in the exponent determines the number of possible bit positions to the right or to the left of the radix point.

Using the number of bits assigned to each field, the largest absolute value of a floating point value appears to be:

$$1.11111111111111111111111 \times 2^{11111111}$$

Exponents containing all zeros and all ones, however, are used to represent special data values. Thus, the usable exponent range is reduced and the decimal range for the entire floating point value is approximately 10^{-45} to 10^{38}.

Floating point numbers with large absolute values have large positive exponents. When overflow occurs, it always occurs within the exponent. Floating point representation also is subject to a related error condition called **underflow**. Very small numbers are represented by negative exponents. Underflow occurs when a negative exponent is too large in absolute value to fit within the bits allocated to store it.

Precision and Truncation Recall that scientific notation, including floating point notation, trades numeric range for accuracy. Accuracy is reduced as the number of digits available to store the mantissa is reduced. The 23-bit mantissa used in floating point format represents approximately seven decimal digits of precision. However, many numbers that may be represented within a computer contain however, more than seven nonzero decimal digits. One example of such a number is the decimal equivalent of the fraction 1/3:

$$1/3 = 0.33333333 \ldots$$

The number of digits to the right of the decimal point is infinite. Only a limited number of mantissa digits are available, however.

Numbers such as 1/3 are stored in floating point format by **truncation**. The numeric value is stored in the mantissa, starting with its most significant bit, until all available bits are consumed. The remaining bits simply are discarded and the mantissa then is truncated. An error or approximation occurs any time a floating point value is truncated. However, the truncated digits are relatively insignificant compared to the significant, or large, value that is stored. Problems can arise when truncated values are used as input to computations. The error introduced by truncation can be magnified as truncated values are used to generate inaccurate results. The error resulting from a long series of computations that starts with truncated inputs can be relatively large.

An added difficulty is that more values have nonterminating representations in the binary system than in the decimal system. For example, the fraction 1/10 is nonterminating in binary notation. The representation of this value in floating point notation is a truncated value. Such problems usually can be avoided through careful programming. As a rule of thumb, programmers reserve floating point calculations for quantities that may vary continuously over wide ranges. Measurements that are made by scientific instruments, for example, may be handled in floating point notation.

When possible, programmers use integer data types to avoid truncation. One common example is financial applications. Because most monetary systems have at least one fractional monetary unit, just as pennies and quarters are fractions of U.S. dollars, novice programmers sometimes assume that monetary amounts should be stored and manipulated as floating point numbers. Truncation errors due to nonterminating representations of tenths and other fractions inevitably will occur. Cumulative errors will mount as truncated numbers, or approximations, are input to subsequent calculations.

Careful programmers use integer arithmetic for accounting and financial applications. To do so, a programmer stores and manipulates monetary amounts in the smallest possible monetary unit, for example, U.S. pennies or Mexican pesos. Small denominations are converted to larger ones only when needed for output, such as printing a dollar amount on a check.

Processing Complexity The difficulty that we encounter when learning to use scientific and/or floating point notation is understandable. The formats are relatively complex—far more complex than integer data formats. This complexity affects both people and computers. Although floating point formats are optimized for processing efficiency, they still require relatively complex processing devices. The simpler two's complement format used for integers requires much less complex computational circuitry.

The difference in complexity of processing circuitry for two's complement and floating point formats translates directly to a speed difference when performing calculations. The magnitude of the difference depends on a number of factors, including the computation being performed and the exact details of the processing circuitry. As a general rule, simple computational operations like addition and subtraction take at least twice as long for real numbers than for integers. The difference is even greater for more complex operations like division.

For this reason and for reasons of accuracy, careful programmers never use a real number when an integer will suffice. This is particularly important for data items that frequently are updated.

Character Data

English and most other languages use alphabetic letters, numerals, punctuation marks, and a variety of other special-purpose symbols (such as $ and &) in their written form. Each of these symbols is a **character**. A sequence of characters that forms a meaningful word, phrase, or other useful grouping of text is a **string**. In most programming languages single characters are surrounded by single quotation marks ('c') and strings are surrounded by double quotation marks ("computer").

Character data cannot be directly represented or processed within a computer because computers are designed to process only digital data (bits). Character data can be represented indirectly by defining a table of correspondence between individual characters and specific numeric values. For example, the integer values 0 through 9 could be used to represent the characters (numerals) '0' through '9'. The uppercase letters 'A' through 'Z' could be represented as the integer values 10 through 36. The lowercase letters 'a' through 'z' could be represented as the integer values 37 through 63. The '$' symbol could be represented as the integer value 64, and so forth.

A set of correspondences or substitutions of numeric values for characters is one method of coding. All coding methods share several important characteristics, including:

▶ All users must use the same table of correspondence.

▶ The values used to represent the data must be capable of being encoded, transmitted, and decoded.

▶ The specific coding method represents a tradeoff among compactness, ease of manipulation, accuracy, range, and standardization.

The following sections describe some of the more common coding methods for character data.

BCD and EBCDIC Binary Coded Decimal (BCD) is a character coding method used by early IBM mainframe computers. Characters are encoded as strings of six bits. Because only six bits are used, the coding method can represent only 64 (2^6) different symbols. The BCD character set includes alphabetic letters in uppercase form only, the numerals 0 through 9, and some punctuation marks.

BCD's compact form was an advantage in the era of limited memory and slow processors. Small bit strings require less memory and simpler processing circuits. BCD also was designed with another processing efficiency in mind. The representation of the symbols '0' through '9' is intentionally the same as the unsigned binary representation of the integer values 0 through 9. This simplifies or eliminates conversion problems between numeric and character data formats, such as when calculating interest on an account and then printing it on an account statement.

When memory and processors were expensive, BCD's limited range was a tolerable characteristic. But as memory and processors became less expensive and more powerful, the simplicity of BCD became less valuable, and larger formats with greater range became more cost effective. BCD also suffered from the fact that it is a proprietary IBM standard never adopted by many other computer manufacturers. BCD is not used in any modern computers or software.

In the 1960s, IBM extended the range of BCD by expanding the format to eight bits and defining additional characters within the expanded table. The extended coding method is **Extended Binary Coded Decimal Interchange Code** (**EBCDIC**). EBCDIC still is used in IBM mainframe computers.

ASCII The **American Standard Code for Information Interchange** (**ASCII**) is a widely used coding method in data communication that has been adopted in the United States. The international equivalent of this character set is **International Alphabet Number 5** (**IA5**), a standard promulgated by the **International Standards Organization** (**ISO**).

ASCII was designed to achieve two primary goals:

► Define a vendor-independent standard for character coding

► Support data communication between computers and peripheral devices

At the time ASCII was defined, IBM controlled most of the world market for computers. For various technical and political reasons, EBCDIC was not chosen as the standard and the ASCII code that was developed is incompatible with EBCDIC. For purposes of backward compatibility, IBM has continued to use EBCDIC in its mainframe computers. Virtually all other computers, including IBM's non-mainframe computers, use ASCII.

Two key characteristics of ASCII arose from its intended use for data communication among computers and peripheral devices:

► A 7-bit format

► The inclusion of device control codes within the coding table

The 7-bit format was chosen because most computers and peripheral devices transmit data in bytes and because parity checking is used widely for detecting transmission errors. Parity checking and other error detection and correction methods are discussed fully in Chapter 8. For now, the important characteristic of parity checking is that it requires one bit per character to be used as a check digit. One bit out of eight is not used as part of the data value, leaving only seven bits for data representation.

The standard version of ASCII used for data communication sometimes is called ASCII-7 to emphasize its 7-bit format. This coding table has 128, or 2^7, defined characters. Computers that use 8-bit bytes are capable of representing 256, or 2^8, different characters. In most computers the ASCII-7 characters are included within an 8-bit character coding table as the first, or lower, 128 table entries. The additional, or upper, 128 entries are defined by the computer manufacturer and typically are used for graphic characters such as line drawing characters, and/or multinational characters such as á, ñ, Ö, and Ω. This encoding method sometimes is called ASCII-8. The term is a misnomer as it implies that the entire table (all 256 entries) is standardized. In fact, only the first 128 entries are defined by the ASCII standard.

Table 3-5 shows portions of the ASCII, EBCDIC, and BCD coding tables.

Table 3-5 ►

Partial listing of ASCII, EBCDIC, and BCD codes

Symbol	ASCII	EBCDIC	BCD
0	0110000	11110000	000000
1	0110001	11110001	000001
2	0110010	11110010	000010
3	0110011	11110011	000011
4	0110100	11110100	000100
5	0110101	11110101	000101
6	0110110	11110110	000110
7	0110111	11110111	000111
8	0111000	11111000	001000
9	0111001	11111001	001001
A	1000001	11000001	010001
B	1000010	11000010	010010
C	1000011	11000011	010011
a	1100001	10000001	
b	1100010	10000010	
c	1100011	10000011	

Device Control When text is printed or displayed on an output device, it often is formatted in particular way. For example, a customer record can be displayed on a video display terminal in a manner similar to a printed form. Text output to a printer normally is formatted into lines and paragraphs as it is in this textbook. Certain text can be highlighted when printed or displayed using methods such as underlining, bold font, or reversed background and foreground color.

Data communication methods in use at the time ASCII was defined were based on the transmission of one character at a time over a single data communication wire. This is serial transmission, as described in Chapter 8. Because only a single wire is used, formatting commands must be sent in the same transmission stream as the text to be printed or displayed. Text and formatting commands are interspersed with formatting commands immediately before or after the text that they modify.

ASCII defines a number of device control codes (see Table 3-6) that can be used for this purpose. Among the simpler codes are carriage return, which moves the print head or insertion point to the beginning of a text line; line feed, which moves the print head or insertion point down one line; and bell, which rings a bell or generates a short sound such as a beep. In ASCII, each of these functions is assigned a numeric code and a short character name, for example, CR for carriage return, LF for line feed, and BEL for bell.

Not all of the device control codes are used for formatting. Some are used for controlling various aspects of the data communication process. For example, ACK is sent to acknowledge correct receipt of data and NAK is sent to indicate that an error has been detected. ASCII defines a total of 33 device control codes that occupy the first 32 table entries (numbered 0 through 31) and the last table entry (number 127).

Including control codes in ASCII was a substantial benefit to users and computer equipment manufacturers. Users were given the flexibility of using peripheral devices from many different manufacturers. To the extent that computer manufacturers and software developers agreed to abide by the ASCII standard, equipment from multiple vendors could be used in a single computer system without compatibility problems. It also standardized the transfer of textual data among computer systems and users.

Software and Hardware Support Because characters usually are represented as integers, there is little or no need for the CPU to provide special support for character processing. Instructions that move and copy integers behave the same whether the content being manipulated is an actual numeric value or an ASCII encoded character. Similarly, an equality, or inequality, comparison instruction that works for actual integers also works for integers that represent characters.

Table 3-6 ▶

ASCII control
codes

Decimal code	Control character	Description
000	NUL	Null
001	SOH	Start of Heading
002	STX	Start of Text
003	ETX	End of Text
004	EOT	End of Transmission
005	ENQ	Enquiry
006	ACK	Acknowledge
007	BEL	Bell
008	BS	Backspace
009	HT	Horizontal Tabulation
010	LF	Line Feed
011	VT	Vertical Tabulation
012	FF	Form Feed
013	CR	Carriage Return
014	SO	Shift Out
015	SI	Shift In
016	DLE	Data Link Escape
017	DC1	Device Control 1
018	DC2	Device Control 2
019	DC3	Device Control 3
020	DC4	Device Control 4
021	NAK	Negative Acknowledge
022	SYN	Synchronous Idle
023	ETB	End of Transmission Block
024	CAN	Cancel
025	EM	End of Medium
026	SUB	Substitute
027	ESC	Escape
028	FS	File Separator
029	GS	Group Separator
030	RS	Record Separator
031	US	Unit Separator
127	DEL	Delete

The results of non-equality comparisons are less straightforward. The assignment of numeric codes to characters follows a specific order called a **collating sequence**. A greater-than comparison for two characters (for example, 'a' less than 'z') returns a result based on the numeric comparison of the corresponding ASCII codes; that is, whether the numeric code for 'a' is less than the numeric code for 'z'. If the character set has an order and if the coding method follows the order, then less-than and greater-than comparisons produce expected results.

Note that using numeric values to represent characters can produce some unexpected or unplanned results. For example, the collating sequence of letters and numerals in ASCII follows the standard alphabetic order for letters and numeric order for numerals, but uppercase and lowercase letters are represented by different codes. This means that an equality comparison between uppercase and lowercase versions of the same letter returns false because 'a' does not equal 'A'. Punctuation symbols also have a specific order in the collating sequence, though there is no general-purpose ordering for such symbols.

The only potential need for CPU support of a separate character data type is so the CPU can detect attempts by a program to perform "illegal" operations on characters. For example, it is acceptable for an "integer add" instruction to be executed with two integer inputs. It makes no sense to execute that instruction with two ASCII codes as input, however, even if their internal coding format is the same as for integers. The CPU can act as an error detection device if characters are treated as a separate data type. However, characters are seldom treated as a distinct data type by the CPU because most CPU designers seek to simplify their designs as much as possible. For the vast majority of CPUs, characters are represented using an integer data coding format and application and/or operating system software is responsible for keeping track of which integers are true numeric values and which are ASCII codes.

ASCII Limitations ASCII's designers could not foresee the long lifetime (approximately 40 years) of the code or the revolutions in I/O device technologies that would occur during that lifetime. Common characteristics of I/O today, such as color, bitmapped graphics, and selectable fonts, were never envisioned by the ASCII designers. Unfortunately, ASCII does not have sufficient range to define a set of control codes large enough to account for the full range of formatting and display capabilities in modern I/O devices.

ASCII also is an English-based coding method. This is not surprising given that at the time ASCII was defined, the United States accounted for most computer use and virtually all computer production. ASCII has a heavy bias toward Western languages in general and American English in particular, which has become a significant limitation as computer use and production have proliferated on a worldwide basis.

Recall that 7-bit ASCII has only 128 table entries, 33 of which are used for device control. Only 95 printable characters can be represented. This is enough for a usable subset of the characters commonly employed in American English text. But that subset does not include any modified Roman characters, such as á and ç, or those from other alphabets, such as Σ.

The International Standards Organization (ISO) partially addressed the problem by defining many different 256-entry tables based on the ASCII model. One of these, called **Latin-1**, contains the ASCII-7 characters in the lower 128 table entries and most of the characters used by Western European languages in the upper 128 table entries. The upper 128 entries sometimes are called the **multinational characters**. The number of available character codes in a 256-entry table, however, still is much too small to represent the full range of printable characters in world languages.

Further complicating matters is the fact that some printed languages are not based on characters in the Western sense. Chinese, Japanese, and Korean written text consists of ideographs, which are pictorial representations of words or concepts. Ideographs are, in turn, composed of individual graphic elements, sometimes called strokes, that number in the thousands. Other written languages present similar though less severe coding problems.

ASCII was never designed to handle the full range of written human language. Despite this limitation it has persisted as a standard for decades because of the dominance of the United States as an economic power and as a designer and manufacturer of computer equipment. A substantial expansion or replacement of ASCII is required to enlarge the standardized character set.

Unicode

The **Unicode** Consortium was founded in 1991 to develop a multilingual character encoding standard encompassing all of the world's written languages. The original members were Apple Computer Corporation and Xerox Corporation, but many computer companies soon joined. The effort was a necessary step to allow software and data to cross international boundaries. Major Unicode standard releases are coordinated with standard 10646 of the International Standards Organization (ISO). The latest standard is Unicode 3.0, published in February, 2000.

Like ASCII, Unicode is a coding table that assigns nonnegative integers to represent individual printable characters. The ISO Latin-1 standard, which includes ASCII-7, is incorporated into Unicode as the first 256 table entries. Thus, ASCII is a subset of Unicode. The primary difference between ASCII and Unicode lies in the number of table entries. Unicode uses 16-bit code providing 65,536 table entries numbered 0 through 65,535.

The additional table entries primarily are used for characters, strokes, and ideograms of languages other than English and its Western European siblings. Unicode includes many other alphabets such as Arabic, Cyrillic, and Hebrew, and thousands of Chinese, Japanese, and Korean ideograms and characters. Some extensions to the ASCII device control codes also are provided. As currently defined, Unicode can represent written text from all modern languages. A total of 49,194 characters are defined in Unicode 3.0. Approximately 6,000 characters are reserved for private use.

Unicode is supported widely in modern software, including most operating systems and word processors since the late 1990s. Because few CPUs directly implement character data types, there is no real need to upgrade processor hardware to support Unicode. Most modern CPUs use 32 bits to represent an integer, so the existing practice of internally coding characters and integers will continue.

The primary impact of Unicode on hardware is in storage and I/O devices. Because the size of the numeric code is double that of ASCII, pure text files are twice as large. This appears to be a problem at first glance, but the impact is reduced by the use of custom word-processing file formats and the ever-declining cost of secondary storage. Most text is not stored as pure ASCII. Instead, it is stored in a format that intermixes text and formatting commands that apply to the text. Because formatting commands also occupy file space, the file size increase experienced when switching from ASCII to Unicode generally is far less than 100 percent.

Devices designed for character I/O traditionally have used ASCII by default and have implemented vendor-specific methods or used older ISO standards to process character sets other than Latin-1. The typical method is to maintain an internal set of alternate coding tables where each table contains a different alphabet or character set. Device-specific commands are used to switch from one table to another. Unicode unifies and standardizes the content of these various tables. It offers a standardized method for processing international character sets. That standard is being widely adopted by I/O device manufacturers. Backward compatibility with ASCII is not a problem because Unicode includes ASCII.

Boolean Data

The **boolean** data type has only two data values—true and false. Most people do not think of boolean values as data items. But the primitive nature of CPU processing requires the ability to store and manipulate boolean values. Recall that processing circuitry acts to transform input signals physically into output signals. If the input signals represent numbers and the processor is performing a computational function, such as addition, then the output signal represents the numerical result. When the processing function is a comparison operation like greater than or equal to, then the output signal represents a boolean result of true or false. This result is stored in a register as is any other processing result and may be used by other instructions as input (a conditional branch or GOTO). The boolean data type potentially is the most concise coding format because only a single bit is required. For example, binary 1 can represent true and 0 can represent false. Most CPU designers seek to minimize the number of different coding formats in use to simplify processor design and implementation. CPUs generally use the coding format for integers (usually two's complement) to represent boolean values. When coded in this manner, the integer value zero corresponds to false and all nonzero values correspond to true.

To conserve memory and storage, programmers sometimes will "pack" many boolean values into a single integer by using each individual bit to represent a separate boolean value. Although this conserves memory, it generally slows program execution due to the complicated instructions required to extract and interpret individual bits from an integer data item.

Memory Addresses

As described in Chapter 2, primary storage is a series of contiguous bytes of storage. Conceptually, each memory byte has a unique identifying or serial number. The identifying numbers are called addresses and their values start with zero and continue in sequence (with no gaps) until the last byte of memory is addressed.

In some CPUs this conceptual view also is the physically implemented method of identifying memory locations. That is, memory bytes are identified by a series of nonnegative integers. This approach to assigning memory addresses is called a **flat memory model**. In CPUs that use a flat memory model, it is logical and typical to use two's complement as the coding format for memory addresses. Memory addresses are stored in the same internal format as integers. The advantage of this approach is that it minimizes the number of different data types and the complexity of processor circuitry.

Some CPU designers choose not to use simple integers as memory addresses in order to maintain backward compatibility with earlier generations of CPUs that didn't use a flat memory model. For example, the Intel Pentium CPU maintains backward compatibility with the 8088 CPU used in the first IBM personal computer. Earlier generations of processors typically used an alternate approach to memory addressing called a **segmented memory model**. In a segmented memory model, primary storage is divided into a series of equal-sized (for example, 64 KB) segments called pages. Pages are identified by sequential nonnegative integers. Each byte also is identified by a sequentially assigned nonnegative integer. Each byte of memory has a two-part address: the first part identifies the page, and the second part identifies the byte within the page.

Segmented memory models are discussed in more detail in Chapter 5. For the moment, we only remain concerned with the fact that a memory address in a segmented memory model consists of two different numbers. Because of this, the data coding method for single integers cannot be used. Instead, a specific coding format for memory addresses must be defined and implemented. For this reason, CPUs with segmented memory models have an extra data type, or coding format, for storing memory addresses. Many flat memory model CPUs code memory addresses using the same method as for unsigned integers.

Technology
Focus

Intel Memory Address Formats

Intel microprocessors have been used in personal computers since the first IBM personal computer was sold in 1981. The original IBM PC used an Intel 8088 microprocessor. Later generations of Intel processors (the 8086, 80286, 80386, 80486, and Pentium family of processors) have all maintained backward compatibility with the 8088. That is, each subsequent processor can execute CPU instructions designed for the 8088. Backward compatibility has been considered an essential ingredient in the success and proliferation of personal computers based on the Intel 80X86 family of CPUs.

Early Intel microprocessors were designed and implemented with speed of processor operation as a primary goal. Because a general-purpose CPU uses memory addresses in nearly every instruction execution, it is important to make memory access as efficient as possible. Processor complexity rises with the number of bits that must be processed simultaneously. Large coding formats for memory addresses and other data types require complicated and relatively slow processing circuitry. Intel designers wanted to minimize the complexity of processor circuitry associated with memory addresses. They also needed to balance the desire for efficient memory processing with the need to maintain a relatively large range of possible memory addresses.

Intel designers made two significant design decisions for the 8088 to achieve this balance:

► A 20-bit size for memory addresses
► A segmented memory model

The 20-bit address format limited usable (addressable) memory to 1 megabyte (MB) (2^{20} addressable bytes). This did not seem like a significant limitation because few computers (including most mainframes) had that much memory at the time. The 20-bit format is divided into two parts—a 4-bit segment identifier and a 16-bit segment offset. The 16-bit segment offset identifies a specific byte within a 64, or 2^{16}, KB memory segment. Because 4 bits are used to represent the segment, there are 16, or 2^4, possible segments.

Intel designers anticipated that most programs would fit within a single 64 KB memory segment. Further, they knew that 16-bit memory addresses could be processed much more efficiently than larger addresses. Therefore, they defined two types of address-processing functions—those that used the 4-bit segment portion of the address and those that ignored it. Memory could be accessed rapidly when the processor ignored the segment identifier. Memory access was much slower when the segment identifier was used. Intel designers assumed that most memory accesses would not require the segment identifier.

Both the 64 KB segment size and the 1 MB total memory limit soon became significant constraints. Although early programs for the IBM personal computer generally were less than 64 KB in size, later versions of those same programs were larger. Because the processor design imposed a performance penalty when the segment identifier was used, later versions ran much more slowly than earlier versions. In addition, both the operating system and the computer hardware were becoming more complex and consuming more memory resources. As a result, computer designers, operating systems designers, and users chafed under the 1 MB limit.

Intel designers first addressed these constraints in the 80286. They increased the segment identifier to 8 bits, increasing total addressable memory to 16 MB. The 80386 went a step further by providing two methods of addressing. The first, called real mode, was a method compatible with that of the 8088. The second, called protected mode, was an entirely new method based on 32-bit memory addresses. Protected mode addressing eliminated the performance penalty for programs larger than 64 KB. The 80486 and Pentium family of microprocessors continue to use both addressing methods.

Although the older design limits were effectively bypassed by protected mode, the need to implement real mode has proved to be a problem in later microprocessors. Implementing two different addressing modes has increased the complexity of processors such as the Pentium. This, in turn, has made it more difficult to implement fast versions of these processors. Intel has struggled to match the performance of other CPUs that implement only a single memory addressing method.

This discussion, and the earlier discussion of ASCII, illustrate a few of the pitfalls for computer designers in choosing and implementing data coding methods and formats. A fundamental characteristic of any design is that it always represents a tradeoff. For Intel memory addresses, the tradeoff was made among processor cost, speed of program execution, and the memory requirements of typical PC software. The tradeoff may have been optimal or, at least, reasonable, given the state of all these variables in 1981. But neither time nor technology stand still. Design decisions have a limited life and that life can be very short for computer technology. Attempts to maintain backward compatibility with previous CPU designs can be difficult, expensive, and may severely limit future design choices.

DATA STRUCTURES

The previous sections outlined the basic data types that a CPU can manipulate directly: integer, real number, character, boolean, and memory address. The data types directly supported by a CPU sometimes are called **primitive data types**, or **machine data types**. It would be difficult to develop programs of any kind if only primitive data types were available. Most application programs need to combine primitive data items to form useful aggregations. A common example is a character string. Most application programs define and use strings, but few CPUs provide instructions that directly manipulate them. Application development is simplified if strings can be defined and manipulated (that is, read, written, and compared) as a single unit instead of manipulated one character at a time.

A **data structure** is a related group of primitive data elements that is organized for some type of common processing. Data structures are defined and manipulated within software. Computer hardware cannot manipulate data structures directly, but must deal with them in terms of their basic components such as integers, floating point numbers, single characters, and so on. Software must translate operations on data structures into an equivalent set of machine instructions that operate on individual primitive data elements. For example, Figure 3-6 shows a comparison of two strings decomposed into comparison operations on each character.

Figure 3-6 ►

Software decomposes operations on data structures into operations on primitive data components

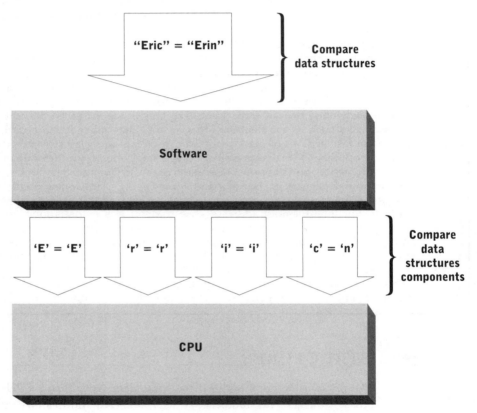

The complexity of data structures is limited only by the imagination and skill of programmers. As a practical matter, certain data structures are useful in a wide variety of situations and are commonly implemented. Examples of such data structures include character strings or arrays, records, and files. System software often provides application services to manipulate these commonly used data structures. For example, an operating system normally provides services for reading and writing to and from files.

Other data structures are less commonly supported by system software. Examples of these include numeric arrays, indexed files, and complex database structures. Indexed files are supported in some, but not all, operating systems.

Numeric arrays are supported within most programming languages but not within most operating systems. Database structures normally are supported by a database management system. Most programming languages support direct manipulation of character strings.

Data structures have an important role in system software development. For example, linked lists commonly are used by operating systems to keep track of memory blocks allocated to programs and disk blocks allocated to files and directories. Indexes are widely used within database management systems to speed search and retrieval operations. Programmers who develop system software generally take an entire course in data structures and efficient methods for manipulating them.

Pointers and Addresses

Whether implemented within system or application software, virtually all data structures make extensive use of pointers and addresses. A **pointer** is a data element that contains the address of another data element. An **address** is the location of some data element within a storage device. Addresses vary in content and representation depending on the storage device being addressed. Secondary storage devices normally are organized as a sequence of data blocks. A block is a group of bytes that is read or written as a unit. For example, floppy disk drives on personal computers usually read and write data in 512-byte blocks. For block-oriented storage devices, an address normally can be represented as an integer containing the sequential position of the block. Integers also can be used to represent the address of an individual byte within a block.

As described earlier, memory addresses can be complex if a segmented memory model is used. For the purpose of discussing data structures, we will assume that a flat memory model is used and that nonnegative integers can be used as memory addresses.

Arrays and Lists

Many sets of data can be considered as lists. A list is a set of related data values. In mathematics, a list is considered unordered. That is, no specific list element is designated as the first, second, or last element. When writing software, a programmer usually prefers to impose some ordering upon the list. For example, a list of the days of the week may be ordered sequentially, starting with Monday.

An **array** is an ordered list in which each element can be referenced by an index to its position. An example of an array for the first five letters of the English alphabet appears in Figure 3-7. Note that the index values are numbered starting at zero, a common (although not universal) practice in computer programming. Although the index values appear in Figure 3-7, they are not stored. Instead, they are inferred from the location of the data value within the storage allocated to the array. In a high-level programming language, individual array elements normally are referenced by the name of the array and the index value. For example, the third letter of the alphabet stored in an array might be referenced as:

alphabet[2]

where alphabet is the name of the array and 2 is the index value (numbered from zero).

Figure 3-7 ►

Array elements in contiguous storage locations

Index	Value
0	A
1	B
2	C
3	D
4	E

Figure 3-8 shows a character array, or string, stored in contiguous, or sequential, memory locations. In this example, each character of the name John Doe is stored within a single byte of memory and the characters are ordered in sequential byte locations starting at byte 1000. An equivalent organization may be used to store the name on a secondary storage device. Access to an individual element of the array can be achieved by using the starting address of the array and the index of the element. For example, if we wish to retrieve the third character in the array we can compute its address as the sum of the address of the first element plus the index value, assuming index values start at zero.

Figure 3-8 ►

Character array in contiguous storage locations

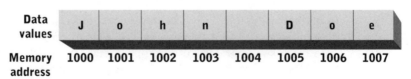

Data values	J	o	h	n		D	o	e
Memory address	1000	1001	1002	1003	1004	1005	1006	1007

Using contiguous storage locations, especially in secondary storage devices, complicates the allocation of storage locations to specific data items or structures. For example, it may be difficult to add new elements to the end of an array because those storage locations already may be allocated to other data items. Because of this, contiguous storage allocation generally is used only for arrays of fixed length.

For a variety of reasons, it may be desirable to store individual elements of the array in widely dispersed storage locations. A **linked list** is a data structure that uses pointers so list elements can be scattered among nonsequential storage locations. Figure 3-9 shows a generic example of a **singly linked list**. Each list element occupies two storage locations. The first location holds the data value of a list element. The second storage location holds the address of the next element value. Figure 3-10 shows a character string stored within a linked list. Note that individual characters are scattered among nonsequential, or noncontiguous, storage locations.

Figure 3-9 ▶

Value and pointer fields of a single linked list

Figure 3-10 ▶

Character array stored in noncontiguous memory locations

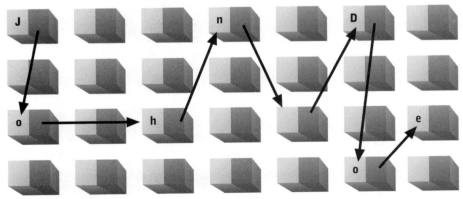

More storage locations are required for a linked list than for an array with equivalent content because both data and pointers must be stored. Using pointers also complicates the task of locating individual array elements. References to specific array elements must be resolved by following the chain of pointers, starting with the first array element. This can be very inefficient if the number of array elements is large. For example, accessing the 1000th element requires reading and following the pointers within the first 999 elements.

Linked lists are easier to expand or shrink than arrays. The procedure to add a new element is as follows (see Figure 3-11):

1. Allocate storage for the new element.

2. Copy the pointer from the element preceding the new element into the pointer field of the new element.

3. Write the address of the new element into the pointer field of the preceding element.

Figure 3-11 ▶

Insertion of a new element into a singly linked list

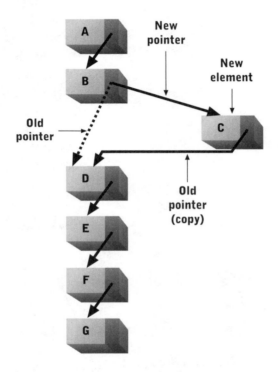

In contrast, insertion of an element into a list stored in contiguous memory can be very time consuming. The procedure is as follows (see Figure 3-12):

1. Allocate a new storage location to the end of the list.

2. For each element past the insertion point, copy the element value to the next storage location, starting with the last element and working backward to the insertion point.

3. Write the new element value in the storage location at the insertion point.

Figure 3-12 ▶

Insertion of a
new element into
an array stored
in contiguous
memory
locations

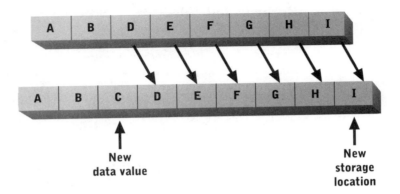

**Before
insertion**

**After
insertion**

**New
data value**

**New
storage
location**

Element insertions near the beginning of the array are highly inefficient due to the large number of required copy operations.

Figure 3-13 depicts a more complicated form of linked list called a **doubly linked list**. Each element of a doubly linked list has two pointers. One pointer points to the next element in the list and the other points to the previous element in the list. The primary advantage of doubly linked lists is that they can be traversed in either direction with equal efficiency. The primary disadvantage is that more pointers must be updated each time an element is inserted into or deleted from the list. In addition, more storage locations are required to hold the extra set of pointers.

Figure 3-13 ▶

A doubly
linked list

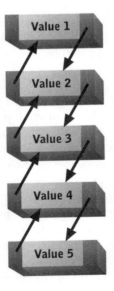

Value 1

Value 2

Value 3

Value 4

Value 5

Records and Files

A **record** is a data structure composed of other data structures or primitive data elements. Records commonly are used as a unit of input and output to files or databases. As an example, consider the following items of data:

Account-Number	Street-Address
Last-Name	City
First-Name	State
Middle-Initial	Zip-Code

This set of data may be the contents of the data structure for a customer record as shown in Figure 3-14. Each component of the record either is a basic data element (for example, Middle-Initial) or another data structure (for example, a character array for Street-Address). To speed input and output, records normally are stored in contiguous storage locations. This restricts the record's array components to a fixed length.

Figure 3-14 ▶

Record data structure

Account-Number	Last-Name	First-Name	Middle-Initial	Street-Address	City	State	Zip-Code

A sequence of records on secondary storage is called a file. A sequence of records stored within main memory normally is called a table, although its structure essentially is the same as a file. Files can be organized in many different ways, the most common being sequential and indexed.

In a sequential file, the records are stored in contiguous storage locations. As with arrays stored in contiguous storage, accessing a specific record is relatively simple. The address of the nth record in a file can be computed as:

$$\text{address-of-first-record} + ((n-1) \times \text{record-size})$$

If the first byte of the first record is at address 1 and the size of a record is 200 bytes, the address of the fourth record is 601.

Sequential files suffer the same problems as contiguous arrays when inserting and deleting records. A copy procedure similar to that shown earlier in Figure 3-11 must be executed to add a record to a file. The procedure is even less efficient for files than for arrays because of the relatively large size of the data, or records, that must be copied.

One method of solving this problem is to use linked lists. With files, the data elements of a linked list are entire records instead of basic data elements. The methods for searching, record insertion, and record deletion essentially are the same as previously described.

Another method of organizing files uses an index. An **index** is an array of pointers to records. The pointers can be ordered in any sequence desired by the user. For example, a file of customer records may be ordered by ascending account number as shown in Figure 3-15.

Figure 3-15 ▶

An indexed file

The advantage of using an index lies in the efficiency of record insertion, deletion, and retrieval. When a record is added to a file, storage is allocated for the record and the data is placed in the storage location. The index then is updated by inserting the address of the new record. The index update is accomplished by the same procedure as an array update. Because the array contains only pointers, it is relatively small in size and fast to update.

Classes and Objects

Up to this point, we have discussed data and programs as two fundamentally different things. Programs contain instructions, which transform data inputs into data outputs when executed. Data items are held within storage devices, moved to CPU registers when needed, transformed into data outputs by executing instructions, and sent to output or storage devices once again. Under this view of computer systems and software behavior, data items are relatively static and programs are the active means for transforming and updating data items.

In the 1980s and 1990s, computer researchers developed an alternative view of computer and software behavior that combined program instructions and data into a single data structure. A **class** is a data structure that contains both traditional, or static, data elements and programs that manipulate that data. The programs in a class are **methods**. A class combines related data items in much the same way as a record, but it extends the record to include methods that manipulate the data items.

Consider the customer record in Figure 3-14 as a starting point for defining a Customer class. The record contains some of the primitive data items and data structures that describe various features of a customer (others, such as account balance, could be added to create a more complete representation of a customer).

To turn the customer record into a Customer class, methods must be added that manipulate or modify data elements in the record. For example, you could add a simple method for updating the content of an individual data element, called Update-First-Name. You could check for legal values, such as that the value supplied for Zip-code is a valid U.S. zip code within the customer's state. Functions such as Print-Mailing-Label, or methods to modify the customer's account balance to reflect transactions, such as Apply-Payment, could be added. One possible set of data elements and methods for the Customer class is shown in Figure 3-16.

Figure 3-16 ▶

A Customer class containing traditional static data elements (darker boxes) and methods (lighter boxes)

An **object** is one instance, or variable, of the class. Each person who is a customer would be represented by one variable or object of the Customer class. Each object could be stored within a storage device, and each object's data elements would occupy a separate part of the storage device.

Viewed as just another data structure, an object differs little from a record. Much as the data element Street-Address could be represented with a character array, a method could be represented as an array of CPU instructions. Methods could be represented in other ways, including linked lists of instructions, or pointers to files containing sequential sets of instructions. In essence, the only new primitive data type required to represent a method within an object is an instruction. As we'll see in the next chapter, instructions, like data, have specific representation and storage formats.

SUMMARY

► Data can be represented in many ways. To be processed by any device, data must be converted from its native format into a form suitable for the processing device. Modern computers represent data as electronic signals and implement processing devices using electronic circuitry. Electronic processing devices exploit the physical laws of electricity. Because the laws of electricity can be stated as mathematical equations, electronic devices can be implemented that use the computational functions embedded within those equations. Other ways of implementing processors, such as mechanics and optics, exploit similar mathematically stated physical laws.

► All data, including nonnumeric data, are represented within a modern computer system as strings of binary digits, or bits. Bits are used because zero and one can be encoded in clearly recognizable electrical states, such as high and low voltage. Binary electrical states can be processed and stored by electrical devices that are reliable and cheap. A handful of data types usually are represented and processed by a CPU, including integers, real numbers, characters, boolean values, and memory addresses.

► Numeric values other than zero and one are represented by combining multiple bits into larger groups, called bit strings, much as multiple decimal digits such as 2, 5, and 9 can be combined to form larger values such as 592. Each bit string has a specific data format and coding method. There are many different data formats and coding methods. CPU designers choose formats and methods that represent the best balance among compactness, ease of manipulation, accuracy, range, and standardization. The optimal balance can vary depending on the type of data and the intended uses for that data.

► Integers have no fractional component and can use relatively simple data formats. Two's complement is the most common integer format, though excess notation sometimes is used. Real numbers have both whole and fractional components and their complex structure requires a complex data format called floating point notation. Floating point notation represents a numeric value as a mantissa multiplied by a positive or negative power of two. A value can have many digits of precision in floating point notation for very large or very small magnitudes, but not both large and small magnitudes at the same time. Floating point formats used in modern computers typically follow a standard promulgated by the IEEE. Floating point format is less accurate and more difficult to process than two's complement format.

► Character data is not inherently numeric. Characters are converted to numbers by means of a coding table. Each character occupies a sequential position within the table. A character is represented by converting it to its integer table position. Characters are extracted from integers by the reverse process. Many different tables are possible. Most modern computers either use the ASCII or Unicode coding tables. ASCII is an older standard geared toward the English language. Unicode is a more modern table that is large enough to encompass all of the world's written languages.

► Boolean values can have only two values, true and false. Memory addresses can be simple or complex numeric values depending on whether the CPU uses a flat or segmented memory model. Flat memory addresses can be represented as a single integer. Segmented memory addresses require multiple integers. Many CPUs also provide double precision numeric data types, which double the number of bits used to store a value.

► Programs often need to define and manipulate data in larger and more complex units than primitive CPU data types. A data structure is a related group of primitive data elements that is organized for some type of common processing. Data structures are defined in software. For a CPU to manipulate a data structure, software must decompose operations on the data structure into an equivalent set of operations on its primitive components. Commonly used data structures include arrays, linked lists, records, tables, files, indices, and objects. Many data structures use pointers, which are stored memory addresses, to link primitive data components.

► In this chapter you've learned the various ways that data is represented within a modern computer system, focusing particularly on the relationship between data representation and the CPU. This chapter and the preceding chapters have introduced basic concepts and terminology of systems architecture. The understanding you've gained in the first three chapters lays the foundation for the more detailed discussion of the hardware implementations of data processing, storage, and communication in upcoming chapters.

Key Terms

address
American Standard Code for
    Information Interchange
    (ASCII)
array
base
Binary Coded Decimal (BCD)
bit
bit string
boolean data type
boolean logic
byte
character
class
collating sequence
data structure
decimal point
double precision
doubly linked list
excess notation

Extended Binary Coded
    Decimal Interchange Code
    (EBCDIC)
flat memory model
floating point notation
hexadecimal notation
high order bit
index
integer
International Alphabet
    Number 5 (IA5)
International Standards
    Organization (ISO)
Latin-1
least significant digit
linked list
long integers
low order bit
machine data type
methods

most significant digit
multinational characters
numeric range
object
octal notation
overflow
pointers
primitive data type
radix
radix point
real number
record
segmented memory model
singly linked list
string
truncation
two's complement
underflow
Unicode

Vocabulary Exercises

1. An element in a(n) _____ contains pointers to both the next and previous
    list elements.

2. _____ notation encodes a real number as a mantissa multiplied by a power
    (exponent) of two.

3. A(n) _____ is an integer stored in double the normal number of bit
    positions.

4. Two's complement notation represents _____ values as ordinary binary
    numbers.

5. Increasing the size (number of bits) of a numeric representation format increases
    the _____ of values that can be represented.

6. Assembly language programs for IBM mainframes commonly use _____
    notation to represent numeric values. Assembly language programs for other com-
    puters commonly use _____ notation to represent numeric values and
    memory addresses.

7. A(n) _____ is a data item composed of multiple primitive data items.

8. In IBM mainframe computers, characters are coded as numbers according to the _____ coding scheme. In most other computers, the _____ character coding scheme is used.

9. A(n) _____ is the address of another data item or structure.

10. In a positional numbering system, the _____ separates digits representing whole number quantities from digits representing fractional quantities.

11. A(n) _____ is an array of characters.

12. Most Intel CPUs implement a(n) _____ where each memory address is represented by two integers.

13. A set of data items that can be accessed in a specified order using a set of pointers is called a(n) _____ .

14. A(n) _____ contains eight _____ .

15. A(n) _____ list stores one pointer with each list element.

16. The result of adding, subtracting, or multiplying two _____ may result in overflow, but never _____ or _____ .

17. A(n) _____ is a sequence of primitive data elements stored in sequential storage locations.

18. A(n) _____ is a group of data elements that usually describe a single entity or event.

19. A(n) _____ data item can contain only the values true or false.

20. A(n) _____ is an array of data items, each of which contains a key value and a pointer to another data item.

21. Many computers implement _____ numeric data types to provide greater accuracy and prevent overflow and underflow.

22. Unlike ASCII and EBCDIC, _____ is a 16-bit character coding table.

23. The _____ is the bit of lowest magnitude within a byte or bit string.

24. The value of a binary signal parameter is compared to a(n) _____ value to determine whether it represents zero or one.

25. _____ occurs when the result of an arithmetic operation exceeds the number of bits available to store it.

26. Within a CPU, _____ arithmetic generally is easier to implement than _____ arithmetic due to a simpler data coding scheme and data manipulation circuitry.

27. Under the _____, memory addresses consist of a single integer.

28. The _____ has defined a character coding table called _____, which combines the ASCII-7 coding table with an additional 128 Western European multinational characters.

29. Portability of floating point data from one computer to another is guaranteed if each computer's CPU represents real numbers using _____ standard floating point notation.

30. The ordering of characters within a coding table is called a(n) _____.

31. A(n) _____ is a data structure that contains both static data and methods.

32. A(n) _____ is one instance or variable of a class.

Review Questions

1. What is the binary representation of the decimal number 10? What is the octal representation? What is the hexadecimal representation?

2. Why is binary data representation and signaling the preferred method of computer hardware implementation?

3. What is excess notation? What is two's complement notation? What criteria determine which notation is used by computer hardware?

4. What is overflow? What is underflow? How can the probability of their occurrence be minimized?

5. How and why are real numbers more difficult to represent and process than integers?

6. Why may a programmer choose to represent a data item in IEEE 64-bit floating point format instead of IEEE 32-bit floating point format? What additional costs are incurred at run time (when the application program executes) as a result using the 64-bit instead of the 32-bit format?

7. Why doesn't a CPU evaluate the expression 'A' = 'a' as true?

8. What are the differences between ASCII and EBCDIC?

9. What primitive data types normally can be represented and processed by a CPU?

10. What is a data structure? List several types of common data structures.

11. What is an address? What is a pointer? For what are they used?

12. How is an array stored in main memory? How is a linked list stored in main memory? What are their comparative advantages and disadvantages? Give an example of data that would be best stored as an array. Give an example of data that would be best stored as a linked list.

13. How does a class differ from other data structures?

Problems and Exercises

1. Develop an algorithm or program to implement the following function:

insert_in_linked_list (element,after_pointer)

The parameter "element" is a data item to be added to a linked list. The parameter "after_pointer" is the address of the element after which the new element will be inserted.

2. Develop an algorithm or program to implement the following function:

insert_in_array (element,position)

The parameter "element" is a data item to be added to the array. The parameter "position" is the array index at which the new element will be inserted. Make sure to account for elements that must be moved over to make room for the new element.

3. Consider the following binary value:

1000 0000 0010 0110 0000 0110 1101 1001

What number (base ten) is represented if the value is assumed to represent a number stored in two's complement notation? Excess notation? IEEE floating point (32-bit) notation?

4. What is the numeric range of a 16-bit two's complement value? A 16-bit excess notation value? A 16-bit unsigned binary value?

Research Problems

1. Choose a commonly used microprocessor such as the Intel Pentium (www.intel.com) or IBM/Motorola PowerPC (www.ibm.com). What data types are supported? How many bits are used to store each data type? How is each data type internally represented?

2. Most personal and office productivity applications, such as word processors, are marketed internationally. To minimize development costs, software producers develop a single version of the program, but have separate configurations for items like menus and error messages that vary from language to language.

Investigate a commonly used development environment for application tools such as Microsoft Developer Studio (www.microsoft.com). What tools and techniques (for example, Unicode data types and string tables) are supported for developing multilingual programs?

3. Object-oriented programming has been adopted widely due to its inherent ability to reuse code. Most application development software provides object libraries and extensive support for complex data structures, including various types of linked lists. Investigate one of these libraries, such as Microsoft Foundation Classes (www.microsoft.com) and the data structure objects provided. What types of data are recommended for use with each data structure object? Which classes contain which data structures, and what methods are provided by the library?

Chapter 4

Processor Technology and Architecture

Chapter Goals

- Describe CPU instruction and execution cycles

- Explain how primitive CPU instructions are combined to form complex processing operations

- Describe the key CPU design features, including instruction format, word size, and clock rate

- Describe the function of general-purpose and special-purpose registers

- Compare and contrast CISC and RISC CPUs

- Describe the principles and limitations of semiconductor-based microprocessors

Chapter 2 gave a brief overview of computer processing, including the function of a processor, general-purpose and special-purpose processors, and the components of a central processing unit (CPU). This chapter explores CPU operation, instructions, components, and implementation (see Figure 4-1). It also takes a look at future trends in processor technology and architecture.

Figure 4-1 ►

Topics covered in this chapter

CPU OPERATION

Recall from Chapter 2 that a CPU has three primary components—the control unit, arithmetic logic unit (ALU), and a set of registers (see Figure 4-2). The control unit moves data and instructions between main memory and registers. The ALU performs all computation and comparison operations. Registers are storage locations that hold inputs and outputs for the ALU.

Figure 4-2 ▶

Components of
the CPU

A complex chain of events occurs when the CPU executes a program. To start, the control unit reads the first instruction from primary storage. The control unit then stores the instruction in a register and, if necessary, reads data inputs from primary storage and also stores them in registers. If the instruction is a computation or comparison instruction, the control unit signals the ALU what function to perform, where the input data is located, and where to store the output data. Instructions to move data to memory, I/O devices, or secondary storage are executed by the control unit itself. When the first instruction has been executed, the next instruction is read and executed. The process continues until the final instruction of the program has been executed.

The actions performed by the CPU can be divided into two groups—the **fetch cycle** (or **instruction cycle**) and the **execution cycle**. During the fetch cycle, data inputs are prepared for transformation into data outputs. During the execution cycle, the transformation takes place and data output is stored. The CPU constantly alternates between instruction and execution cycles. Figure 4-3 shows the flow between instruction and execution cycles denoted by solid arrows and data and instruction movement denoted by dashed arrows.

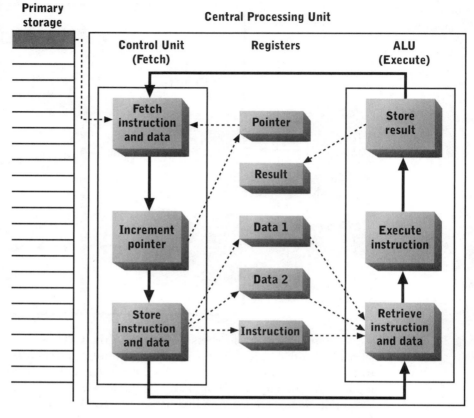

Figure 4-3 ▶

Control and data flow during the fetch and execution cycles

During the fetch cycle, the control unit:

▶ Fetches an instruction from primary storage

▶ Increments a pointer to the location of the next instruction

▶ Separates the instruction into components—the instruction code (or number) and the data inputs to the instruction

▶ Stores each component in a separate register

During the execution cycle, the ALU:

▶ Retrieves the instruction code from a register

▶ Retrieves data inputs from registers

▶ Passes data inputs through internal circuits to perform the addition, subtraction, or other data transformation

▶ Stores the result in a register

At the conclusion of the execution cycle a new fetch cycle is started. The control unit keeps track of the next program instruction location by incrementing a pointer after each fetch. The second program instruction is retrieved during the second fetch cycle, the third instruction is retrieved during the third fetch cycle, and so forth.

INSTRUCTIONS AND INSTRUCTION SETS

An **instruction** is a command to the CPU to perform one of its primitive processing functions on specific data inputs. It is the lowest-level command that software can direct a processor to perform. As stored in memory, an instruction is merely a bit string. The bit string logically is divided into a number of components. The first group of bits represents the unique binary number of the instruction, commonly called the **op code**. Subsequent groups of bits hold the input values for the instruction, called **operands**. The content of an operand can represent a data item (such as an integer value) or the location of a data item (such as a memory address, a register address, or the address of a secondary storage or I/O device).

Figure 4-4 ▶

An instruction containing one op code and two operands

Operand 1

Op code

Operand 2

Physically, an instruction directs the CPU to route electrical signals representing data input(s) through a predefined set of processing circuits that implement the desired function. Data inputs are accessed from storage or extracted directly from the operands, and stored in one or more registers. For computation and logic functions, the ALU is directed to access these registers and send the corresponding electrical signals through the appropriate processing circuitry. This circuitry transforms the input signals into output signals representing the processing result. This result is stored in a register in preparation for movement to a storage device or I/O device, or for use as input to another instruction.

Some instructions are executed by the control unit without assistance from the ALU. Instructions for moving or copying data, as well as some simple CPU functions like halting or restarting the CPU, are handled by the control unit alone.

In general, instructions that require transforming data inputs into new outputs are executed by the ALU. All other instructions are executed by the control unit.

The collection of instructions that a CPU can process is called the CPU's **instruction set**. Instruction sets vary among CPUs in the following ways:

- Size of the instruction set
- Size of individual instructions, op codes, and operands
- Supported data types
- Number and complexity of processing operations implemented as individual instructions

Instruction set variations reflect differences in design philosophy, processor fabrication technology, class of computer system, and type of application software. CPU cost and speed depend on these design parameters.

The full range of processing operations expected of a modern computer can be implemented with approximately one dozen instructions. Such an instruction set can perform all of the computation, comparison, data movement, and branching functions for integer and boolean data types. Computation and comparison of real numbers can be accomplished by software with complex sequences of integer instructions operating separately on the whole and fractional parts. Small instruction sets were common in early CPUs and microprocessors. Details of this minimal instruction set are described in the following sections.

Data Movement

A **MOVE** instruction copies data bits to storage locations. MOVE can copy data among any combination of registers and primary storage locations. Data transfers from main memory into registers usually are called **load** operations. Data transfers from registers into primary storage usually are called **store** operations.

MOVE tests the bit values in the source location and places copies of those values in the destination location. The former bit values in the destination are overwritten. At the completion of the MOVE, both sending and receiving locations hold identical copies. The name "move" is a misnomer because the data content of the source location is unchanged.

MOVE also is used to access storage and I/O devices. An input or storage device writes to a specific memory location and the CPU retrieves the input by reading that memory location and copying its value into a register. Similarly, data is output or stored by writing to a predefined memory address or range of addresses. The output or storage device continually monitors the content of its assigned memory address(es) and reads newly written data for storage or output.

Data Transformations

The most primitive data transformation instructions are based on boolean logic:

- ► NOT
- ► AND
- ► OR
- ► XOR

These four boolean instructions and the ADD and SHIFT instructions (which are discussed in detail in the following sections) are the basic building blocks of all numeric comparisons and computations. They are summarized in Table 4-1.

Table 4-1 ►

Primitive data transformation instructions

Instruction	Function
NOT	Each result bit is the opposite of the operand bit
AND	Each result bit is true if both operand bits are true
OR	Each result bit is true if either or both operand bits are true
XOR	Each result bit is true if either, but not both, operand bits are true
ADD	Result is the arithmetic sum of operands
SHIFT	Move all bit values left or right as specified by operand

NOT   A **NOT** instruction transforms the boolean value true (1) into false (0) and the value false into true. The rules that define the output of NOT on single bit inputs are:

$$\text{NOT } 0 = 1$$
$$\text{NOT } 1 = 0$$

With bit strings, NOT treats each bit in the bit string as a separate boolean value. For example, executing NOT on the input 10001011 produces the "opposite" result—01110100. Note that NOT has only one data input, whereas all other boolean instructions have two.

AND   An **AND** instruction generates the result true if both of its data inputs are true. The following rules define the result of AND on single bit data inputs:

$$0 \text{ AND } 0 = 0$$
$$1 \text{ AND } 0 = 0$$
$$0 \text{ AND } 1 = 0$$
$$1 \text{ AND } 1 = 1$$

The result of AND with two bit string inputs is shown in the following example:

```
        10001011
AND     11101100
        10001000
```

OR   There are two types of OR operations in boolean logic. An **inclusive OR** instruction (the word inclusive usually is omitted) generates the value true if either or both data inputs are true. The rules that define the output of inclusive OR on single bit inputs are:

```
0 OR 0 = 0
1 OR 0 = 1
0 OR 1 = 1
1 OR 1 = 1
```

The result of inclusive OR with two bit string inputs is shown in the following example:

```
       10001011
OR     11101100
       11101111
```

An **exclusive OR,** also called **XOR,** instruction generates the value true if either, but not both, data inputs are true. The rules that define the output of XOR on single bit inputs are:

```
0 XOR 0 = 0
1 XOR 0 = 1
0 XOR 1 = 1
1 XOR 1 = 0
```

Note that if either operand is one the result is the complement of the other operand. Specific bits within a bit string can be inverted by XORing the bit string with a string containing zeros in all positions except the positions to be negated. For example, the following XOR inverts only the right four bit values and leaves the left four bit values unchanged:

```
        10001011
XOR     00001111
        10000100
```

Every bit in a bit string can be inverted by XORing with a string of ones:

```
        10001011
XOR     11111111
        01110100
```

Note that XORing any input with a string of ones produces the same result as executing NOT. Thus, NOT isn't required in a minimal instruction set.

ADD An **ADD** instruction accepts two numeric inputs and produces their arithmetic sum. For example:

$$
\begin{array}{r}
10001011 \\
\text{ADD} \quad \underline{00001111} \\
10011010
\end{array}
$$

This shows the binary addition of two bit strings. Note that the mechanics of the addition operation are the same regardless of what the bit strings represent. In this example, if the bit strings represent unsigned binary numbers, then the operation is:

$$139_{10} + 15_{10} = 154_{10}$$

If the bit strings represent signed integers in two's complement format, then the operation is:

$$-117_{10} + 15_{10} = -102_{10}$$

Binary addition does not work for complex data types such as floating point and double precision numbers. If complex numeric data types are supported by the CPU, then a separate ADD instruction must be implemented for each type.

SHIFT The effect of a **SHIFT** instruction is shown in Figure 4-5. In Figure 4-5(a), the value 01101011 occupies an 8-bit storage location. Bit strings can be shifted to the right or left, and the number of positions shifted may be greater than one. Typically, a second operand is used to hold an integer value that indicates the number of bit positions by which the value will be shifted. Positive or negative values of this operand can be used to indicate shifting to the left or right.

Figure 4-5(b) shows the result of shifting the value two positions to the right. The resulting value is 00011010. In this case, the values in the two least significant positions of the original string have been dropped and the vacant bit positions are filled with zeroes.

Figure 4-5 is an example of a **logical SHIFT**. Logical SHIFT typically is used to extract a single bit from a bit string. Figure 4-6 shows how shifting an 8-bit value (Figure 4-6(a)) to the left by four bits (b) and then to the right by seven bits, creates a result (c) with the third bit of the original string in the rightmost position. Since all other bit positions contain zeros, the entire bit string can be interpreted as true or false. Shifting an 8-bit two's complement value to the right by seven positions is a simple way to extract and test the sign bit.

Figure 4-5 ►

Original data
byte (a) shifted
two bits to the
right (b)

Figure 4-6 ►

Extracting a
single bit with
logical SHIFT
instructions

An **arithmetic SHIFT** instruction performs multiplication or division, as illustrated in Figure 4-7. If a bit string contains an unsigned binary number (Figure 4-7(a)), then shifting to the left by one bit (b) multiplies the value by two, and shifting the original bit (a) to the right by two bits (c) divides by four. Arithmetic SHIFT instructions are more complex when applied to two's complement values because bits must be shifted "around" the sign bit. Most CPUs provide a separate arithmetic SHIFT instruction that preserves the sign bit of a two's complement value.

Figure 4-7 ▶

Multiplying and dividing binary values using SHIFT instructions

Binary values

Decimal values

(a) Original byte — 40

(b) Arithmetic shift one position to left — 80

(a) Original byte — 40

(c) Arithmetic shift two positions to right — 10

Sequence Control

Sequence control operations alter the flow of instruction execution in a program. These operations include:

▶ Unconditional branch

▶ Conditional branch

▶ Halt

Branch Operations   A BRANCH (or JUMP) instruction causes the processor to depart from sequential instruction order. Recall that the control unit fetches the next instruction from memory at the conclusion of each execution cycle. The control unit consults a register to determine where the instruction is located in primary storage. BRANCH has one operand containing the memory address of the next instruction. BRANCH actually is implemented as a MOVE instruction. The BRANCH operand simply is loaded into the register that the control unit uses to fetch the next instruction.

In an **unconditional BRANCH,** the processor always departs from the normal execution sequence. In a **conditional BRANCH,** the BRANCH occurs only if a specified condition is met, such as the equivalence of two numeric variables. The condition must be evaluated and the boolean result stored in a register. The conditional BRANCH instruction checks the content of that register and only branches if the value contained there is true.

HALT   A **HALT** operation suspends the normal flow of instruction execution in the current program. In some CPUs, it causes the CPU to cease all operations. In others it causes a BRANCH to a predetermined memory address. A portion of the operating system typically is loaded at this address, effectively transferring control to the operating system and terminating the previously executing program.

Complex Processing Operations

Complex processing operations can be performed by combining the simpler operations described previously. For example, subtraction can be implemented as complementary addition. That is, the operation A - B can be implemented as A + (-B). As described in Chapter 3, a negative two's complement value can be derived from its positive counterpart by taking the complement of the positive value and adding one. A bit string's complement can be generated by XORing it with a string of binary one digits.

For example, the complement of 0011 (3_{10}), represented as a two's complement value can be derived as:

$$XOR(0011,1111) + 0001 = 1100 + 0001 = 1101 = -3_{10}$$

This result then can be added to implement a subtraction operation. For example, the result of subtracting 0011 from 0111 can be calculated as:

$$7_{10} - 3_{10} = ADD(ADD(XOR(0011,1111),0001),0111)$$

$$= ADD(ADD(1100,0001),0111)$$

$$= ADD(1101,0111)$$

$$= 10100$$

Because 4-bit values are used, the result of 10100 is truncated from the left, resulting in a value of 0001.

Comparison operations can be implemented similar to subtraction. A comparison operation generates a boolean output value. Typically, an integer value of zero is interpreted as false and any nonzero value is interpreted as true. The comparison A ≠ B can be implemented by generating the complement of B and adding it to A. If the two numbers are equal, the result of the addition will be a string of zeros (interpreted as false). An equality comparison can be implemented by negating the boolean result of an inequality comparison.

Greater than and less than comparisons also can be performed with subtraction followed by extraction of the sign bit. For the condition A < B, subtracting B from A will generate a negative result if the condition is true. In two's complement notation, a negative value always will have a one in the leftmost position (that is, the sign bit). SHIFT can be used to extract the sign bit from the remainder of the value. For example, the two's complement value 10000111 is a negative number.

The sign bit may be extracted by shifting the value 7 bits to the right, resulting in the string 00000001. The SHIFT result may be interpreted as a boolean value (nonzero, or true, in this case).

For example, the comparison:

$$0111 < 0011$$

can be evaluated as:

$$SHIFT(ADD(0111,ADD(XOR(0011,1111),0001)),0011)$$

$$SHIFT(ADD(0111,ADD(1100,0001)),0011)$$

$$SHIFT(ADD(0111,1101),0011)$$

$$SHIFT(0100,0011)$$

$$0000$$

The second operand of the SHIFT instruction is a binary number representing the direction and number of bit positions to shift (+3, or right 3 in this example). The result is zero and is interpreted as the boolean value false.

A Short Programming Example[1]

Consider the following high-level programming language statement:

```
IF (BALANCE < 100) THEN
        BALANCE = BALANCE - 5
ENDIF
```

Such a computation may be used in a program that applies a monthly service charge to checking or savings accounts with a balance below a certain minimum. A program that implements this computation using only the previously defined low level CPU instructions is shown in Figure 4-8. Table 4-2 shows the register contents after each instruction is executed when the account balance is $64. An 8-bit two's complement is the assumed coding format for all numeric data.

---

[1] The reader may wish to review the discussion of two's complement notation in Chapter 3 before studying these examples.

Figure 4-8 ▶

A simple
program using
primitive CPU
instructions

```
1     MOVE      M1, R1           ' load BALANCE
2     MOVE      M2, R2           ' load minimum balance 100₁₀
3     MOVE      M3, R3           ' load service charge 5₁₀
4     MOVE      M4, R4           ' load constant 1₁₀
5     MOVE      M5, R5           ' load constant 7₁₀
6     NOT       R2, R2           ' start < comparison
7     ADD       R2, R4, R2       '
8     ADD       R1, R2, R0       '
9     SHIFT     R0, R5           ' end < comparison
10    XOR       R0, R4, R0       ' invert comparison result
11    BRANCH    R0, 16           ' branch if comparison false
12    NOT       R3, R3           ' start negation of service charge
13    ADD       R3, R4, R3       ' end negation of service charge
14    ADD       R1, R3, R0       ' subtract service charge from balance
15    MOVE      R0, M1           ' store new balance
16    HALT                       ' terminate program
```

Table 4-2 ▶

Register contents after executing each instruction in Figure 4-8 when account balance in M1 is $64

Instruction	R0	R1	R2	R3	R4	R5
1		01000000				
2			01100100			
3				00000101		
4					00000001	
5						00000111
6			10011011			
7			10011100			
8	11011100					
9	00000001					
10	00000000					
11						
12				11111010		
13				11111011		
14	00111011					
15						
16						

Instructions 1 through 5 load the account balance, minimum balance, service charge, and needed binary constants from memory locations M1 through

M5. A less than comparison is performed in instructions 6 through 9. The right side of the comparison is converted to a negative value by executing a NOT instruction (instruction 6) and adding 1 to the result (instruction 7). The result is added to the account balance (instruction 8) and the sum is shifted seven places to the right to extract the sign bit (instruction 9). At this point register R0 holds the boolean result of the less than comparison.

For this example, all BRANCH instructions are assumed to be conditional on the content of a register. The BRANCH is taken if the register holds a boolean true value, and otherwise ignored. To jump beyond the code that implements the service charge if the account balance is above the minimum, we must invert the boolean result of the condition prior to branching. Instruction 10 inverts the sign bit stored in the rightmost bit of R0 by XORing it against 00000001 (stored in R5). The conditional BRANCH then is executed. Because the original sign bit was one, the inverted value is zero. Thus the BRANCH is ignored and processing proceeds with instruction 12.

Instructions 12 through 14 subtract the service charge stored in register R3 from the account balance. Instructions 12 and 13 convert the positive value to a negative value, and instruction 14 adds it to the account balance. Instruction 15 saves the new balance in memory. Instruction 16 halts execution of the program.

Instruction Set Extensions

The instructions described to this point are sufficient for a simple general-purpose processor. All of the more complex functions normally associated with computer processing can be implemented by combining those primitive building blocks. Most CPUs provide a much larger set of instructions, however, including advanced computation operations such as multiplication and division, negation of two's complement values (NOT followed by ADD 1), and testing a sign bit and other single-bit manipulation functions. Such instructions sometimes are referred to as **complex instructions** because they represent combinations of primitive processing operations.

Complex instructions represent a tradeoff between processor complexity and programming simplicity. For example, consider the three-step process of NOT - ADD - ADD used to implement subtraction in Figure 4-8, instructions 12 through 14. Because subtraction is a commonly performed operation, most CPU instruction sets include one or more subtraction instructions. This complicates the CPU, because there is extra processing circuitry for the additional instruction, but simplifies machine language programs.

Complex instructions also represent a tradeoff between CPU complexity and program execution speed. Multistep instruction sequences execute faster if they are executed within hardware as a single instruction, avoiding the overhead of fetching multiple instructions and accessing intermediate results in registers.

Other efficiencies also may be realized by hardwiring the steps together. However, these efficiencies have limits, as described later.

Additional instructions are required when new data types are added. For example, most CPUs provide instructions for adding, subtracting, multiplying, and dividing integers. If double-precision integers and floating point numbers are supported, additional computation instructions must be included for those data types. It is not unusual to see a half dozen different ADD instructions in an instruction set, to support integers and real numbers in single and double precision, as well as signed and unsigned "short" integers.

Some instruction sets also include instructions that combine data transformations with data movement. The primitive instructions in Figure 4-8 use registers for both data input and data output. In some CPUs, data transformation instructions allow one or more operands to be a memory address. In essence, this combines data movement with data transformation (AND, OR, XOR, ADD, etc.).

INSTRUCTION FORMAT

Recall that an instruction consists of an op code (instruction number) and zero or more operands (representing data values or storage locations). An **instruction format** is a template that specifies the number of operands and the position and length of the op code and operands. Instruction formats vary among CPUs in many ways, including op code size, meaning of specific op code values, data types used as operands, and the length and coding format of each type of operand. The term instruction format erroneously implies that a single format is used by each CPU. Because instructions vary in the number and type of operands they require, there generally are multiple instruction formats used by a CPU. Different formats represent various combinations of operand types.

Figure 4-9 shows an instruction format consisting of an op code and three operands with a total length of 20 bits. The sample format is typical of instructions that use register inputs and outputs. The op code occupies the first 8 bits. Most CPUs represent the op code as an unsigned binary number. Thus, an 8-bit op code allows for 256 possible instructions numbered 0 through 255. Each operand is a 4-bit unsigned binary number identifying one of 16 possible registers.

Figure 4-9 ▶

An instruction
format with
three register
operands

Figure 4-10 shows another instruction format for load and store instructions in a CPU that uses segmented memory. The format contains three operands and is 32 bits in length. The first two operands store register numbers. The first register contains the data value to be moved for a store instruction, or to hold the value being moved from memory, for a load instruction. The second register contains a partial memory address.

Figure 4-10 ▶

An instruction
format that uses
segmented
memory

Recall from the previous chapter that some CPUs use a segmented memory addressing scheme in which a memory address consists of two parts—a memory segment identifier and a segment offset. For segmented memory schemes, the second operand identifies a register that holds the segment identifier, and the third operand is the segment offset. The control unit uses the contents of the register identified in the second operand and the content of the third operand to construct a complete memory address.

Instruction Length

Instruction formats within a single CPU can be fixed or variable in length. **Fixed length instructions** simplify the instruction-fetching process implemented within the control unit. If the instruction format is of a fixed length, then the amount by which the instruction pointer must be incremented after each fetch is a constant. This increment is the length of an instruction.

It appears that a CPU that uses the two instruction formats shown earlier in Figure 4-9 and Figure 4-10 is using variable length instructions because the formats are not of equal length. This is not necessarily the case. Instruction format lengths can be equalized by "padding" the shorter instruction formats with trailing zero bits. For example, the format in Figure 4-9 could be padded with 12 zero bits to increase its length to match the format in Figure 4-10. The extra bits simply are ignored by the CPU when processing instructions that use the padded format.

With a **variable length instruction**, the amount by which the instruction pointer is incremented after a fetch is the length of the most recently fetched instruction. The control unit must check the op code of each fetched instruction to determine the proper incrementation value. Conceptually, you might think of a table with op code values on the left and format lengths on the right. The control unit would look up the op code in the table and extract the appropriate value to add to the instruction pointer.

Variable length instructions also complicate the instruction fetching operation because the number of bytes to be fetched is not known in advance. One method of addressing this uncertainty is always to fetch the number of bytes in the longest instruction format, but this can result in many unnecessary memory accesses. Another method that can be used is to fetch only the number of bytes in the shortest format. Then the op code is extracted and used as before, to determine the length of the instruction and the number of additional bytes to fetch, if any. In either case, extra computer resources are used in the fetch operation that would not have been used if the instruction length were fixed and known before the fetch.

Although fixed length instructions and fields simplify control unit functions, they do so at the expense of efficient memory use. Some instructions have no operands and others have one, two, or three operands. If fixed length instructions are used, then the instruction length must be the length of the longest instruction, for example, an instruction with two or three large operands. Smaller instructions stored in memory must be padded with empty bit positions to extend their length to the maximum. The memory used to pad short instructions is wasted, and programs require more memory during execution and more time to load from secondary storage.

Reduced Instruction Set Computing

Reduced Instruction Set Computing (RISC) is a philosophy of processor and computer system design that first was employed in the late 1980s. The primary architectural feature of a RISC processor is the absence of some, but not all, complex instructions in the instruction set. In particular, instructions that combine data transformation and data movement operations usually are avoided. For example, a non-RISC processor might provide a data transformation instruction of the form:

Transform(Address1,Address2,Address3)

Address1 and Address2 are the memory addresses of data inputs, and Address3 is the memory address to which the result will be stored. Depending on the processor implementation, this instruction may require as little as one or as many as four execution cycles to complete. In a RISC processor, transformation operations always use register inputs and outputs. The single complex instruction shown above would require four separate RISC instructions:

MOVE(Address1,R1)
MOVE(Address2,R2)
Transform(R1,R2,R3)
MOVE(R3, Address3)

Each instruction would be loaded and executed independently, consuming at least four execution cycles.

Although the lack of many complex instructions is the primary distinguishing feature of a RISC processor, other differences also are typical. These include the use of fixed length instructions, short instruction length, and a large number of general-purpose registers. To describe these differences fully, we must first contrast RISC with its opposite design philosophy—**Complex Instruction Set Computing (CISC)**.

Complex instruction sets were developed because early computers had limited memory and processing power. Memory was very expensive and many computers barely had enough to hold an entire application program in memory. To compensate for limited memory, many CPU designers used complex instructions that would do more work per instruction. As a result, programs required less memory, and complex operations executed more quickly.

For example, assume that a floating point addition operation, implemented as a single complex instruction, can be executed in one processor cycle. Assume further that an equivalent operation can be performed with five integer math instructions, each of which requires one processor cycle. In this example, a direct implementation of the complex floating point instruction saves four processor cycles each time it is executed.

If complex instructions are so beneficial, then why would anyone want to eliminate them? The most important reason is that the benefits of complex instructions are subject to the law of diminishing returns. Each complex instruction provides benefits if measured in isolation. But as more and more complex instructions are added, the instruction set as a whole becomes bloated. This leads to costs that affect the entire instruction set.

A bloated instruction set is a problem in two ways. First, it complicates the control unit because there are more instructions to interpret and usually more instruction formats with which to deal. Also, a large set of complex instructions often goes hand in hand with variable length instruction formats. This adds even more complexity to the job of fetching and decoding instructions, which in turn increases fetching and decoding time. Because every simple or complex instruction must be fetched before it is executed, a performance penalty is applied to every instruction, even the simple ones.

The second problem arising from a bloated instruction set is microprocessor size. As discussed later in this chapter, speed improvements in microprocessors primarily are achieved by shrinking them. The simpler the processor, the easier the task of shrinking it and the more reliably smaller versions can be fabricated. Because CISC processors are much more complex than RISC processors, they are more difficult to fabricate reliably in smaller sizes. A RISC processor is designed to trade the size and complexity of the instruction set for raw speed in fetching and executing instructions. RISC design follows a "less is more" strategy and extends the instruction set only when the benefits are very high.

The primary disadvantage of RISC CPUs is the extra memory required for program storage and execution. Because many complex instructions are not present, programs must use multiple simple instructions in their place. The equivalent sequences of simple instructions occupy more storage than their complex equivalents and increase program memory requirements. Although this is a disadvantage for RISC, it is not one that results in significantly higher memory cost because memory cost has fallen more rapidly than CPU cost.

RISC processors also are relatively inefficient at executing programs that do many of the functions for which complex instructions are designed, for example, transforming data items stored in memory and immediately storing the results back to memory. Although this potentially is a disadvantage, detailed studies of typical program behavior have shown that many complex instructions are used infrequently. Typical programs spend most of their time executing relatively primitive instructions such as load, store, addition, and comparison. In most cases, the speed advantage of complex instructions is not realized frequently enough to make up for the performance penalty applied to every instruction in a more complex instruction set.

The choice of CISC versus RISC highlights the multiple factors that must be balanced by a computer hardware designer. CISC and RISC are not fundamentally different strategies; they are simply different points along a design continuum representing tradeoffs among design factors. The economics and technology of program design and behavior, processor fabrication, and memory cost made CISC an optimal solution in the early days of computing. As memory became cheaper and it became possible to build smaller and faster CPUs, the optimal tradeoff shifted toward RISC. Twenty more years of technology changes may shift the optimal tradeoff back toward CISC or toward a completely different approach to CPU design and implementation.

CLOCK RATE

The system clock is a digital circuit that generates timing pulses, or signals, and transmits the pulses to other devices within the computer. The clock generally is an entirely separate device with a dedicated communication line that is monitored by all devices within the computer system. All actions, especially the instruction and execution cycles of the CPU, are timed according to this clock. Storage and I/O devices are timed by the clock signal, and all devices within a computer system coordinate their activities with the system clock.

The frequency at which the system clock generates timing pulses is the **clock rate** of the system. Each "tick" of the clock is a **clock cycle**. CPU and computer system clock rates are expressed in **hertz (Hz)**. One hertz corresponds to one clock cycle per second. Modern CPUs and computer systems typically use clocks that generate millions or billions of timing pulses per second. The frequency of these clocks is measured in **megahertz (MHz)**, meaning millions of cycles per second, or **gigahertz (GHz)**, billions of cycles per second.

The inverse of the clock rate is called the CPU **cycle time**. In most CPUs, the cycle time is the time required to fetch and execute the simplest instruction in the instruction set. For example, assume that the CPU clock rate is 1 GHz and that NOT is the simplest and fastest instruction. The time required to fetch and execute that instruction can be computed as the inverse of the clock rate:

$$\text{cycle time} = \frac{1}{\text{clock rate}} = \frac{1}{1,000,000,000} = 0.000000001 \text{ second} = 1 \text{ nanosecond}$$

Clock rate and cycle time are important CPU performance measures. However, they tell only part of the performance story for a CPU or computer system. CPU clock rate frequently is misinterpreted by equating it with:

► Instruction execution rate

► Overall computer system performance

From the perspective of program execution speed, the most important CPU performance consideration is the rate at which instructions are executed. That rate generally is stated in units called **millions of instructions per second (MIPS)**. MIPS generally are assumed to measure CPU performance when manipulating single precision integers. CPU performance when manipulating single precision floating point numbers is measured in **millions of floating point operations per second** (**MFLOPS**).

If all instructions, regardless of function or data type, were executed within a single clock cycle, then clock rate, MIPS, and MFLOPS would be equivalent. Although the time required to fetch an instruction is the same for all instructions, the time required to execute instructions varies. Instructions for simple processing functions such as boolean functions, equality comparison, and integer addition execute quickly. Instructions for more complex functions like multiplication and division execute more slowly.

Most complex instructions require more than one clock cycle to execute. Execution of a complex instruction continues through one or more successive fetch and execute cycles. A new instruction fetch does not begin until the beginning of the next clock cycle after the previous instruction completes execution. Complex integer computations typically require two to four clock cycles and complex floating point computations typically require three to ten clock cycles.

The number of instructions executed in a given time interval depends on the mix of simple and complex instructions in a program. Assume, for example, that a program executes 100 million integer instructions. Also assume that 50 percent are simple instructions requiring a single clock cycle and 50 percent are complex instructions requiring an average of three clock cycles. The average program instruction requires two clock cycles, in this case, $(1 \times 0.50) + (3 \times 0.50)$, and the CPU MIPS rate is 50 percent of the clock rate when executing this program. Different mixes of simple and complex instructions and different mixes of integer and floating point data will result in different ratios of clock rate to MIPS and MFLOPS. For all but the most trivial programs, MIPS and MFLOPS will be much smaller than the CPU clock rate.

The previous MIPS calculation assumes that nothing hinders the CPU in fetching and executing instructions. Unfortunately, the CPU relies heavily on other devices in the computer system to keep it supplied with instructions and data. These devices are slower than the CPU and are a potential source of delay. For example, main memory typically is two to ten times slower than the processor. That is, the time required to complete a main memory read or write operation typically is two to ten CPU clock cycles. Secondary storage accesses are thousands or millions of times slower than the CPU. The CPU may be idle while waiting for access to storage and I/O devices. Each clock cycle that the CPU spends waiting for a slower device is called a **wait state**.

Unfortunately, a CPU can spend much of its time in wait states, during which no instructions are being executed. The effective MIPS rate of a computer system can be much less than the MIPS rate of the CPU measured in isolation because of the delays imposed by waiting for storage and I/O devices. Various methods of minimizing wait states are discussed in Chapter 6.

Business
Focus

RISC or CISC for the Desktop?

With the notable exception of Intel and its clones, most CPU developers have abandoned CISC in favor of RISC. Computer system manufacturers that heavily or exclusively use RISC CPUs include IBM and Sun Microsystems. The only CISC-based computers produced by these companies are those directly aimed at the IBM PC-compatible market and some mainframe computers.

Most application software for the desktop market is sold in binary form. That is, source code instructions already have been translated into executable code (binary CPU instructions). Binary application software depends on the CPU instruction set for which it was compiled. Any new CPU that executes an older binary application program must implement a compatible instruction set for the program, sometimes referred to as **binary compatibility**. Newer Intel CPUs have implemented binary compatibility by preserving the instruction set of older Intel processors. Users can upgrade CPUs and computers while preserving their investment in application software.

A CPU that does not directly implement an older instruction set must emulate its function to run binary software designed for that older instruction set. Emulation is accomplished by translating CPU instructions for the older processor into equivalent instructions for the new processor. Translation is performed during program execution as each instruction is fetched. Pentium CPU clones, such as those manufactured by AMD, combine a RISC processor with hardware emulation to translate Intel 80x86 instructions into "native" RISC instructions.

Emulating older instruction sets within newer CPUs provides backward compatibility, but at the expense of CPU complexity. In theory, the added complexity slows performance and limits increases in clock rate. In practice, Intel processors have appeared immune to this fate to this point. Although Intel clock rates lag slightly behind many RISC designs, there has not been a significant performance difference. Intel has kept pace by investing heavily in processor design and world-class fabrication facilities. In essence, Intel has kept up with RISC designs by staying one step ahead in the fabrication process. The huge profits made from the IBM PC-compatible boom have made this possible.

RISC processors generally operate at higher clock rates than do their CISC counterparts. This difference is apparent to a user when executing software designed for the RISC processor, but much of the advantage is lost when executing older software due to the cost of emulating older instruction sets. A newer RISC processor provides little or no performance improvement if it must emulate an older CISC instruction set. The lack of a clear performance advantage has left purchasers of desktop computers with little reason to switch to newer RISC CPUs until recently.

Newer operating systems such as Windows 2000 and Linux support CPUs from a variety of manufacturers. Operating systems designed for Intel CPUs can execute on Intel CPUs (such as the Pentium III) and their clones (such as the AMD Athlon). Other versions can execute on RISC CPUs, including the Sun SPARC and Compaq Alpha CPUs. The new order being ushered in by newer operating systems and RISC competitors of Intel has complicated the task of acquiring PC-compatible desktop computers. Until recently there was no real choice in CPUs for PCs. You simply purchased the fastest and most current Intel CPU that you could afford. Now a purchaser is presented with the choice of Intel CPUs, RISC-based Intel clones, and true RISC CPUs with high clock rates. How do you decide which is best?

Questions

1. You are asked by a family member to recommend a replacement for his aging PC-compatible. He has a large inventory of software, including games, entertainment, and personal productivity programs. Most of this software is less than 3 years old, but some of it is as much as 10 years old. He has heard from a friend that substantial cost savings can be achieved by NOT buying a computer system with an Intel processor. What is your recommendation to him?

2. You are a system administrator for a small building contractor with 50 employees. Your user base includes clerical, managerial, and technical personnel, including architects and draftsmen. The company has just obtained a large capital infusion in preparation for a substantial business modernization and expansion. You have been asked to prepare a proposal to replace all of the company's desktop computers over the next two years. Should you seriously consider alternatives to Intel-based PC-compatibles?

CPU REGISTERS

Registers play two primary roles in CPU operation. First, they provide a "scratchpad" for the currently executing program, holding data that is needed quickly or frequently by the program. Second, registers store information about the currently executing program and about the status of the CPU, for example, the address of the next program instruction or error messages and signals from external devices.

General-Purpose Registers

General-purpose registers are used only by the currently executing program. General-purpose registers typically are used to hold intermediate results or to hold data values, such as loop counters or array indices, that will be used frequently. Register accesses are fast because registers are implemented within the CPU. In contrast, storing and retrieving from primary storage is much slower. Using registers to store data needed immediately or frequently increases program execution speed. In general, increasing the number of general-purpose registers increases the execution speed of a program.

Adding registers only increases execution speed up to a point. There are a limited number of intermediate results or frequently used data items in any given process or program, so CPU designers try to find the optimal tradeoff among the number of general-purpose registers, the extent to which those registers will be used by a typical process, and the cost of implementing those registers. As the cost of producing individual registers has decreased, their number has increased. Current CPUs typically provide several dozen general-purpose registers. They also can have several dozen other special-purpose registers.

Special-Purpose Registers

Every processor has a number of registers that are used by the CPU for specific purposes. Some of the more important special-purpose registers are:

► Instruction register

► Instruction pointer

► Program status word

When the control unit fetches an instruction from memory, it stores it in the **instruction register**. The control unit then extracts the op code and operands from the instruction and performs any additional data movement operations that are needed to prepare for execution. The process of extracting the op code and operands, loading data inputs, and signaling the ALU is called instruction **decoding**.

The **instruction pointer** also can be called the **program counter**. Recall that the CPU constantly alternates between the instruction (fetch and decode) and execution (data movement and/or transformation) cycles. At the end of each execution cycle, the control unit starts the next fetch cycle by retrieving the next instruction from memory. The instruction retrieved is the instruction at the memory address in the instruction pointer. The instruction pointer is incremented by the control unit either during or immediately after each fetch cycle.

The CPU deviates from sequential execution only if a BRANCH instruction is executed. A BRANCH is implemented by overwriting the value of the instruction pointer with the address of the instruction to which the branch is directed. An unconditional BRANCH instruction actually is a MOVE from the branch operand, which contains the branch address, to the instruction pointer.

The **program status word** (**PSW**) contains data that describes the status of the CPU and the currently executing program. Each bit within the PSW is a separate boolean variable, sometimes called a **flag**, that represents one data item. The content and meaning of the flags varies widely from one CPU to another. In general, flags have three primary uses:

► Store the result of a comparison operation

► Control conditional branch execution

► Indicate actual or potential error conditions

The sample program shown earlier in Figure 4-8 performs comparison using the XOR, ADD, and SHIFT instructions. The result was stored in a general-purpose register and was interpreted as a boolean value. This method was used because the instruction set was limited.

Most CPUs provide one or more COMPARE instructions. COMPARE takes two operands and determines whether the first is less than, equal to, or greater than the second. Because there are three possible conditions, the result cannot be stored in a single boolean variable. Most CPUs use two PSW flags to store the result of a COMPARE. One flag is set to true if the operands are equal and the other flag indicates whether the first operand is greater than or less than the second. If the first flag is true then the second flag is ignored.

To implement program branches based on a COMPARE result, two additional conditional BRANCH instructions are provided. One conditional BRANCH is based on the equality flag and the other is based on the less than or greater than flag. Using COMPARE with related conditional BRANCH instructions simplifies machine language programs, speeding up their execution.

Other PSW flags represent status conditions resulting from instruction execution by the ALU. Conditions such as overflow, underflow, or an attempt to perform an undefined operation such as dividing by zero are each represented by a PSW flag. These flags are tested by the control unit after each execution cycle to determine if an error has occurred. They also can be tested by an operating system or other software to determine appropriate error messages and corrective actions.

WORD SIZE

A **word** is a unit of data that contains a fixed number of bytes or bits. A word can be defined loosely as the amount of data that a CPU processes at one time. Depending on the CPU, processing may include arithmetic, logic, fetch, store, and copy operations. Word size normally matches the size of general-purpose registers. Word size is a fundamental CPU design decision with implications for most other computer system components.

In general, a CPU with a large word size can perform a given amount of work faster than a CPU with a small word size. As an example, consider the addition or comparison of two 64-bit double precision integers. A processor with a 64-bit word size adds or compares by executing a single instruction because the registers that hold the operands and the ALU circuitry are 64 bits wide.

Now consider manipulating 64-bit data in a CPU with a 32-bit word size. Because the operands are larger than the word size, they must be partitioned and the operation(s) carried out on the pieces. For example, in a comparison operation, the CPU compares the first 32 bits of the operands and then, in a second execution cycle, compares the second 32 bits. The process is inefficient because multiple operands are loaded from or stored to memory, and multiple instructions are executed to accomplish what logically is a single operation.

Because of these inefficiencies, a 64-bit CPU usually is more than twice as fast as a 32-bit processor when processing 64-bit data values. Inefficiencies are compounded as the complexity of the operation increases. For example, complex calculation such as division and exponentiation may be four or five times slower on a 32-bit processor than on a 64-bit processor.

CPU word size also has implications for the design of the system bus. Maximum CPU performance is achieved when the bus width is at least as large as the CPU word size. If the width is smaller, every load and store operation requires multiple transfers to or from primary storage. For example, moving 8 bytes of character data to contiguous memory locations requires two separate data movement operations on a 32-bit bus, even if the word size of the processor is 64 bits. Similarly, fetching a 64-bit instruction requires two separate transfers.

The physical implementation of memory is likewise affected by word size. Although the storage capacity of memory always is measured in bytes, data movement between memory and the CPU generally is performed in units that are factors of the word size. For a 64-bit CPU, memory should be organized to read or write 8 contiguous bytes in a single access operation. Lesser capabilities will cause CPU wait states.

As with many other CPU design parameters, increases in word size are subject to the law of diminishing returns. Two primary issues are the usefulness of the extra bits and their cost. The extra bits are useful if the larger word is less than or equal to the size of data items that normally are manipulated by programs.

Consider the early days of microprocessors. In the 1970s, most microprocessors had an 8-bit word size. This limited the range of a one-word two's complement integer to -128 to +127. Little useful processing can be performed on such a restricted set of numbers. Storage for multiple words, two words for a "short" integer and four words for a "normal" 32-bit integer was needed by most programs. Multiword storage required complex and inefficient programs to implement even addition and subtraction. Dozens of instructions were required to perform simple processing operations on pieces of data and to combine these results into a complete result.

As processor technology and fabrication improved, word size increased. Improvements in fabrication were needed because doubling word size requires at least double, typically 2.5 to 3 times, the number of electrical components within the CPU. In the late 1970s and early 1980s, 16-bit processors appeared, 32-bit processors appeared in the mid to late 1980s, and 64-bit processors appeared in the 1990s.

The added benefit of increased word size, however, drops off sharply past 64 bits. Recall from the previous chapter that integers typically are stored in 32-bit two's complement format and that real numbers typically are stored in either 32-bit or 64-bit IEEE floating point format. Of what use are the extra 32 bits of a 64-bit CPU when adding together two 32-bit data items? They are of no use at all. The extra 32 bits simply store zeros that are carried through the computational process and provide benefit only when manipulating 64-bit data items. But many current word processors, financial applications, and other programs never manipulate 64-bit data items. Only a few types of application programs can use the additional bits. Examples include numerical-processing applications that need very high precision such as navigation and numerical simulations of complex phenomena, some database- and text-processing applications, and programs that manipulate continuous streams of audio or video data. For other programs, the increased computational power of the larger word size is wasted.

The waste of extra word size wouldn't be a major concern if cost were not an issue. Doubling word size, however, generally increases the number of CPU components by 2.5 to 3 times. This increase, in turn, increases CPU cost and complicates microprocessor fabrication. Costs tend to rise at nonlinear rates when approaching the limits of current fabrication technology. Until recently, fabrication technology limits made 64-bit CPUs an expensive luxury—generally used only in larger and more expensive computer systems. Now technology has reached the point where placing tens of millions of transistors on a single chip is cost effective. The forthcoming generation of Intel CPUs (called IA-64 or Itanium®) will feature a 64-bit word size. Internet and multimedia applications will need that larger word size.

Technology
Focus

Intel Pentium Processor Family

The Pentium processor is a direct descendent of the Intel 8086 and 8088 processors used in the original IBM personal computer. It maintains backward compatibility with those processors in all areas, including instruction set, data types, and memory addressing. The Pentium processor was introduced in 1993. The Pentium has been updated several times through generations named the Pentium Pro, Pentium MMX (MMX stands for MultiMedia eXtensions), Pentium II, Pentium III, and Pentium Xeon. This discussion will concentrate on features common to the entire Pentium family, the MMX, and streaming multimedia extensions.

Data Types

The Pentium supports a rich variety of data types (see Table 4-3). For purposes of describing data types, a word is considered to be 16 bits in length (as it was in the

8086/8088). Data types are of five sizes including 8 (byte), 16 (word), 32 (double-word), 64 (quadword), and 128 bits. Two variable length data types also are defined.

Table 4-3 ►

Pentium data type summary

Data Type	Length (Bits)	Coding Format
Unsigned integer	8, 16, or 32	Ordinary binary
Signed integer	8, 16, or 32	Two's complement
Real	32, 64, or 80	IEEE floating point standards
Bit field	32	32 binary digits
Bit string	Variable up to 2^{32} bits	Ordinary binary
Byte string	Variable up to 2^{32} bytes	Ordinary binary
MMX (integer)	8×8, 4×16, 2×32, 1×64	Packed two's complement
MMX (real)	4×32	Packed IEEE 32-bit floating point

No special data types are provided for boolean or character data, which are assumed to use appropriate integer data types. A bit field data type consists of up to 32 individually accessible bits. Two variable length data types also are defined. A bit string is a sequence of 1 to 2^{32} bits. This data type commonly is used for large binary objects such as compressed audio or video files. A byte string consists of 1 to 2^{32} bytes. Elements of a byte string can be bytes such as ASCII characters, words such as Unicode characters, or doublewords.

Four 64-bit MMX data types are provided. Three of these are packed data types in which multiple integers are "packed" within a single 64-bit unit. The packed integers may be 8, 16, or 32 bits in length. The fourth data type is a quadword (64-bit) integer. Packed data types allow MMX instructions to operate on each component of a packed data type in parallel. An MMX instruction can execute the same instruction on eight data inputs at once. Streaming multimedia instructions can use all of the MMX data types and an additional 128-bit data type containing four 32-bit floating point numbers.

Four different memory address formats are provided. Two of these are segmented memory addresses that are backward compatible with earlier 80x86 processors. The third is a 32-bit flat memory model address compatible with the 80386 and later processors. The fourth is a 36-bit flat address model that was introduced with the Pentium Pro. Segmented memory addresses are offset with respect to predefined segment registers. Six different segment registers are available and are assumed to point to specific portions of program address space. Flat 32-bit and 36-bit addresses are the preferred format for newly developed programs.

Instruction Set

The Pentium instruction set contains 28 system instructions, 92 floating point instructions, 52 MMX instructions, 61 streaming multimedia instructions, and 164 integer, logical, and other instructions. This is one of the largest instruction sets among modern microprocessors. As expected, such a large instruction set has many instruction formats of varying length. Instructions can have from zero to three operands. Legal operands include words, registers, and 32-bit memory addresses.

A detailed discussion of the Pentium instruction set fills an entire volume of several hundred pages of Intel documentation. The large number of instructions represents both the rich variety of supported data types and the desire to implement as many processing functions as possible in hardware. This clearly places the Pentium in the CISC camp of CPU design.

Word Size

There is no clear definition of word size for the Pentium. For backward compatibility with the 8088 and 8086 microprocessors, a word is defined as 16 bits. All registers in the integer ALUs are 32 bits in length, the upper and lower 16 bits can be accessed separately for backward compatibility. The two internal processing paths of the integer ALUs also are 32 bits wide. The external bus interface is 64 bits wide and the on-chip memory cache interface is 128 bits wide.

The Pentium has a separate floating point ALU with its own registers and internal processing paths. Registers within this ALU are 80 bits wide as are the internal processing paths, which allows single instruction operations to be performed on the largest floating point data type. Although this is a powerful floating point processing capability, it seldom is used by most application programs.

Clock Rate

The original Pentium debuted at a clock rate of 60 MHz. Clock rate for this processor eventually reached 200 MHz. The fastest Pentium III processor at this writing has a 1.13 GHz clock rate and 100 and 133 MHz interfaces to the system bus.

THE PHYSICAL CPU

A modern CPU is a complex system of interconnected electrical switches. Early CPUs typically contained several hundred to a few thousand switches. Modern CPUs contain millions of switches. In this section, we first will look at how these switches perform basic processing functions, and then at how these switches and circuits have been implemented physically.

Switches and Gates

The basic building blocks of computer processing circuits are electronic **gates**. Gates are composed of one or more electrical switches wired together to apply one or more boolean transformations to individual binary electrical signals, or bits. Basic processing functions on binary digits are performed with the logical functions AND, OR, XOR, and NOT. Figure 4-11 shows the electrical component symbols for NOT, AND, OR, XOR, and NAND (NOT AND). A NOT gate also is called a signal inverter, because it transforms a value of zero or one into its inverse or opposite.

Figure 4-11 ►

Electrical
component
symbols for
signal inverter or
NOT gate (a);
AND gate (b);
OR gate (c);
XOR gate (d);
and NAND
gate (e)

(a)

(b)

(c)

(d)

(e)

Processing circuits for more complex operations are constructed by combining logic gates. One example is the XOR gate. Although it is represented by a single symbol as shown in Figure 4-11(d), it actually is constructed by combining NOT, AND, and OR gates.

Figure 4-12 shows the components of half adder and full adder circuits for single bit inputs. Addition circuits for multiple bit inputs use half adder and full adder circuits as building blocks. More complex processing functions require more complicated arrangements of gates. The variety of processing functions implemented and the large number of bits that must be processed in parallel account for the extremely large number of gates and switches within a CPU.

Figure 4-12 ►

Circuit diagrams
for half adder
(a) and full
adder (b)

(a)

(b)

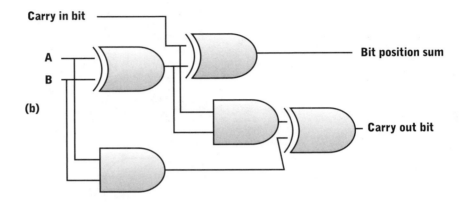

Electrical Properties

Gates and the wires that connect them carry and transform electrical signals representing binary ones and zeros. The construction of these switches and their connections is important in determining the CPU's speed and reliability. The speed and reliability of a CPU are affected not only by the materials used in its fabrication, but also by the properties of electricity such as conductivity, resistance, and heat.

Conductivity Electrical current is the flow of electrons from one place or device to another. An electron requires a sufficient energy input to excite it to move. Once excited, electrons move from place to place using molecules as stepping stones. One type of connection that allows electrons to travel from one place to another is a **wire**. Although the switches used in a CPU number in the millions, there is a much larger number of wires connecting those switches. The wires are as important as the switches in determining CPU speed and reliability.

The ability of an element or substance to enable electron flow is called **conductivity**. Substances that electrons can flow through are called **conductors**. Electrons travel through a perfect conductor with no loss of energy. With less than perfect conduction, energy is lost as electrons pass through. If enough energy is lost, the electrons cease to move and the flow of electrical current is halted.

Resistance   The loss of electrical power that occurs within a conductor is called **resistance**. A perfect conductor would have no resistance. Unfortunately, all substances have some degree of resistance. Conductors with relatively low resistance include some well-known and valuable metals such as silver, gold, and platinum. Fortunately, some cheaper materials such as copper serve nearly as well.

The laws of physics tell us that energy is never really lost but merely converted from one form to another. Electrical energy isn't really lost due to resistance. Instead, it is converted to heat. The amount of generated heat depends on the amount of electrical power being transmitted and the resistance of the conductor. Higher power and/or higher resistance increase the amount of generated heat.

Heat   Heat has two negative effects on electrical conductivity. The first is physical damage to the conductor. To reduce heat and avoid physical destruction of a wire, we must use a very low resistance wire, reduce the power input, or increase the diameter of the wire.

The second negative effect of heat is that it changes the inherent resistance of the conductor. The resistance of most materials increases as their temperature increases. To keep operating temperature and resistance low, some method must be provided to remove or dissipate heat from the wire. This can be accomplished in many ways. The simplest way is to provide a cushion of moving air around the wire. Heat migrates from the surface of the wire or its coating to the air, and then is transported away by fans or through ventilation openings.

A **heat sink** can be used to improve heat dissipation (see Figure 4-13). A heat sink is an object specifically designed to absorb heat and rapidly dissipate it via air or water movement. A heat sink is placed in direct physical contact with an electrical device to dissipate heat from the device. A heat sink exposes a large surface area to the moving air to allow more rapid dissipation.

Figure 4-13 ▶

A heat sink attached to a surface-mounted microprocessor

Speed and Circuit Length To perform processing, electrons must move through the wires and switches of a processing circuit. Electrons move at the (constant) speed of light, which is approximately 180,000 miles per second. There is a fundamental relationship between circuit length and processing speed. The time required to perform a processing operation is a function of the length of the circuit and the speed of light.

Because circuit length is the only variable, the path to faster processing is clear: reduce circuit length. Shorter circuits require smaller and narrower switches, and shorter wires. Miniaturization has been the primary basis for CPU speed and clock rate improvement since the first electrical computer. The first IBM personal computer used an Intel 8088 microprocessor with a clock rate of 4.77 MHz. As of this writing, Intel microprocessors have clock rates as high as 1.13 GHz. Some RISC microprocessors have achieved even greater clock rates.

Processor Fabrication

Basic electrical design tradeoffs apply to CPU design as well. Reliable, efficient, and cost-effective electrical circuits must balance power requirements, resistance, heat, size, and cost. The earliest computers were constructed with ordinary copper wire and vacuum tube switches. They were unreliable due to the large amounts of heat generated by the vacuum tubes. They also were quite large, typically filling an entire room with processing circuitry less powerful than that found today in a cheap calculator. Improvements in materials and fabrication techniques have vastly increased the performance and reliability of processors.

Transistors and Integrated Circuits   In 1947, researchers at Bell Laboratories discovered a class of materials called **semiconductors**. The conductivity of these materials varies in response to the electrical inputs applied. **Transistors** are made of semiconductor material that has been treated, or doped, with chemical impurities to enhance the semiconducting effects. Silicon and germanium are basic elements with resistance characteristics that can be tailored through the use of chemicals called dopants.

In the early 1960s, techniques were developed to fabricate miniature electronic circuits from multiple layers of metals, oxides, and semiconductor materials. This new technology made it possible to implement several transistors and their interconnections on a single chip to form an **integrated circuit** (**IC**). Integrated circuits reduced manufacturing cost per circuit because many chips could be manufactured in a single sheet, or wafer. Combining multiple gates on a single chip also reduced the manufacturing cost per gate and created a compact, modular, and reliable package. As fabrication techniques improved, it became possible to put hundreds, then thousands, and today, millions of electrical devices on a single chip. The term **microchip** was coined to refer to this new class of electronic devices.

Microchips and Microprocessors   A **microprocessor** is a microchip that contains all of the circuits and connections that implement a CPU. The first microprocessor (see Figure 4-14) was designed by Ted Hoff of Intel and introduced in 1971. Microprocessors ushered in a new era in computer design and manufacture. The most important part of a computer system now could be produced and purchased as a single package. Computer system design was simplified because computer designers no longer needed to construct processors from smaller components. Microprocessors also opened an era of standardization as a small number of microprocessors became widely used. The personal computer revolution would not have been possible without standardized microprocessors.

Figure 4-14 ▶

The Intel 4004
microprocessor
containing 2,300
transistors

Current Technology Capabilities and Limitations

Gordon Moore, an Intel founder, made an observation during a speech in 1965 that has come to be known as **Moore's Law**. Moore observed that the rate of increase in transistor density on microchips had increased steadily, roughly doubling every 18 to 24 months. He further observed that each doubling was achieved with no increase in unit cost. Moore's Law has proven surprisingly durable. Figure 4-15 shows that increases in transistor density of Intel microprocessors have followed Moore's Law. Software developers and users have come to expect CPU performance to double every couple of years at no price increase.

Figure 4-15 ▶

Increases in transistor density for Intel microprocessors

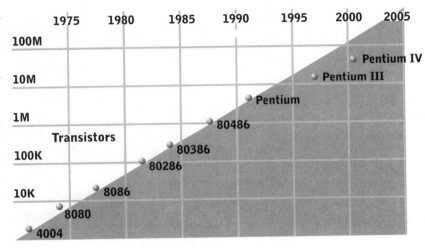

Arthur Rock, a venture capitalist, made a short addendum to Moore's Law. **Rock's Law** states that the cost of fabrication facilities for the latest generation of chips doubles every four years. This law also has proven quite durable. A fabrication facility using the latest production processes currently costs 5 to 10 billion dollars. Improvements in chip fabrication have been achieved by ever more expensive and exotic techniques. The added expense hasn't led to price increases because newer fabrication methods produce larger numbers of chips and the demand for chips exceeds the production capacity of any single factory.

Current fabrication technology is capable of squeezing over 28 million transistors onto a wafer of silicon less than one square centimeter in size (see Figure 4-16). The process starts with a wafer, which is a flat disk of pure silicon. A thin layer of conductive or semiconductive material is spread over the surface of the wafer. The material then is exposed to a laser or X-ray focused through a patterned map similar to an overhead transparency. An etching chemical then is applied which removes the portion of the layer that was exposed. The portion that remains has a very specific pattern, or map, corresponding to one layer of the microprocessor circuitry. Additional layers are added, then etched, gradually building up complete transistors and wires.

The wafer contains hundreds of microprocessors, which are cut apart, encased in individual carriers, and tested. It is not unusual for 70 to 80 percent of the microprocessors to fail initial testing. Testing failures can be the result of impurities in materials, contamination of layers as they are laid down on the wafer, errors in etching, and variations in any number of production process parameters. Those that survive initial tests then are subjected to further tests at successively higher clock rates. A microprocessor that fails to perform reliably at a higher clock rate still may be usable and rated for sale at a lower clock rate.

Increases in the number of circuits and packing density on a chip have been achieved by shrinking the size of transistors and the wires connecting them, and by increasing the number of layers added to the wafer. Shrinking the size of transistors and wires is accomplished by an etching process that leaves thinner and thinner lines of conductive and semiconductive material on the wafer, while increasing the number of layers puts additional components one on top of another on the chip.

Increased packing density has some problems, however. As the size of the devices and wiring decreases, the ability to transmit electrical power is reduced. This requires a reduction in normal operating voltage (currently as low as 1.8 volts, which is down from 5 volts a few years ago. It also makes the circuits far more susceptible to damage from voltage surges and from static electricity.

Electrical resistance also is a problem because wires and devices must have uniform resistance properties throughout the chip. A slight increase in resistance in one part of the chip may prevent sufficient electrical power from reaching another part of the chip. Uniform resistance requires very high and uniform purity of the conductive and semiconductive material in the chip. Resistance also can cause problems due to heat. Because a chip is a sealed package, there is no way to circulate air or any other substance among its circuits. Heat must migrate through the chip carrier to be dissipated by a fan or heat sink. Heat generation per circuit is minimized by their small size and low resistance materials. But each device and wire does generate heat. Higher clock rates generate more heat as more electrons are forced through the circuits. Inadequate or barely adequate cooling can shorten a microprocessor's operating life substantially.

FUTURE TRENDS

Will Moore's Law continue to hold for the foreseeable future? Will transistor size and cost continue to shrink at the present rate? Many industry experts don't think so. They predict that a technological "wall" will be reached between 2010 and 2020 when the limits of current microprocessor design and fabrication will be reached.

Beyond a certain point, further miniaturization will be more difficult to achieve because of the nature of the etching process and the limits of semiconducting materials. The width of wires and devices depends on the etching beam wavelength, but there are limits to our ability to generate and focus short wavelength lasers and X-rays. The width of chip wiring also is rapidly approaching the molecular width of the materials from which the wires are constructed; a wire can't be less than one molecule wide.

Subsequent improvements will require fundamentally different approaches to microprocessor design and fabrication. These approaches may be based on a number of technologies including optics and quantum physics. None of these technologies has yet been proven on a commercial scale.

Optical Processing

In an electrical computer, computation is performed by forcing electrons to flow through millions of electrical switches and wires. Optical computing could eliminate interconnection and simplify fabrication problems because photon pathways can cross without interfering with one another. Although fiber-optic wires usually are used to connect optical devices, it is possible to connect them without wires. If the sending and receiving devices are aligned precisely, then light signals sent between them can travel through free space. Eliminating wires would vastly improve fabrication cost and reliability.

Commercially available and affordable optical computers are at least two decades from being a reality. As long as semiconductors continue to get faster and cheaper, there isn't sufficient economic incentive to develop optical technologies. There is economic incentive to pursue specialized processing applications in the areas of telecommunications and networking. Practical optical processors probably will appear first as dedicated communication controllers, and later evolve into full-fledged computer processors.

Electro-Optical Processing

Electro-optical transistors hold promise as the interface between semiconductor processors and purely optical memory and storage devices. One material that shows great promise is **gallium arsenide**. Like silicon and other materials, it can be used to fabricate conventional semiconductors. Unlike silicon, it has both optical and electrical properties. Gallium arsenide can be used to create transistors with optical controls; a gallium arsenide transistor can enable electron flow when an optical input is applied.

Gallium arsenide transistors already are used in telecommunications switching equipment. They also are used in some supercomputers so electrical CPUs can interface directly with an optical bus. They provide sufficient data transfer capacity to supply dozens or hundreds of processors with instructions and data. Technologies such as holographic storage have the potential to deliver gigabytes of data in a matter of seconds. Optical communication channels will be required to provide processor access to all of that data.

Quantum Processing

Quantum computing was introduced as a Technology Focus section in Chapter 2. Quantum computing uses quantum states to simultaneously encode two bits, called a qubit, and uses quantum processing devices to perform computations. In theory, quantum computing is well-suited to solving problems that require massive amounts of computation, such as factoring large numbers and transmitting private encryption keys.

Current computers solve many different types of problems, only a few of which require massive computation. As currently envisioned, quantum computing provides no benefit for the other problem classes as compared to conventional electrical processors. Quantum processors may never replace conventional processors in the vast majority of computer systems, or perhaps they will develop in ways we can't imagine. Much as Bell Labs researchers could not have imagined 1 GHz microprocessors when they developed the first transistor, current researchers probably can't foresee the quantum devices that will be built later in the twenty-first century.

SUMMARY

▶ The CPU continuously alternates between the instruction, or fetch, cycle and the execution cycle. During the fetch cycle, the control unit fetches an instruction from memory, separates the op code and operands, stores the operands in registers, and increments a pointer to the next instruction. During the execution cycle, either the control unit or the ALU executes the instruction. The ALU executes the instruction if it is an arithmetic or logical operation. The control unit executes all other instruction types.

▶ Primitive CPU instructions can be classified into three types: data movement, data transformation, and sequence control. Data movement instructions copy data among registers, primary storage, secondary storage, and I/O devices. Data transformation instructions implement simple boolean operations (NOT, AND, OR, and XOR), addition (ADD), and bit manipulation (SHIFT). Sequence control instructions control the next instruction to be fetched or executed. More complex processing operations are implemented by appropriate sequences of primitive instructions.

▶ An instruction format is a template describing the op code position and length and the position, type, and length of each operand. Most CPUs support multiple instruction formats. CPUs can support fixed length or variable length instruction formats. Fixed length formats are simpler for the control unit to fetch and decode. Variable length instructions use primary and secondary storage more efficiently. RISC CPUs use fixed length instructions and generally avoid complex instructions, particularly those that combine data movement and data transformation. RISC CPUs are simpler than CISC CPUs and generally operate at higher clock rates. They are less efficient, however, than CISC CPUs when performing some complex operations.

▶ The CPU clock rate is the number of instruction and execution cycles potentially available in a fixed time interval. Typical CPUs have clock rates measured in hundreds or thousands of MHz (1000 MHz = 1 GHz). A CPU generally executes fewer instructions than are implied by its clock rate. The rate of actual or average instruction execution is measured in MIPS or MFLOPS. Many instructions, especially floating point operations, require multiple execution cycles to complete. The CPU also may incur wait states pending storage or I/O device accesses.

▶ CPU registers are of two types—general and special purpose. General-purpose registers are used by programs to hold intermediate results and frequently needed data items. General-purpose register accesses don't incur wait states because they are implemented as part of the CPU. Within limits, increasing the number of general-purpose registers decreases program execution time. Special-purpose registers are used by the CPU itself to track processor and program status. Important special-purpose registers include the instruction pointer and program status word.

▶ Word size is the number of bits that a CPU can process simultaneously. Within limits, CPU efficiency increases with word size because inefficient piece by piece operations on large data items are avoided. For optimal performance, other computer system components such as the system bus should match or exceed CPU word size.

▶ CPUs are electrical devices implemented as silicon-based microprocessors. Like all electrical devices, they are subject to basic laws and limitations of electricity including speed as a function of circuit length, resistance, and heat generation. Microprocessors use small circuit size, low-resistance materials, and heat dissipation to ensure fast and reliable operation. Microprocessors are fabricated using expensive processes based on X-ray or laser etching and chemical deposition. Those processes, and semiconductors themselves, are approaching fundamental physical limits that will stop further improvements due to miniaturization. Candidate technologies that may move performance beyond semiconductor limitations include optical processing, hybrid optical-electrical processing, and quantum processing.

▶ Now that we've described the inner workings of computer processors, we need to move on to the other computer system components. In the next chapter you will learn the characteristics and inner function of primary and secondary storage devices, including magnetic and optical storage technology—the bases of all modern storage devices.

ADD
AND
arithmetic SHIFT
binary compatibility
BRANCH
clock cycle
clock rate
Complex Instruction Set
    Computing (CISC)
complex instructions
conditional BRANCH
conductivity
conductors
cycle time
decoding
exclusive OR (XOR)
execution cycle
fetch cycle
fixed length instructions
flag
gallium arsenide
gates

gigahertz (GHz)
HALT
heat sink
hertz (Hz)
inclusive OR
instruction
instruction cycle
instruction format
instruction pointer
instruction register
instruction set
integrated circuit (IC)
JUMP
load
logical SHIFT
megahertz (MHz)
microchip
microprocessor
millions of floating point
    operations per second
    (MFLOPS)

millions of instructions per
    second (MIPS)
Moore's Law
MOVE
NOT
op code
operands
program counter
program status word (PSW)
Reduced Instruction Set
    Computing (RISC)
resistance
Rock's Law
semiconductor
SHIFT
store
transistor
unconditional BRANCH
variable length instruction
wait state
wire
word

Vocabulary Exercises

1. The _____ time of a processor is one divided by the clock rate (in Hz).

2. A CPU typically implements multiple _____ to account for differences among instructions in the number and type of operands.

3. _____ generates heat within electrical devices.

4. _____ is a semiconducting material with optical properties.

5. A(n) _____ is an electrical switch built of semiconducting materials.

6. A(n) _____ improves heat dissipation by providing a thermal mass and a large thermal transfer surface.

7. One _____ is one cycle per second.

8. Applying a(n) _____ OR transformation to input bit values one and one generates true. Applying a(n) _____ OR transformation to those same inputs generates false.

9. When first fetched from memory, an instruction is placed in the _____ and then _____ to extract its components.

10. Using _____ instructions simplifies the process of instruction fetching and decoding.

11. A(n) _____ is an electrical circuit that implements a boolean or other primitive processing function on single bit inputs.

12. A microchip containing all of the components of a CPU is called a(n) _____

13. A(n) _____ instruction transforms the bit pairs 1/1, 1/0, and 0/1 into 1.

14. The address of the next instruction to be fetched by the CPU is held in the _____ .

15. The contents of a memory location are copied to a register while performing a(n) _____ operation.

16. A(n) _____ or _____ contains multiple transistors or gates in a single sealed package.

17. A(n) _____ instruction always alters the instruction execution sequence. A(n) _____ instruction alters the instruction execution sequence only if a specified condition is true.

18. A(n) _____ processor does not directly implement complex instructions.

19. A(n) _____ instruction copies data from one memory location to another.

20. The CPU incurs one or more _____ when it is idle, pending the completion of an operation by another device within the computer system.

21. A(n) _____ is the number of bits processed by the CPU in a single instruction. It also describes the size of a single register.

22. In many CPUs, a register called the _____ stores condition codes, including those representing processing errors and the results of comparison operations.

23. The components of an instruction are its _____ , and one or more _____ .

24. Two one-bit values generate a one result value when a(n) _____ instruction is executed. All other input pairs generate a zero result value.

25. A(n) _____ CPU typically uses variable length instructions and has a large instruction set.

26. A(n) _____ operation transforms a zero-bit value to one and a one-bit value to zero.

27. _____ predicts that transistor density and processor power will double every two years or less.

Review Questions

1. Describe the operation of MOVE. Why is the name "move" a misnomer?

2. Why does program execution speed generally increase as the number of general-purpose registers increases?

3. What are special-purpose registers? Give three examples of special-purpose registers and explain how each is used.

4. What are the advantages and disadvantages of fixed length instructions as compared to variable length instructions? Which type generally are used in a RISC processor? Which type are generally used in a CISC processor?

5. Define word size. What are the advantages and disadvantages of increasing word size?

6. What characteristics of the CPU and of primary storage should be balanced to obtain maximum system performance?

7. How does a RISC processor differ from a CISC processor? Is one processor type better than the other? Why or why not?

8. What factor(s) account for the dramatic improvements in microprocessor clock rates since their invention in the early 1970s?

9. What potential advantages are provided by optical processors as compared to electrical processors?

Problems and Exercises

1. Develop a program consisting of primitive CPU instructions to implement the following procedure:

```
integer i,a;

i=0;
while (i < 10) do
  a=i*2;
  i=i+1;
endwhile
```

2. Assume that a microprocessor has a cycle time of four nanoseconds. What is the processor clock rate? Also, assume that the fetch cycle is 40 percent of the processor cycle time. What memory access speed is required to implement load operations with zero wait states? What memory access speed is required to implement load operations with two wait states?

3. Processor R is a RISC processor with a 1.2 GHz clock rate. The average instruction requires two cycles to complete, assuming zero wait state memory accesses.

Processor C is a CISC processor with a 750 MHz clock rate. The average simple instruction requires 2 cycles to complete, assuming zero wait state memory accesses. The average complex instruction requires 4 cycles to complete, assuming zero wait state memory accesses. Processor R can't directly implement the complex processing instructions of Processor C. Executing an equivalent set of simple instructions requires an average of 10 cycles to complete, assuming zero wait state memory accesses.

Program S executes 90 percent simple instructions and 10 percent complex instructions. Program C executes 70 percent simple instructions and 30 percent complex instructions. Which processor will execute program S more quickly? Which processor will execute program C more quickly? At what percentage of complex instructions will the performance of the two processors be equal?

4. Assume the existence of a CPU with a 1.2 GHz clock rate. The instruction and execution cycles are each 50 percent of the clock cycle. The average instruction requires 2 nanoseconds to complete execution. Main memory access speed for a single instruction is 8 nanoseconds on average. What is the expected average MIPS rate for this CPU?

Research Problems

1. Investigate the instruction set and architectural features of a modern RISC processor such as the NEC VR5000 (http://www.necel.com) or IBM Power PC (www.chips.ibm.com). In what ways does it differ from the architecture of the Intel Pentium processor family?

2. A few companies produce microprocessors that execute CPU instructions designed for the Intel Pentium processor family. These companies include VIA (www.cyrix.com) and AMD (www.amd.com). Investigate their current product offerings. Do they offer true Pentium compatibility? Is their performance comparable to that of Intel processors?

3. IBM (www.ibm.com) recently has developed a semiconductor fabrication process that uses copper instead of aluminum to implement transistor interconnections. Investigate the advantages of chips fabricated with the new process. Will the new process extend the lifetime of semiconductor microprocessor technology? Why or why not?

Chapter 5

Data Storage Technology

Chapter Goals

- Describe the distinguishing characteristics of primary and secondary storage

- Describe the devices used to implement primary storage

- Describe the memory allocation schemes

- Compare and contrast secondary storage technology alternatives

- Describe factors that determine storage device performance

- Choose appropriate secondary storage technologies and devices

In Chapter 2, you were briefly introduced to the topic of storage, including the role of storage in a computer system and the differences between primary and secondary storage. In this chapter, you will explore storage devices and their underlying technologies in depth (see Figure 5-1). The chapter outlines characteristics common to all storage devices, and then explains the technology, strengths, and weaknesses of primary storage (RAM and ROM) and secondary storage (magnetic and optical).

Figure 5-1 ►

Topics covered in this chapter

Storage

Speed, volatility, access
Random access memory
Memory organization
Magnetic technology
Optical technology
Tape storage devices
Disk storage devices

STORAGE DEVICE CHARACTERISTICS

A storage device generally consists of a read/write mechanism and a storage medium. The **storage medium** is the device or substance that actually holds data. The **read/write mechanism** is the device used to read or write data to and from the storage medium. A **device controller** usually provides the interface between the storage device and system bus. Device controllers are discussed in detail in Chapter 6.

In some storage devices, the read/write mechanism and storage medium are a single unit using the same technology. For example, most types of primary storage use electrical circuits implemented with semiconductors for both the read/write

---

mechanism and storage medium. In other storage devices, the read/write mechanism and the storage medium use fundamentally different technologies. For example, tape drives use an electro-mechanical device for the read/write mechanism and a magnetic storage medium composed of polymers and metal oxides.

In some cases, such as tape and diskette storage devices, the storage medium can be removed. This allows data and programs to be transported physically and backup copies to be created and stored at multiple locations. In theory, the capacity of a removable media storage device is infinite because many media can share the same read/write mechanism.

A typical computer system has many storage devices (see Figure 5-2). Multiple devices are required because no one device or technology can cost-effectively meet all storage needs for a single computer system or user. Different devices must be chosen for different storage purposes because there is no single storage device that is best in all characteristics. The primary characteristics that distinguish storage devices are:

► Speed
► Volatility
► Access method
► Portability
► Cost and capacity

Each characteristic is discussed in detail in the sections that follow.

Speed

Speed is the most important characteristic that differentiates primary and secondary storage. Primary storage extends the limited capacity of CPU registers. The CPU continually moves data and instructions between registers and primary storage. CPU instruction and execution cycles are extremely short, a few nanoseconds at most.

If primary storage uses devices that cannot be read or written within a single CPU cycle, then the CPU must wait for the read/write operation to complete. Wait states reduce the performance of the CPU and the entire computer system. To minimize wait states, all or part of primary storage is implemented with the fastest available storage devices. With current technology, primary storage speed is typically greater than secondary storage speed by a factor of 10^5 or more.

Figure 5-2 ▶

Primary and
secondary
storage and their
component
devices

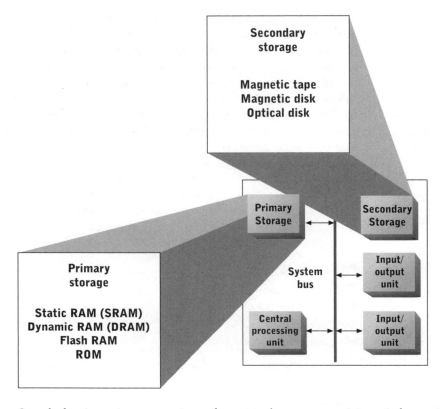

Speed also is an important issue for secondary storage. Many information system applications require access to large databases, to support ongoing processing in a customer order query system. Program response time in such systems depends on secondary storage access speed. Secondary storage access speed also affects overall computer performance in other ways. Starting program execution generally requires reading the corresponding executable code from secondary storage and copying it to primary storage. The delay between a user request for program execution and the first prompt for user input depends on the speed of both primary and secondary storage.

Storage device speed is called access time. **Access time** is the time required to execute one read or write operation completely. Access time is assumed to be the same for both reading and writing unless otherwise stated. Access time for some storage devices such as random access memory (RAM) is the same regardless of which storage location is accessed. For other storage devices such as disks, access time varies with storage location. For such devices, access time typically is expressed as an average called **average access time**. Access times of primary storage devices generally are expressed in nanoseconds (billionths of a second).

Access times for secondary storage devices typically are expressed in milliseconds (thousandths of a second).

By itself, access time is an incomplete measure of data access speed. A complete measure of data access speed consists of access time and the unit of data transfer to/from the storage device. Data transfer units vary from one storage device to another. The data transfer unit for primary storage devices is usually a word. Depending on the CPU, a word may represent 2, 4, 8, or more bytes, 4 and 8 bytes being the most common word sizes. Data transfer unit size sometimes is used as an adjective, as in "32-bit memory."

Secondary storage devices read or write data in units larger than words. **Block** is a generic term that describes secondary storage data transfer units. Block size normally is stated in bytes. It also varies widely among storage devices and can even vary within a single storage device. By itself, the term block has little significance and requires a specific definition of **block size**.

The term **sector** is used to describe the data transfer unit for magnetic and optical disk drives. As with blocks, sector size generally is stated in bytes and can vary from one device to another. A 512-byte block size is a widely used, but not universal, data transfer unit for magnetic disks.

A storage device's **data transfer rate** is computed by dividing 1 by the access time (expressed in seconds) and multiplying the result by the unit of data transfer (expressed in bytes). For example, the data transfer rate for a primary storage device with 15 nanosecond access time and a 32-bit word data transfer unit can be computed as:

$$\frac{1 \text{ second}}{15 \text{ nanoseconds}} \times 32 \text{ bits} = \frac{1}{0.000000015} \times 4 \text{ bytes} \approx 266{,}666{,}667 \text{ bytes/second}$$

The same formula can be used to compute a sustained data transfer rate using average access time for devices with variable access times.

Volatility

A storage device or medium is **non-volatile** if it holds data without loss over long periods of time. A storage device or medium is **volatile** if it cannot reliably hold data for long periods. Primary storage devices are generally volatile. Secondary storage devices are generally non-volatile.

Volatility is actually a matter of degree and conditions. For example, RAM is non-volatile as long as external power is continually supplied. But it generally is considered a volatile storage device because continuous power cannot be guaranteed under normal operating conditions. Magnetic tape and disk are considered non-volatile storage media. But data held on magnetic media is often lost after a few years due to natural magnetic decay and other factors. Data stored on non-volatile media may be lost due to obsolescence of compatible read/write devices, for example, 12-inch laser disks.

Access Method

The physical structure of a storage device's read/write mechanism and storage medium determines the way(s) in which data can be accessed. Three broad classes of access are recognized:

► Serial access

► Random access

► Parallel access

A single device can use multiple access methods.

Serial Access   A **serial access** storage device stores and retrieves data items in a linear, or sequential, order. Magnetic tape is the only widely used form of serial access storage. Data is written to a linear storage medium in a specific order. Once written, data can be read back only in that same order. For example, to play the last song on a cassette or digital audio tape requires playing or fast forwarding past all of the songs that precede it. Data storage on tapes also is read and written in linear order. If there are n data items stored on the tape, then the nth data item can be accessed only after the preceding $n-1$ data items have been accessed.

Serial access time depends on the current position of the read/write mechanism and on the position of the desired data item within the storage medium. If both positions are known, then access time can be computed as the difference between the current and target positions multiplied by the amount of time required to move from one position to the next.

Serial access devices are not used for frequently accessed data due to their inherently inefficient access method. Even when data items are accessed in the same order as written, serial access devices are much slower than other forms of storage. Serial access storage primarily is used to hold backup copies of data stored on other storage devices, for example, in a weekly tape backup of user disk files.

Random Access   A random **access device**, also called a **direct access device**, is not restricted to any specific order when accessing data. Rather, it can directly access any desired data item stored on the storage medium. All primary storage devices and disk storage devices are random access devices.

Access time for a random access storage device may or may not be a constant. It is a constant for most primary storage devices. It is not a constant for disk storage because of the physical, or spatial, relationships of the read/write mechanism, storage media, and storage locations on the media. This issue is discussed fully in a later section.

Parallel Access   A **parallel access** device is capable of simultaneously accessing multiple storage locations. Although RAM is considered to be a random access storage device, it also is a parallel access device. The confusion arises from differences in the definition of the unit of data access. If one considers the unit of data access to be a bit, then access is parallel. That is, random access memory circuitry can access all of the bits within a byte or word at the same time.

Parallel access also can be implemented by subdividing data items and storing the component pieces on multiple storage devices. For example, some operating systems can store file content on several disk drives. Different parts of the same file can be read at the same time by sending parallel commands to each disk drive.

Portability

Data can be made portable by storing it on a removable storage medium or device. For example, compact disc and tape storage are portable because the storage media are removable and their storage formats are standardized. External disk drives are portable if they can be added or removed easily from a computer system and transported without damage to the device or its data content.

Portable secondary storage devices and devices with removable storage media typically have slower access speed than permanently installed devices and those with non-removable media. High-speed access requires tight control of environmental factors. For example, high-speed access for magnetic disks is achieved, in part, by sealed enclosures for the storage media, thus minimizing or eliminating dust and air density variations. Removable storage media come in contact with a larger variety of damaging environmental influences than do isolated media. Portable storage devices typically sacrifice access speed to compensate for the lack of environmental control.

Cost and Capacity

Each of the storage device attributes described so far is directly related to device cost (see Table 5-1). An increase in speed, permanence, or portability generally comes at increased cost if all other factors are held constant. Cost also increases as an access method moves from serial to random to parallel. Primary storage generally is expensive compared to secondary storage due to its high speed and its combination of parallel and random access methods.

Secondary storage devices usually are thought of as having much greater capacity than primary storage devices. In fact, the capacity difference between primary and secondary storage in most computer systems results from a compromise between cost and other device characteristics. If cost were not a factor, most users would opt for non-volatile semiconductor-based memory devices instead of disk drives to implement secondary storage. However, most users need gigabytes of secondary storage and that much non-volatile semiconductor-based memory costs more than most users can afford. So users sacrifice speed and parallel access to gain sufficient capacity at an acceptable cost.

Table 5-1 ▶

Storage device
characteristics
and their rela-
tionship to cost

Characteristic	Description	Cost
Speed	Time required to read or write a bit, byte, or larger unit of data	Cost increases as speed increases
Volatility	Ability to hold data indefinitely, particularly in the absence of external power	For devices of similar type, cost decreases as volatility increases
Access method	Can be serial, random, or parallel; parallel devices also are serial or random access	Serial is the least expensive; random is more expensive than serial; parallel access is more expensive than non-parallel access
Portability	Ability to easily remove and reinstall the storage media from the device or the device from the computer	For devices of similar type, portability increases cost; if all other characteristics are held constant
Capacity	Maximum data quantity held by the device or storage medium	Cost usually increases in direct proportion to capacity

Memory-Storage Hierarchy

A typical computer system has a variety of primary and secondary storage devices. A small amount of high speed RAM is typically used within the CPU. Slower RAM is used to implement the bulk of primary storage. Multiple secondary storage devices generally are used. One or more magnetic disk drives are usually complemented by an optical disk drive and at least one form of removable magnetic storage.

The range of storage devices used within a single computer system forms a memory-storage hierarchy, as illustrated in Figure 5-3. Cost and access speed generally decrease as one moves down the hierarchy. Due to lower cost, capacity tends to increase as one moves down the hierarchy. A computer system designer or purchaser attempts to find an optimal mix of cost and performance for a particular purpose.

Hardware and software technology can be used to increase performance by substituting one form of storage for another. The exact mechanisms by which this may be accomplished are varied and complex. For example, memory capacity can be increased by supplementing primary storage with secondary storage using a technique known as virtual memory management. Memory access speed can be improved by using relatively small amounts of very high-speed RAM using a technique called caching. A similar technique can improve access speed to disk storage devices. These and other performance-enhancing storage substitutions are discussed in Chapter 6.

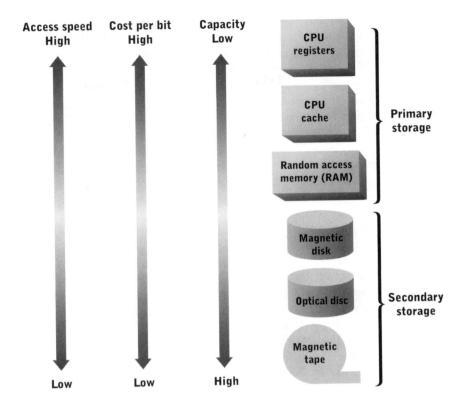

Figure 5-3

Comparison of storage devices in terms of cost and access speed

PRIMARY STORAGE DEVICES

The critical performance characteristics of primary storage devices are access speed and data transfer unit size. Primary storage devices must match CPU speed and word size closely to avoid wait states. The evolution of memory devices has mirrored closely the evolution of processing devices. Technologies applied to the construction of processors have, in general, been applied simultaneously to the construction of primary storage devices.

Storing Electrical Signals

Data are represented within the CPU as digital electrical signals. Digital electrical signals also are the basis of data transmission among all devices attached to the system bus. Any storage device or controller must accept electrical signals as input for a write operation and generate electrical signals as output for a read operation.

Electrical power can be stored directly by various devices including batteries and capacitors. Unfortunately, there is a tradeoff between access speed and volatility. Batteries are quite slow to accept and/or regenerate electrical current. They also lose their ability to accept charge with repeated use. Capacitors can charge or discharge with much greater speed than batteries. However, capacitors lose their charge at a fairly rapid rate and require a fresh injection of electrical current at regular intervals (hundreds or thousands of times per second).

An electrical signal can be stored indirectly by using its energy to alter the state of a device such as a mechanical switch, or a substance such as a metal. An inverse process is used to regenerate an equivalent electrical signal. For example, electrical current can generate a magnetic field. The strength of the magnetic field can induce a permanent magnetic charge in a nearby metallic compound. The stored magnetic charge then can be used to generate an electrical signal equivalent to the one used to create the original magnetic charge. Magnetic polarity, which either is positive or negative, can be used to represent the values zero and one.

Early computers implemented primary storage as rings of ferrous material (iron and iron compounds). The rings are embedded in a two-dimensional wire mesh. The rings are called cores and the technology is called **core memory**. Signals sent along the wires induce magnetic charges in the metallic rings. The polarity of the charge depends on the direction in which current flows through the wires.

Modern computers use memory implemented with semiconductors. Basic types of semiconductor memory include random access memory (RAM) and read-only memory (ROM). There are many variations of each memory type, described in the following sections.

Random Access Memory

Random access memory (RAM) is a generic term describing primary storage devices with the following characteristics:

► Microchip implementation using semiconductors

► Ability to read and write with equal speed

► Random access to stored bytes, words, or larger data units

RAM is fabricated in the same manner as microprocessors. One might assume that microprocessor clock rates are well matched to RAM access speeds. Unfortunately, this is not the case for many reasons, including:

► Reading and writing many bits in parallel requires additional circuitry.

► When RAM and microprocessors are on separate chips, delays are encountered when moving data from one chip to another.

There are two basic RAM types and several variations on each. **Static RAM (SRAM)** is implemented entirely with transistors. The basic unit of storage is a flip-flop circuit (see Figure 5-4). Each flip-flop circuit contains two transistors for bit storage. Additional transistors (typically two or four) are used to perform read and write operations. A flip-flop circuit is an electrical switch that remembers its last position. One position represents zero and the other represents one. Flip-flop circuits require a continuous supply of electrical power to maintain their positions. SRAM is volatile unless a continuous supply of power can be guaranteed.

Figure 5-4 ►

A flip-flop circuit composed of two NAND gates: the basic component of SRAM and CPU registers

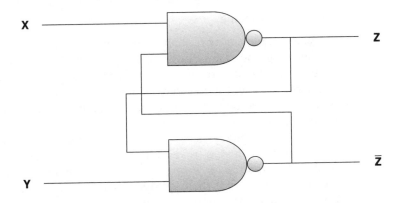

Dynamic RAM (DRAM) uses transistors and capacitors. The capacitors are the *dynamic* element. DRAM circuitry is less complex than SRAM circuitry. However, because the capacitors quickly lose their charge, they require a fresh infusion of power thousands of times per second. DRAM chips include circuitry that automatically performs refresh operations. Each refresh operation is called a **refresh cycle**. Unfortunately, a DRAM chip cannot perform a refresh operation at the same time it performs a read or write operation.

SRAM is more expensive to fabricate than DRAM due to its more complex circuitry. The current cost difference is approximately 10:1. Although simpler circuitry usually translates to faster speed, DRAM is slower than SRAM due to its required refresh cycles. With current fabrication technology, best-case access times are 50 nanoseconds for DRAM and 5 nanoseconds for SRAM (a 10:1 difference). Improvements in fabrication technology can decrease both access times but cannot change the relative performance difference.

Note that neither RAM type can match current microprocessor clock rates, which range from 500 MHz to 1.5 GHz at this writing. For zero wait state memory accesses, those clock rates require the following memory access speeds:

$$\frac{1}{1.5 \text{ GHz}} = \frac{1}{1{,}500{,}000{,}000} = 0.00000000067 = 0.67 \text{ nanoseconds}$$

$$\frac{1}{500 \text{ MHz}} = \frac{1}{500{,}000{,}000} = 0.000000002 = 2 \text{ nanoseconds}$$

The fastest DRAM is at least 25 times slower than modern microprocessors. SRAM is at least 2.5 times slower.

A number of technologies are used to bridge the performance gap between memory and microprocessors:

► Read-ahead memory access
► Synchronous read operations
► On-chip caching

Computer memory contains more than just SRAM or DRAM circuits. Memory circuits are grouped in large, for example 64 MB, memory modules; each module contains additional circuitry to implement read and write operations to random locations. Read and write circuitry must be activated before data can be accessed. Activating the read/write circuitry consumes time, which in turn lengthens the time required to access a memory location.

Programs usually access instructions and data items sequentially. Read-ahead memory access takes advantage of this tendency by activating the read/write circuitry for location $n+1$ during or immediately after an access to memory location n. If the CPU subsequently issues an access command for location $n+1$ then the memory module already has performed part of the work needed to complete the access.

Synchronous DRAM (SDRAM) is a read-ahead RAM that uses the same clock pulse as the system bus. Read and write operations are broken into a series of simple steps and each step can be completed in one bus clock cycle. Several clock cycles are required to complete a single random access. But if memory is being written or read sequentially, then accesses occur once per system bus clock tick (for example, 100 million times per second) after the first access has completed. Current SDRAM can match bus clock rates up to 266 MHz.

Cached DRAM, also called **enhanced DRAM (EDRAM)**, uses a small amount of SRAM in each DRAM device. When a read request for a specific word is received, that word and several words that follow are retrieved. The extra words are stored in the SRAM cache in anticipation of future sequential read requests. If those requests are made, the data already are waiting in the SRAM cache and can be more quickly accessed. Other uses of caching are discussed in Chapter 6.

Ferroelectric RAM embeds iron or iron compounds within a microchip to store bits in much the same manner as old-fashioned core memory. Read and write operations are implemented similar to conventional DRAM or SRAM at comparable speeds. Fabricating ferroelectric RAM chips has proved difficult and costly but the technology is very young.

Read-Only Memory

Read-only memory (ROM) is a random access memory device that can store data permanently or semipermanently. ROM commonly is used to store programs such as computer system boot subroutines like the system BIOS. These instructions can be loaded at high speed into main memory from ROM for execution by the processor. Instructions that reside in ROM are called **firmware**.

ROM technology has evolved through several stages, only two of which are currently in use. An **electronically erasable programmable read-only memory (EEPROM)** can be programmed, erased, and reprogrammed by signals sent from an external control source, such as a CPU. The primary drawback of EEPROM technology is that read and erase operations require relatively high voltages and repeated cycles to cause permanent change.

A newer form of EEPROM called **flash memory**, sometimes erroneously called **flash RAM**, can be erased and rewritten much more quickly. It is competitive with DRAM in storage density (capacity) and read performance. Unfortunately, flash memory tends to wear out after 100,000 or more write operations, making it unsuitable for primary storage. Flash memory is used for programs and data that aren't frequently updated, such as computer bootstrap programs and configuration data, and in some consumer products such as digital cameras.

Memory Packaging

Memory packaging is similar to microprocessor packaging. Memory circuits are embedded within microchips and groups of chips are packed on a small circuit board that can be installed or removed easily from a computer system. Early RAM and ROM circuits were packaged in **dual in-line packages (DIPs)**, as shown in Figure 5-5. Installing a DIP on a printed circuit board is a tedious and precise operation. Also, single DIPs mounted on the board surface occupy a relatively large portion of the total surface area.

Figure 5-5 ►

Dual in-line package (DIP) chip and single in-line memory module (SIMM)

Dual in-line package (DIP)

Single in-line memory module (SIMM)

In the late 1980s, memory manufacturers adopted the **single in-line memory module (SIMM)** as a standard RAM package (see Figure 5-5). Each SIMM incorporates multiple DIPs on a tiny printed circuit board. The edge of the circuit board has a row of electrical contacts and the entire package is designed to lock into a SIMM slot on a computer motherboard. The **double in-line memory module (DIMM)** is a newer packaging standard. It is essentially a double-sided SIMM with memory DIPs and electrical contacts on both sides of the module.

Current microprocessors include a small amount of on-chip memory; this technology is fully described in Chapter 6. As fabrication techniques improve, the amount of memory that can be packaged with the CPU on a single chip will grow. The logical extension of this trend is a CPU and all of its primary storage implemented on a single chip. This virtually would eliminate the gap that currently exists between microprocessor clock rates and memory access speeds.

CPU MEMORY ACCESS

Although main memory currently is not implemented as a part of the CPU, the need for the CPU to load instructions and data from memory and to store processing results requires a close coordination between both devices. Specifically, the physical organization of memory, the organization of programs and data within memory, and the method(s) of referencing specific memory locations are critical design issues for both primary storage devices and processors.

Physical Memory Organization

The main memory of any computer can be regarded as a sequence of contiguous, or adjacent, memory cells, as shown in Figure 5-6. Addresses of these memory locations are assigned sequentially so that available addresses proceed from zero (low memory) to the maximum available address (high memory).

Figure 5-6 ▶

Sequential physical organization of memory cells

Data values and instructions generally require multiple bytes of storage. For example, an integer typically requires 4 consecutive bytes, or 32 bits, of storage.

When written in a program, the bits and bytes of a numeric value typically are ordered from highest position weight on the left to lowest position weight on the right. If a 32-bit integer is written as:

$$00110010\ 11010011\ 01001100\ 01101001$$

it is assumed that the left position weight is the largest (2^{31}) and the right position is the smallest (2^0). When considered as a byte sequence, the leftmost byte is called the **most significant byte** and the rightmost byte is called the **least significant byte.**

In many CPU and memory architectures, the least significant byte of a data item is placed at the lower memory address. For example, the 32-bit word $02468ACE_{16}$ would be stored in memory with CE at the lowest memory address, 8A at the next address, 46 at the address after that, and 02 at the highest address. Other CPU and memory architectures store bytes in the reverse order. **Big endian** describes architectures that store the most significant byte at the lowest memory address. **Little endian** describes architectures that store the least significant byte at the lowest memory address.

The **addressable memory** of a CPU or computer system is the highest numbered storage byte that can be represented. Addressable memory usually is determined by the number of bits used to represent an address. For example, if 32 bits are used to represent a memory address, then addressable memory is 2^{32} = 4,294,967,296 bytes, or 4 GB. A computer system's **physical memory** is the actual number of memory bytes that physically are installed in the machine. Physical memory is usually less than addressable memory.

Memory Allocation and Addressing

Memory allocation describes the assignment of specific memory addresses to system software, application programs, and data. The simple allocation scheme shown in Figure 5-7 supports an operating system and a single program stored in contiguous memory blocks. The operating system occupies the lowest memory block and the program occupies the next block of memory. Memory allocation diagrams such as Figure 5-7 show physical memory as an array of addresses with the lowest memory address at the bottom and the highest memory address at the top.

In Figure 5-7, the operating system occupies the lowest range of memory addresses and the first instruction of the program is one address higher than the last byte of the operating system. The difference between the first address in physical memory and the address of the first program instruction is called the program's offset. In Figure 5-7, the offset is $3000_{16} - 0 = 3000_{16}$.

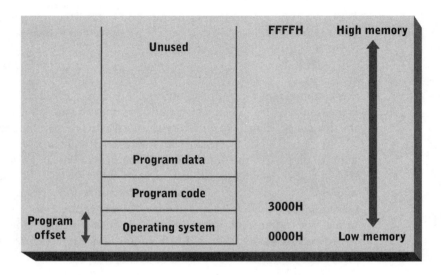

Figure 5-7 ►

A simple memory allocation scheme

Unused

FFFFH   High memory

Program data

Program code

3000H

Program offset

Operating system

0000H   Low memory

A programmer explicitly could include the program offset in any memory address operands when writing or compiling a program. For example, assume that the program offset is 100,000 bytes. If the tenth program instruction is a BRANCH to the first instruction, then the tenth instruction's operand must be 100,000. Note that the programmer must know the program offset to specify correctly any memory address operands in BRANCH, MOVE, and other instructions. The term **absolute addressing** describes memory address operands that refer to actual physical memory locations.

Generally, a programmer doesn't know exactly where a program will be loaded in memory when it is executed, or the exact value of the offset. Even if the programmer did know the offset when writing the program, the offset value might later change for many reasons including:

► Upgrading the operating system, thus changing its size and the program offset

► Reconfiguring the operating system, such as allocating more I/O buffers or adding new device drivers

► Loading and executing multiple programs

If a program uses absolute addressing then it must be rewritten every time the program offset is changed.

Because a program's offset generally can't be known in advance, programs are written as though their first instruction is loaded in memory address 0. The CPU automatically converts memory address operands into physical memory addresses as the program executes. To do so, the operating system calculates and stores the program offset in a register when the program first is loaded into memory. During program execution, the CPU automatically adds the program offset to all memory address operands before accessing memory. This method of automatically computing physical memory addresses is called **indirect addressing** or **relative addressing**. The register that holds the offset value is called an **offset register**.

CPUs with segmented memory have multiple offset registers called **segment registers**. Each offset register points to a different memory segment. Using multiple offset registers is advantageous when memory segments are relatively small. For example, the Intel 8088 used in the original IBM PC used sixteen 64 KB segments and four segment registers. A program could address up to 256 KB of memory at any one time by storing the offset value for a different segment in each segment register (see Figure 5-8). Later generations of Intel microprocessors have implemented a flat memory model, but also have maintained the segmented model and the four segment registers for backward compatibility.

Figure 5-8 ▶

The segmented memory architecture of the Intel 8088 processor; a reference to a specific memory address consists of a segment register and a 16-bit offset value (for example, CS + 1F40)

Memory addressing and allocation schemes become much more complex when multiple programs are loaded into memory. Figure 5-9 shows an operating system and two programs stored in memory simultaneously. Note that the starting address and offset value of each program is different (3000_{16} for Program 1, 9000_{16} for Program 2). The operating system stores the offset value for each program currently loaded into memory. Before CPU control is passed from one program to another the operating system loads the appropriate offset value(s) into the offset register(s).

Figure 5-9 ▶

Memory
allocation for
multiple
programs

MAGNETIC STORAGE

Magnetic storage devices exploit the duality of magnetism and electricity. That is, electrical current can be used to generate a magnetic field and a magnetic field can be used to generate electricity. A magnetic storage device converts electrical signals into magnetic charges, captures the magnetic charge on a storage medium, and later uses the stored magnetic charge to generate an electrical current. The polarity of the magnetic charge represents the bit values zero and one.

Figure 5-10 illustrates a simple magnetic storage device. A wire is coiled around a metallic **read/write head**. To perform a write operation, an electrical current is passed through the wire, which generates a magnetic field across the gap in the read/write head. The direction of current flow through the wire determines the polarity of the field, that is, the position of the positive and negative poles of the magnetic field. Reversing the direction of the current reverses the polarity.

A magnetic storage medium is placed next to the gap. The portion of the medium surface closest to the magnetic field is permanently charged. The stored charge polarity is identical to the polarity of the generated magnetic field. The strength of the stored charge is directly proportional to the strength of the magnetic field. The strength of the magnetic field is determined by several factors including the number of coils in the wire, the strength of the electrical current, and the mass of the metallic read/write head. The storage medium must be constructed of or coated with a substance capable of accepting and storing a magnetic charge, such as a metallic compound.

Figure 5-10 ▶

Principles of
magnetic data
storage

A read operation is the inverse of a write operation. The portion of the storage medium with the desired data is placed near the gap of the read/write head. The stored charge radiates a small magnetic field. When the magnetic field comes in contact with the read/write head, a small electrical current is induced in the read/write head and the wire wrapped around it. The current direction depends on the polarity of the stored charge. Electrical switches at either end of the coiled wire detect the direction of current flow and sense either a zero or one bit value.

Magnetic storage devices must control or compensate for some undesirable characteristics of magnetism and magnetic storage media including:

▶ Magnetic decay

▶ Magnetic leakage

▶ Minimum threshold current for read operations

▶ Storage medium coercivity

▶ Long-term storage medium integrity

A magnetic storage device must carefully balance all of these factors to achieve cost-effective and reliable storage. Different storage device requirements, such as speed vs. cost, dictate different trade-offs. Table 5-2 summarizes the factors and measures that protect data against loss.

Table 5-2 ▶

Factors that
lead to data
loss in magnetic
storage devices

Factor	Description
Magnetic decay	Natural charge decay over time—data must be written at higher power than the read threshold to avoid data loss
Magnetic leakage	Cancellation of adjacent charges of opposite polarity and migration of charge to nearby areas—data must be written at higher power than the read threshold to avoid data loss
Coercivity	Ability of storage medium to hold charge—medium must be of sufficient mass and coercivity to hold charges strong enough to counteract decay and leakage
Recording density	Coercible material per bit decreases as recording density increases—higher recording density makes stored data more susceptible to loss due to decay and leakage
Media integrity	Stability of coercible material and its attachment to substrate—physical stress and temperature/humidity extremes must be avoided to prevent loss of coercible material

Magnetic Decay and Leakage

The tendency of magnetically charged particles to lose their charge over time is called **magnetic decay**. Magnetic decay is constant over time and proportional to the power of the charge. The electrical switches used in a read operation require a minimum, or threshold, current level. Because induced current power is proportional to magnetic charge strength, a successful read operation requires magnetic charge above a certain threshold.

Over a sufficient length of time, magnetic decay will cause the stored charge power to fall below the threshold required for a successful read operation. At that point, the data content of the storage medium effectively is lost. This phenomenon is the primary reason that data stored on diskettes usually is unreadable after a few years. To minimize the potential for such data loss, data bits must be written with sufficient charge to compensate for decay. Magnetic storage devices write data at a substantially higher charge power than the read threshold, thus ensuring long-term readability. But there are limits to how much charge a read/write head can generate and how much the storage medium can hold.

The strength of individual bit charges also can decrease due to **magnetic leakage** from adjacent bits. This phenomenon can be heard clearly in audio recordings made on analog magnetic tape (refer to Problem 3 at the end of this chapter). Any charged area continuously generates a magnetic field that may affect nearby bit areas. If the polarity of adjacent bits is opposite then their magnetic fields will tend to cancel out the charge of both areas. The strength of both charges will fall below the read threshold.

Storage Density

Coercivity is the ability of a substance or magnetic storage medium to accept and hold magnetic charge. Coercivity varies widely among elements and compounds. In general, metals and metallic compounds offer the highest coercivity at reasonable cost. For any specific material, coercivity is directly proportional to mass. In magnetic storage media, mass is a function of the surface area in which a bit is stored, the thickness of the medium or its coercible coating, and the density of the chargeable material within the medium or coating. Larger, thicker, and/or more dense areas can hold more charge due to their higher mass of coercible material.

Most users want to store as much data as possible on a storage medium. The simplest way to do this is to reduce the surface area used to store a single bit value, thus increasing the total number of bits that can be stored on the media. For example, the storage density of a two-dimensional medium can be quadrupled by halving the length and width of each bit area (see Figure 5-11). The amount of surface area allocated to a bit is referred to as the **recording density**. Recording density typically is expressed in bits, bytes, or tracks per inch, abbreviated as bpi, Bpi, and tpi, respectively.

Figure 5-11 ▶

Storage density is a function of the length and width of an individual bit area (a); density can be quadrupled by halving the length and width of bit areas (b)

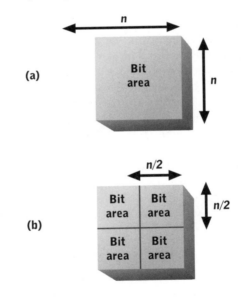

Some magnetic storage devices such as hard disks have fixed recording density but others such as tape have variable recording density. Because reductions in area per bit reduce coercible mass per bit, the problems of magnetic decay and leakage increase as recording density increases. Designers and purchasers of magnetic media and devices must find a suitable balance between high recording density and the reliability of the media, especially over relatively long periods of time.

Media Integrity

The integrity of a magnetic storage medium depends on the nature of its construction and the environmental factors to which it is subjected. Some magnetic storage media, such as tape and floppy disk, use thin coatings of coercible material layered over a plastic or other substrate. Age and environmental stress can loosen the bond between the coating and substrate causing the coating to wear away. Physical stress on the media from fast forwarding and rewinding tape can accelerate the process. Temperature and humidity extremes also can accelerate the process.

Loss of coercible coating represents a loss of strength in stored magnetic charges. As with magnetic leakage and decay, data becomes unreadable when the remaining charge falls below the threshold of readability. To extend the life of magnetic media they must be protected from physical abuse and temperature and humidity extremes. Removable magnetic media sometimes are stored in climate-controlled vaults to ensure long-term media integrity.

Magnetic Tape

A **magnetic tape** is a ribbon of plastic with a coercible (usually metallic oxide) surface coating. Tapes are mounted in a tape drive for reading and writing. The **tape drive** contains motors that wind and unwind the tape. It also contains one or more read/write heads and related circuitry. Tape drives are relatively slow serial access devices. Tapes primarily are used to make backup copies of data stored on faster secondary storage devices and to physically transport large data sets.

Tapes and tape drives come in a variety of shapes and sizes. Older tape drives used wide tapes of up to 1 inch wound on large open reels of up to 10 inches. Modern tape drives use tapes permanently mounted in plastic cassettes or cartridges (see Figure 5-12). When a cassette is inserted into a tape drive, an access door automatically is opened and the read/write mechanism is positioned around or next to the exposed tape. Tape widths have shrunk over time because improvements in technology have increased recording density.

Tapes compound the magnetic leakage problem by winding the tape upon itself. Leakage can occur from adjacent bit positions on the same area of the tape as well as from the layer of tape wound above or below on the reel. Tapes also are susceptible to problems arising from stretching, friction, and temperature variations. As a tape is wound and unwound it tends to stretch. If enough stretching occurs, the distance between bit positions can be altered, making individual bits difficult or impossible to read.

There are two basic approaches to recording data onto a tape surface. **Linear recording** places bits along parallel tracks that run along the entire length of the tape, as shown in Figure 5-13(a). Recording density depends on bit spacing within each track and on the number of tracks. **Helical scanning**, as shown in Figure 5-13(b) reads and writes data to or from a tape by rotating the read/write head at an angle to the tape and moving from tape edge to tape edge. This method requires much more complex read/write mechanisms.

Figure 5-12 ▶

Components of a typical cassette or cartridge tape

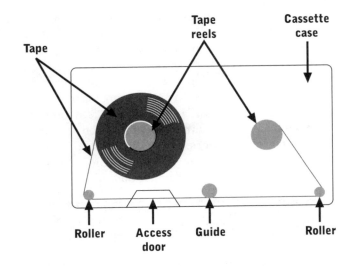

Tape

Tape reels

Cassette case

Roller

Access door

Guide

Roller

Figure 5-13 ▶

Data recorded in linear parallel tracks (a) and with helical scanning (b)

(a)

(b)

Helical scanning originally was developed for video recording, such as VHS tapes. For equal cost drives, helical scanning usually can pack more data into the same tape area than linear recording. The primary disadvantages of helical scanning are relatively slow tape motor speeds and direct contact of the read/write heads with the tape. Searching for specific data on a tape is slower with helical scanning drives due to slower tape motor speeds. Helically scanned tapes are subject to wear due to friction as the tape passes over the read/write head. Linear recording drives usually maintain a small gap between the read/write head and the tape to eliminate friction.

Recording data to tape at high speed and density requires precise alignment of the tape with internal drive components. Reliable tape storage requires careful control of environmental factors such as heat, dust, and dirt. Environmental control and precise alignment can be implemented within the cartridge, drive, or both.

High precision cartridges are relatively expensive, but cheaper drive components can be used. If high precision and environmental control are designed into the tape drive then its price increases but cheaper cartridges can be used. If an application calls for high capacity, speed, and reliability, then both the cartridge and the drive must employ relatively expensive components.

There are at least five competing magnetic tape technology/format combinations and several variations on each (see Table 5-3). The tape storage industry has had relatively little cooperation on standards in the last decade. Modern standards are proprietary and subject to licensing fees. For the consumer, the result has been a maze of competing technologies with lots of hype, lots of misinformation, and little compatibility.

Table 5-3 ►

Comparison of modern tape formats

Format Family	Standard	Uncompressed Capacity (GB)	Manufacturers
DDS (DAT)	DDS-1	2	Hewlett-Packard
	DDS-2	4	Sony
	DDS-3	12	
	DDS-4	20	
	DDS-5 (proposed)	40	
AIT	AIT-1	25 or 35	Sony
	AIT-2	50	
Mammoth	Mammoth 1	20	Exabyte
	Mammoth 2	60	
	Mammoth 3 (proposed)	200	
DLT	DLT4000	20	Quantum
	DLT7000	35	
	DLT8000	40	
	Super DLTtape 1	80 or 110	
	Super DLTtape 2 (proposed)	Not yet specified	
	Super DLTtape 3 (proposed)	Not yet specified	
	Super DLTtape 4 (proposed)	Not yet specified	
LTO	Accelis-1	25	Hewlett-Packard
	Accelis-2 (proposed)	50	IBM
	Accelis-3 (proposed)	100	Seagate
	Accelis-4 (proposed)	200	
	Ultrium-1	100	
	Ultrium-2 (proposed)	200	
	Ultrium-3 (proposed)	400	
	Ultrium-4 (proposed)	800	

Data Storage Technology

Technology
Focus
Magnetic Tape Formats and Standards

Prior to the mid-1990s, there were several widely used tape storage standards. Mainframe tape drives typically used 0.5- or 1-inch open reel tape. IBM set standards for mainframe tapes because it made the majority of mainframe computers. Tape drives for smaller computers either followed IBM's standards or the open standards of the **Quarter Inch Committee (QIC)**, as shown in Table 5-4. In those days, IBM's dominance helped to create an "us vs. IBM" mentality that perhaps explains the cooperative spirit that led to the QIC.

Table 5-4 ►

Sample QIC tape format specifications

Format	Year	Cartridge Size (Inches)	Capacity (GB)	Tracks	Recording Density (bpi)
QIC-80	1988	4 x 6	.08	28	14,700
QIC-120	1991	4 x 6	0.125	15	10,000
QIC-525	1992	4 x 6	0.525	26	20,000
QIC-2100	1993	4 x 6	2.1	30	50,800
QIC-3095	1995	3.25 x 2.5	4	72	67,733
QIC-3220	1997	3.25 x 2.5	10	108	106,400

The explosion in the market for PCs, small servers, and minicomputers in the 1980s and 1990s led to a corresponding surge in the tape market. Many companies entered the business and many new technologies were developed. To gain a competitive edge, many vendors closely guarded their technology. Cooperation became the exception and when companies did cooperate, they created proprietary standards and charged substantial licensing fees for other companies to produce compatible media and drives. Technology quickly outpaced the QIC standards and no vendors were willing to contribute proprietary technology to update them.

The **Digital Data Storage (DDS)** standards were developed by Hewlett-Packard and Sony based on an earlier technology called **Digital Audio Tape (DAT)**. DDS tapes are 4 mm wide and DDS drives use helical scanning. DDS tapes are relatively cheap, but they aren't designed for precise alignment. DDS drives compensate for cheap tape cassettes with elaborate technology for tape and environmental control, making the drives relatively expensive. DDS storage is used widely in small servers.

Digital Linear Tape (DLT) is a standard developed by Quantum. DLT tape is 0.5 inches wide and the cartridge contains only one reel. The other reel is located in the drive, which uses an elaborate mechanism to grab a plastic tape leader from the cartridge and wind it onto the takeup reel. Longer tape lengths and higher capacity per tape can be achieved with the single reel cartridge. DLT records in parallel linear tracks, one at a time, from one end of the tape to another. At the end of the tape, the drive motors reverse direction and the next track is recorded. The end-to-end track format has advantages when searching for specific data but it requires many passes to fill the tape completely.

Sony and Exabyte have attempted to breathe new life into DAT. Sony's standard is called **Advanced Intelligent Tape (AIT)** and Exabyte's is called **Mammoth**. Both standards use helical scanning on 8 mm tapes, and employ a more expensive and precisely manufactured tape cartridge. Both standards also improve the tape drive technology to pack more data onto a single tape. AIT includes a small RAM cache within each cartridge.

Linear Tape Open (LTO) is a standard developed by Hewlett-Packard, IBM, and Seagate. Despite its name and claims, LTO is a proprietary standard. Actually, LTO is two standards—Accelis and Ultirum. LTO uses linear recording and employs a number of technology improvements in tape cartridges, coercible materials, read/write heads, and tape control. All of this technology comes at a substantial price; LTO primarily is designed for enterprise-wide servers supporting mission-critical applications. The Accelis standard maximizes retrieval speed at the expense of per tape capacity. Some Accelis devices can retrieve and load one tape quickly from a library of dozens or hundreds. The Ultrium standard maximizes capacity at the expense of access speed. It is designed to back up large data sets such as data warehouses and on-line transaction-processing databases.

Magnetic Disk

Magnetic disk media are flat, circular **platters** with metallic coatings that are rotated beneath read/write heads (see Figure 5-14). Multiple platters can be mounted and rotated with a common spindle. Data normally are recorded on both sides of a platter. A **track** is one concentric circle of a platter, or the surface area that passes under a read/write head when its position is fixed. A sector is a fractional portion of a track (see Figure 5-15). The number of sectors per track ranges from eight for low-density floppy disks to several dozen for high-capacity hard disks. A single sector, or block, usually holds 512 bytes, which also is the normal data transfer unit to or from the drive.

Figure 5-14 ▶

Primary components of a typical disk drive

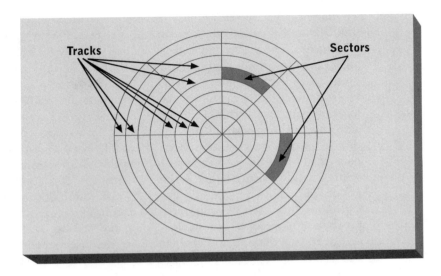

There is a read/write head for each side of each platter. Read/write heads are mounted on the end of an **access arm**. Access arms of equal length are attached to a positioning servo. The servo moves the access arms so that the read/write heads can be positioned anywhere between the outermost track, which is closest to the outer edge of the platter, and the innermost track, which is closest to the center of the disk. Read/write heads normally do not make contact with the recording surface of a platter. Rather, the head floats above the platter on a thin cushion of air as the disk spins.

A **hard disk** is a magnetic disk media with a rigid metal base, or substrate. Typical platter size for hard disks is between three and five inches in diameter. Multiple platters are spun at relatively high speeds of up to 10,000 revolutions per minute. Drive capacity depends on the number of platters, platter size, and recording density. Typical hard disk drive capacities at this writing range from 5 to 100 GB. Multiple hard drives often are enclosed in a single storage cabinet and referred to as a **drive array**. The term drive array also can be used to describe arrays of optical disk drives.

A **floppy disk**, or **diskette**, uses a base of flexible or rigid plastic material. Modern diskettes range from 2.5 to 3.5 inches in diameter. A diskette contains a single flexible plastic platter coated with an iron or other metallic compound on both sides. The platter is mounted permanently in a plastic case that can be removed from a diskette drive. The case has an access door that is closed when the diskette is removed from the drive and opened automatically when the diskette is inserted into the drive. Diskettes are spun at lower speeds than hard disks. Diskette capacities range from less than 1 MB to as much as 5 MB.

Larger capacity removable magnetic disks are similar externally to floppy disks. Examples include Iomega Jaz and Zip disks. Platter sizes are usually 3.5 or 5.25 inches. The platter has a dense coating of coercible material. Spin rates are higher than for floppies but not as high as for hard disks. Storage capacity ranges up to 2 GB.

When originally manufactured, all magnetic disks are blank. Although the location of tracks within a platter is fixed, sector locations within a track are not. One of the operations performed when formatting a disk is to fix the location of sectors within tracks. This is accomplished by writing synchronization data at the beginning of each sector. Synchronization information also can be written within a sector either during formatting or during subsequent write operations. The read/write circuitry uses synchronization information to compensate for minor variations in rotation speed and other factors that might disturb the precise timing needed for reliable reading and writing.

Read or write operations are made in response to a command that identifies the platter, track within the platter, and sector within the track. Controller circuitry switches activate the read/write head for the platter. The positioning servo moves the access arms so they are positioned over the track. The controller circuitry then waits for the appropriate sector to rotate underneath the read/write head. As the sector starts to rotate beneath the read/write head, the read or write operation is started. The operation is completed just as the last portion of the sector passes the read/write head. The access arms typically are left in their last position until the next read or write command is received.

Disk access time depends on several factors including:

► Time required to switch among read/write heads
► Time required to position the read/write heads
► Rotational delay

Disk drives generally share one set of read/write circuits among all of the read/write heads. The read/write circuitry must be electronically attached, or switched, to the appropriate read/write head before a sector can be accessed. The time required to perform this switching is no more than a few nanoseconds and is referred to as **head-to-head switching time**. Switching occurs in sequence among the read/write heads and multiple switching operations may be required to activate a desired head. For example, if there are 10 read/write heads, switching from the first head to the last requires 9 separate switching operations (first to second, second, to third, and so forth). Head-to-head switching time is the least significant component of access time (delay).

The positioning servo that moves the access arms and read/write heads is a mechanical device. It is essentially a bidirectional motor that can be started and stopped very quickly to position the read/write heads precisely over a desired track. The time required to move from one track to another is called

track-to-track seek time. Track-to-track seek time typically is measured in milliseconds. It is an average number due to variations in the time required to start, operate, and stop the positioning servo.

Once the read/write heads are positioned over the desired track, the controller waits for the desired sector to rotate beneath the heads. The controller keeps track of the current sector position by continually scanning the sector synchronization information written on the disk during formatting. When the appropriate sector is sensed, the read or write operation is started. The time that the disk controller must wait for the proper sector to rotate beneath the heads is called **rotational delay**. Rotational delay depends on platter rotation speed. High spin rates such as 10,000 RPM yield relatively low rotational delay.

Both track-to-track seek time and rotational delay vary from one read or write operation to another. For example, if two consecutive read requests access adjacent sectors on the same track and platter, then rotational delay and track-to-track seek time for the second read operation will be zero. If the second read access is to the sector preceding the first, then rotational delay will be high; almost a full platter rotation. If one read request references a sector on the outermost track and the other request references the innermost track, then track-to-track seek time will be high.

The amount of track-to-track seek time and rotational delay cannot be known in advance because the location of the next sector to be accessed also can't be known in advance. Access time for a disk drive is not a constant. Instead, it usually is stated using a variety of performance numbers including raw times for head-to-head switching and rotation speed. The most important performance numbers are average access time, **sequential access time**, and **sustained data transfer rate**.

Average access time is computed by assuming that two consecutive accesses are sent to random locations. Given a large number of random accesses, the expected value of head-to-head switching time will correspond to switching over half the number of recording surfaces. The expected value of track-to-track seek time will correspond to movement over half the tracks. The expected value of rotational delay will correspond to one-half of a platter rotation. If head-to-head (HTH) switching time is 5 nanoseconds, there are 10 read/write heads, and track-to-track (TTT) seek time is 5 microseconds, 1,000 tracks are on each recording surface, and the platters are rotated at 7,500 RPM, the average access delay is computed as:

average access delay = HTH switch + TTT seek + rotational delay

$$= (5 \text{ nanoseconds} \times \frac{10}{2}) + (\frac{1000}{2} \times 5 \text{ microseconds}) + \left(\frac{60 \text{ seconds}}{7500 \text{ RPM}} \div 2 \right)$$

$$= .000000025 + (500 \times .000005) + (.008 \div 2) \text{ seconds}$$

$$= .000000025 + .0025 + .004 \text{ seconds}$$

$$\approx 6.5 \text{ milliseconds}$$

Average access time is the sum of average access delay and the time required to read a single sector. The time required to read a single sector depends entirely on the rotational speed of the disk and the number of sectors per track. If the previous figures are assumed and there are 24 sectors per track, the average access time is computed as:

$$\text{average access time} = \text{average access delay} + \left(\frac{60 \text{ seconds}}{7500 \text{ RPM}} \div 24 \right)$$

$$= .0065 + (.008 \div 24)$$

$$\approx 6.83 \text{ milliseconds}$$

Sequential access time is the time required to read the second of two adjacent sectors on the same track and platter. Head-to-head switching time, track-to-track seek time, and rotational delay are all zero for such an access. The only component of access time is the time required to read a single sector, that is, the second half of the above equation, or 0.33 milliseconds in that example.

Because sequential access time is so much faster than average access time, disk performance is improved dramatically if related data are stored in sequential sectors. For example, loading a program from disk to memory is relatively fast if the entire program is stored in sequential locations on the same track. If the program won't fit in a single track, then performance is maximized by equivalently positioned tracks on multiple platters.

If portions of the program are scattered around the disk in random sectors, then performance is reduced due to switching read/write heads, positioning the access arm, and waiting for the desired sectors to rotate beneath the heads. Note that the performance improvement resulting from storing programs and files in sequential locations is realized only if those files are accessed sequentially. If file contents are accessed in random order, then sequential storage will not provide significant performance improvements.

File contents tend to become scattered, or fragmented, in many nonsequential sectors over time. Fragmentation is a normal by-product of adding, deleting, modifying, and moving files and directories. Most operating systems provide a utility program that performs an operation called **disk defragmentation**. A disk defragmentation utility reorganizes disk content so that all of the portions of an individual file are stored in sequential sectors, tracks, and platters.

Effective disk drive performance also depends on the speed at which data can be moved through the disk controller circuitry to/from memory or the CPU. This circuitry typically is not a communication bottleneck for a single disk drive. A summary performance number that combines the physical aspects of data access with the electronic aspects of data movement is the disk drive data transfer rate. As with access times, there are generally two different numbers stated. Maximum data transfer rate is the fastest rate that the drive can support and assumes sequential access to sectors. Because mechanical delays are minimal with sequential accesses, this number is quite high, typically several megabytes per second. The maximum data transfer rate for the previously described performance figures, assuming no controller delays, is:

$$\frac{1}{0.00033 \text{ seconds}} \times 512 \text{ bytes per sector} \approx 1.5 \text{ megabytes per second}$$

Sustained data transfer rate is computed with much less optimistic assumptions about data location. The most straightforward method is to use average access time. For the previous example, sustained data transfer rate is:

$$\frac{1}{0.00683 \text{ seconds}} \times 512 \text{ bytes per sector} \approx 75 \text{ kilobytes per second}$$

Because operating systems typically allocate disk space to files sequentially, the sustained data transfer rate is generally much higher than the above computation would imply. Table 5-5 provides performance statistics for various hard disk drive models from Seagate Technology, Inc.

Table 5-5 ►

Hard disk drive performance statistics

Model	Platters	Cylinders	Capacity (GB)	Rotation Speed (RPM)	Avg. Access Time (Milliseconds)
ST39175	3	11721	9.1	7200	10.9
ST318416	3	14384	18.4	7200	10
ST336704	6	14100	36.7	10,000	8.2
ST173404	12	14100	73.4	10,000	8.6

Technology
Focus

Whither the Floppy Disk?

Floppy disk drives have been a standard component of nearly every personal computer ever manufactured. They also commonly were used for archiving and transporting data from minicomputers and mainframes during the 1970s and early 1980s. Their attractiveness, particularly in 3.5-inch hard case format, is obvious. They offer cheap removable storage that can fit in a shirt pocket. Until recently, floppy disks were the technology of choice for data transfer, data backup, and the distribution of both software and data on personal computers. Data transport via floppy disk sometimes is called SneakerNet, because carrying floppies between computers was a popular way to share data before networks were widely implemented.

Table 5-6 shows specifications for the floppy disks used in IBM-compatible personal computers since 1981. Storage capacity has increased by approximately 20 times during the 10-year period of 1981 through 1991. The 5.25-inch flexible format is virtually obsolete and was replaced by the hard shell 3.5-inch format in use today. Note that no format or capacity improvements are listed after 1991. Why are there no improvements after that date? Has floppy disk technology peaked out?

Table 5-6 ▶

IBM-compatible PC floppy disk formats

Year	Capacity (KB)	Platter Size (Inches)	Number of Sides	Tracks per Side	Sectors per Track
1981	160	5.25	1	40	8
1982	180	5.25	1	40	9
1982	320	5.25	2	40	8
1983	360	5.25	2	40	9
1985	1200	5.25	2	80	15
1985	720	3.5	2	80	9
1987	1440	3.5	2	80	18
1991	2880	3.5	2	80	36

The primary reason that no standard format improvements have been adopted since 1991 is that IBM no longer defines the standard for IBM-compatible personal computers. Intel and Microsoft have been the source of many new standards since the late 1980s, but they have been largely silent on the issue of improved floppy disk drives. There has been no authority willing and able to define any new floppy disk formats.

Removable magnetic storage technology, however, has continued to evolve since 1991. Iomega Corporation has been a leader in removable magnetic media technology since the mid-1980s and is responsible for many advances since that time. Its Bernoulli Box products commonly were used as backup devices and second magnetic disk drives although they never became standard equipment on a PC. The Bernoulli Box products have evolved into two newer products lines—the Zip drive and the Jaz drive.

The Zip drive is a removable magnetic media drive that uses 250 MB, 3.5-inch cartridges. Average access time is 29 milliseconds with a maximum sustained data transfer rate of 2.4 MB per second. The Jaz drive also uses 3.5-inch removable magnetic cartridges. It increases all of the statistics of the Zip drive, including the cost. Average access time is 10 milliseconds for reading and 12 milliseconds for writing, with a rotational speed of 5,400 RPM, and a maximum sustained data transfer rate of approximately 8 MB per second. Jaz specifications rival those of many magnetic hard disks.

The Jaz drive is too expensive to be considered the "next floppy drive standard" but the Zip drive is more affordable. Iomega has attempted to make the Zip the next floppy standard by producing them in large quantities and selling Zip drives relatively cheaply. Iomega has signed license agreements for drive manufacture and installation with many PC manufacturers. It also has entered production agreements with several magnetic media manufacturers to meet the growing demand for disk cartridges. The Zip drive appears at present to be the closest thing to a new floppy disk standard.

OPTICAL MASS STORAGE DEVICES

Optical storage came of age in the 1990s. Its primary advantages, as compared to magnetic storage, are higher recording density and longer data life. Typical optical recording densities are at least 10 times greater than for magnetic storage devices. Higher density is achieved with tightly focused lasers that can access a very small storage medium area. Magnetic fields cannot be focused on as small an area without overwriting surrounding bit positions.

Most optical storage media can retain data for decades. Optical storage is not subject to problems of magnetic decay and leakage. Optical storage is popular due to standardized and relatively inexpensive storage media. The compact disc (CD) format used for musical recordings also is supported by most optical storage devices. The cost to manufacture CDs is considerably less per bit than for removable magnetic storage technologies.

Optical storage devices store bit values as variations in light reflection (see Figure 5-16). The storage medium is a surface of highly reflective material. The read mechanism consists of a low-power laser and a photoelectric cell. The laser is focused onto one bit location of the storage medium at a specific angle. The photoelectric cell is positioned at a complementary angle to intercept reflected laser light. A highly reflective surface spot will cause the photoelectric cell to generate a detectable electrical current. A poorly reflective surface spot will not reflect enough light to cause the photoelectric cell to fire. The current, or lack of current, produced by the photoelectric cell is interpreted as a binary zero or one.

There are multiple ways to implement bit areas with low reflectivity. In Figure 5-16, low reflection is a result of a concave "dent" or pit in the reflective layer. Other ways of achieving low reflectivity bit areas include burned areas, dyes, and substances that change from crystalline to amorphous states.

An optimal surface material could be rapidly and easily changed from highly to poorly reflective and then back again an unlimited number of times. The search for such a material and a reliable and efficient method of altering its reflectivity has proven problematic. Some materials change quickly in one direction but not another, in a way similar to the properties of many automatic sunglass materials. The ability of most materials to change reflectivity degrades over time and/or with repeated use. There are several different optical write technologies currently in use or under development and there is currently no clearly dominant technology.

All current optical storage devices use a disk storage medium. (Note that in current usage, optical storage media often are called *discs* while magnetic storage media are called *disks*.) Because the recording surface is a disk, there are many similarities to magnetic disk storage including the use of a read/write mechanism mounted on an access arm, a spinning disk, and performance limitations due to rotational delay and movement of the access arm.

Figure 5-16 ▶

Optical disc read
operations for a
one bit (a) and a
zero bit (b)

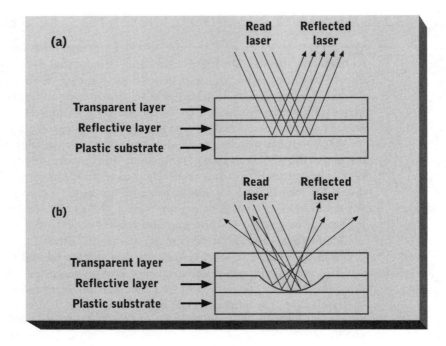

Although magnetic and optical devices both use disk technology, their performance is quite different. Optical disc storage is slower for a number of reasons, including the use of removable media, the use of a single spiral track, and the inherent complexity of an optical write operation. The performance gap has narrowed considerably in recent years and is approximately 10 to 1 at present. Magnetic hard disks typically have access times under 10 milliseconds. Read access time for optical drives is generally greater than 100 milliseconds. Write access time is typically two or three times longer than read access time. The gap is expected to shrink as optical technology matures and as magnetic technology reaches its physical limits.

Magnetic and optical storage are not direct competitors at present because they have very different cost/performance tradeoffs. Magnetic hard disks are still the cost performance leader for general-purpose on-line secondary storage. Optical storage is favored for read-only storage with low performance requirements and for applications with extremely high capacity requirements. The two technologies will compete more directly as optical access times improve and as optical write technology develops and matures.

Optical storage devices are available currently in a wide variety of storage formats and write technologies. Some of the more popular include:

► CD-ROM
► WORM (CD-R)
► Magneto-optical
► CD-RW
► DVD

Table 5-7 summarizes each technology. The technologies and media formats are discussed in detail below.

Table 5-7 ►

Technologies and storage formats for optical and magneto-optical storage

Technology/Format	Writable?	Description
CD-ROM	No	Adaptation of musical compact disc technology, 650 MB capacity
CD-R	One time only	CD-ROM format with a dye reflective layer that can be altered by a low-power laser
CD-RW	Yes	CD-ROM format with a phase change reflective layer, can be written up to 1000 times
Magneto-optical	Yes	Combination of optical and magnetic technology, expensive and outdated, rapidly giving way to CDs and DVDs
DVD-ROM	No	Adaptation of DVD video technology, similar to CD-ROM but more advanced, 4.7 GB capacity
DVD-R	One time only	DVD-ROM format, same basic technology as CD-R with performance and capacity improvements
DVD-RAM	Yes	Format similar but not completely compatible with DVD-ROM, updated CD-RW phase change technology with performance and capacity improvements
DVD-RW	Yes	A competitor to DVD-RAM
DVD+RW	Yes	A competitor to DVD-RAM

CD-ROM

Compact disc technology originally was developed by Sony and Philips for storing and distributing music in a format known as **compact disc digital audio (CD-DA)**. **Compact disc read-only memory (CD-ROM)** is compatible with CD-DA but includes additional formatting information to store directory and file information. Both CD-DA and CD-ROM are read-only storage technologies and formats because data permanently are embedded in the disc during manufacture.

A compact disc is 120 millimeters (approximately 4.75 inches) in diameter, is made largely of durable polycarbonate, and holds approximately 650 MB. Bit values are represented as flat areas called *lands* and concave dents called *pits* in the reflective layer. Data are recorded in a single continuous track that spirals outward from the center of the disc. Error correction data written along the track at regular intervals, thus helping the CD drive read mechanism to recover from errors introduced by dust, dirt, and scratches.

CD-ROM is a popular medium for distributing software and large data sets such as maps, census data, telephone directories, and other large databases. Its standardized format, high density, and cheap manufacturing cost make it well suited for these purposes. Its primary drawbacks are that discs cannot be rewritten, and their capacity is limited to 650 MB.

WORM and CD-R

A **write once read many (WORM)** disc is manufactured with all bit areas in a highly reflective state. As with CD-ROM, when a bit area state is changed to low reflectivity, the process is irreversible. WORM devices use high-powered lasers to burn holes in the reflective layer. This process proved too slow, costly, and cumbersome compared to newer technologies, but "burning a CD" still is used to describe the process of writing an optical disc.

Storage device and medium manufacturers eventually standardized a cheaper and faster write method and medium called compact disc–read (**CD-R**). CD-R drives use a laser that can be switched between high and low power and a laser-sensitive dye embedded in the CD-R disc. The dye is stable when scanned at low power during a read operation, but changes its reflective properties when scanned at higher power during a write operation.

CD-R discs and drives are relatively cheap. CD-R drives can read standard CDs and most computer CD readers can read CD-R discs. CD-R drives and discs commonly are used in several application areas including:

► Creating music CDs on home computers
► Backing up data from other storage devices
► Creating permanent archives of large data sets, such as photo albums
► Manufacturing small quantities of identical CDs for distribution, such as parts catalogs and conference proceedings

Magneto-Optical

A **magneto-optical** drive uses a laser and reflected light to sense bit values. The variations in reflectivity of the disk surface are achieved by applying a magnetic charge to the bit area. Reading is based on the polarity of the reflected laser light, not its strength. The magnetic charge shifts the polarity of the reflected laser light. The direction of the shift is determined by the polarity of the magnetic charge. This polarity shift is known as the Kerr effect.

The disk surface is a material that is resistant to magnetic charge at normal temperatures. At higher temperatures, approximately 150 degrees Celsius, the material can be charged by a magnetic field in much the same way as purely magnetic media. To bring the material to a chargeable temperature, a high-power laser is focused on a bit position. A magnetic write head positioned below the media surface then generates the appropriate charge. Because the laser beam focuses on a single bit position, only that bit position is raised to a sufficient temperature. Surrounding bits are unaffected by the magnetic field from the write head because they are cooler.

Magneto-optical technology peaked in the mid-1990s. Few new magneto-optical drives have been sold since then due to the introduction of cheaper CD-RW drives. Magneto-optical technology still has advantages over CD-RW in access speed and capacity, but those advantages are rapidly fading as phase-change technology matures and as newer phase-change formats such as DVD-RAM are being developed and deployed. There are many magneto-drives still in use by organizations that used them to create large collections of archived data. The technology probably will not completely fade away for several more years.

Phase-Change Optical Discs

Phase-change optical technology allows optical storage media to be written non-destructively. Phase-change technology is based on materials that can change state easily from non-crystalline, or amorphous, to crystalline, and then back again. The reflective characteristics of the materials are quite different in their amorphous and crystalline states. The difference is less than with traditional CDs, but sufficient to be detected by newer optical scanning technologies.

The reflective layer of a phase-change disc is a compound of tellurium, selenium, and tin. It changes from an amorphous state to a crystalline state when heated to a precise temperature. Heating the material to its melting point changes it back to an amorphous state. The melting point is relatively low so high-power lasers aren't required. However, multiple passes usually are required to generate sufficient heat, so write times are substantially longer than read times. The reflective layer loses its ability to change state with repeated heating and cooling. Current CD-RW media wear out after approximately 1000 write operations.

Compact disc–read/write (**CD-RW**) is a standard for phase-change optical discs that are the same size as standard CDs. Most CD-RW drives can read and write CD-R and CD-ROM discs. Unfortunately, older and some lower-priced newer CD-R and CD-ROM drives cannot read CD-RW discs because they are substantially less reflective than CD-R and CD-ROM discs. Access time is comparable to CD-ROM and CD-R drives, typically between 100 and 150 milliseconds.

DVD-ROM

DVD is an acronym for both **digital video disc** and **digital versatile disc** (the two terms are equivalent). DVD originally was developed by a consortium of companies as a standard format for distributing movies and other audiovisual content. Like CD-ROM, **DVD-ROM** is an adaptation of DVD in which DVD disks can store computer data. Most DVD-ROM drives also can read CD and CD-R discs. Most newer DVD-ROM drives are able to read CD-RW discs.

DVD capacity has improved on several CD and CD-RW technologies, including:

► Increased track and bit density due to smaller wavelength lasers and more precise mechanical control
► Improved error correction
► Multiple recording sides and layers

DVD technology now can store between 4 and 12 times as much data as a CD on the same size disc.

There are several DVD formats, some of which still are under development. DVD-ROM is the current standard format for read-only computer data. A single-sided, single-layer DVD-ROM disc holds 4.7 GB. The DVD-ROM standard also defines discs with capacities of 8.5 GB (double-sided) and 17 GB (double-sided, two recording layers per side). Virtually all DVD, DVD-R, DVD-RW, and DVD-RAM drives can read DVD-ROM discs. Also, most DVD drives can read all CD formats.

DVD discs have more densely packed data and are spun at higher speed than CDs. Sustained data transfer rates are several times those of CD formats. However, because they use similar disc access technology, DVD access times approximate those of CD formats (100 to 150 milliseconds). Access times should improve substantially as the technology matures.

DVD-R drives and media are similar in most ways to CD-R drives and media. First-generation DVD-R discs had a 3.9 MB capacity, which was later expanded to 4.7 MB to ensure compatibility with DVD-ROM.

There are three competing standards for rewritable DVDs: DVD-RAM, DVD-RW, and DVD+RW. **DVD-RAM** devices were released over a year earlier than DVD-RW and DVD+RW devices. DVD-RAM capacity is approximately 2.4 GB for single-sided discs and 4.8 GB for double-sided discs. DVD-RAM discs cannot be read in current DVD-ROM and home DVD players. DVD+RW discs hold 3 GB per side. The DVD+RW standard has the backing of several large companies such as Hewlett-Packard, Philips, and Sony, so it may be the most viable of the DVD-RAM competitors.

Business
Focus

A Modern SneakerNet?

Computer Technology Publishing (CTP) is a medium-sized publisher of high school, trade school, and college textbooks, software manuals, and computer-oriented "how to" books. The company catalog currently has over 300 titles in active distribution and normally has between 50 and 100 titles under development.

CTP has approximately 50 full-time employees and a single office building in Los Angeles. This workforce is much too small to do all of the work associated with its large catalog. CTP contracts much of its development and production work, such as copyediting, content reviews, and illustrations. All printing and distribution is subcontracted to a larger publisher.

Many copyeditors, graphic artists, and other development and production personnel are independent contractors working out of their homes. Others are employees of small companies with 10 or fewer employees. All of the authors and reviewers are independent contractors. The contractors, authors, and reviewers are scattered throughout North America.

Development projects for new and revised titles require a great deal of communication among authors, editors, reviewers, product managers, and other production staff. A manuscript typically is shipped dozens of times during development. Until recently, all of this communication was performed by sending paper via an overnight mail service. Over the last few years CTP has standardized the software used for development, using Microsoft Word, Adobe Illustrator, and a handful of other software products for most projects. CTP would like to go one step further and transfer manuscripts and other information in electronic form.

This would be a relatively easy task if CTP had a network with all of its workers directly attached. But the large number of contract employees and the wide variation in their computer configurations makes this impossible. An additional problem is the size of the files comprising a typical title, typically a few hundred megabytes to a few gigabytes. The company does maintain a Web site and electronic mail links. Some data currently are moved by these means but many authors, reviewers, and contract employees lack high-speed Internet connections. CTP still relies heavily on overnight delivery of manuscripts.

CTP is considering adopting a standard storage medium for electronic data exchange. Although this wouldn't eliminate the need for all overnight deliveries, it would reduce the production and consumption of paper. It also would reduce overnight shipping costs due to reduced weight of the shipment. CTP has decided on the following requirements for their standard storage medium:

▶ A removable cartridge that is small, light, and rugged—something that is easily and inexpensively mailed without damage to itself or the data

▶ Reliable hardware and media—drives should last at least five years, data should be 100 percent readable for at least five years

▶ A long-lived and widely available format—a drive that is widely available now and a cartridge that is likely to be readable by drives purchased in 3 to 5 years

▶ Inexpensive drive—$500 maximum, $250 or less preferred; inexpensive media—10 cents per megabyte maximum, 2 cents or less preferred

SUMMARY

► A typical computer system has primary and secondary storage devices. Primary storage devices support the immediate execution of programs, and secondary storage devices provide long-term storage to support future program execution. The characteristics of speed, volatility, access method, portability, cost, and capacity vary among storage devices, forming a memory-storage hierarchy. Primary storage tends to be at the top of the hierarchy with faster access speeds and higher costs per bit of storage, while secondary storage tends to be in the lower portion of the hierarchy, with slower access speeds and lower costs.

► The critical performance characteristics of primary storage devices are their access speed and the number of bits that can be accessed in a single read or write operation. Modern computers use memory implemented with semiconductors. Basic types of memory that are built from semiconductor microchips include random access memory (RAM) and read-only memory (ROM). The primary RAM types are static and dynamic. Static RAM (SRAM) is implemented entirely with transistors. Dynamic RAM (DRAM) is implemented using transistors and capacitors. The more complex circuitry of SRAM is more expensive to implement but provides faster access times. Neither type of RAM can match current microprocessor clock rates.

► Programs generally are created as though they occupied contiguous primary storage locations starting at the first location. Programs are rarely physically located in low memory due to system software or other programs that may occupy those locations. Indirect addressing reconciles the difference between where a program "thinks" it is located and where it actually is located. The actual address of a program's first instruction is stored in an offset register and this address is added to each address reference made by the program. Indirect addressing allows multiple programs to reside in memory and allows programs to be moved during execution.

► Magnetic secondary storage devices store data bits as magnetic charges. Magnetic tapes are ribbons of plastic coated with a coercible coating. Data is written to (or read from) tapes by passing it over the read/write head of a tape drive. Magnetic disks are platters coated with coercible material. Platters are rotated within a disk drive and read/write heads access data at various locations on the platter(s). Magnetic disk drives are random access devices because the read/write head can be freely moved directly to any location on a disk platter.

▶ Optical discs store data bits as variations in light reflection. An optical disc drive reads data bits by shining a laser beam onto a small disc location. High and low reflections of the laser are interpreted as ones and zeros. The cost per bit of optical storage is less than magnetic storage, at the expense of slower access speed. Types of optical discs (and drives) include various forms of CDs and DVDs. CD-ROM and DVD-ROM are written during manufacture. CD-R and DVD-R are manufactured blank and can be written to once. CD-RW and various types of rewritable DVDs use phase-change technology to write and rewrite the media.

Now that CPU, primary storage, and secondary storage have been covered in some detail, it's time to turn our attention to how those devices are integrated in a computer system. Chapter 6 describes how processing, I/O, and storage devices interact, and discusses methods of improving overall computer system performance. Chapter 7 describes various types of input/output devices.

Key Terms

absolute addressing
access arm
access time
addressable memory
Advanced Intelligent Tape (AIT)
average access time
big endian
block
block size
CD-DA (compact disc digital audio)
CD-R
CD-ROM (compact disc read-only memory)
CD-RW
coercivity
core memory
data transfer rate
device controller
Digital Audio Tape (DAT)
Digital Data Storage (DDS)

Digital Linear Tape (DLT)
direct access
disk defragmentation
diskette
double in-line memory module (DIMM)
drive array
dual in-line packages (DIPs)
DVD
DVD-RAM
DVD-ROM
dynamic RAM (DRAM)
electronically erasable programmable read-only memory (EEPROM)
enhanced DRAM (EDRAM)
ferroelectric RAM
firmware
flash memory
flash RAM
floppy disk

hard disk
head-to-head switching time
helical scanning
indirect addressing
least significant byte
linear recording
Linear Tape Open (LTO)
little endian
magnetic decay
magnetic leakage
magnetic tape
magneto-optical
Mammoth
memory allocation
most significant byte
non-volatile
offset register
parallel access
physical memory
platters
Quarter Inch Committee (QIC)

random access
random access memory (RAM)
read-only memory (ROM)
read/write head
read/write mechanism
recording density
refresh cycle
relative addressing
rotational delay

sector
segment registers
sequential access time
serial access
single in-line memory module
   (SIMM)
static RAM (SRAM)
storage medium
sustained data transfer rate

synchronous DRAM
   (SDRAM)
tape drive
track
track-to-track seek time
volatile
write once read many
   (WORM)

Vocabulary Exercises

1. _____ RAM requires frequent _____ to maintain its data content.

2. _____ rate is the speed at which data can be moved to/from a storage device over a communication channel.

3. Two standard optical storage media that are written only during manufacture are called _____ and _____ .

4. _____ , _____ , and _____ are competing standards for rewritable DVD discs.

5. The _____ of a hard disk drive generates or responds to a magnetic field.

6. Data stored on magnetic media for long periods of time may be lost due to _____ and _____ .

7. A(n) _____ stores data in magnetically charged areas on a rigid platter.

8. The _____ endian storage format places the _____ byte of a word in the lowest memory address. The _____ endian storage format places the _____ byte of a word in the highest memory address.

9. The contents of most forms of RAM are _____ , making them unsuitable for long-term data storage.

10. _____ is an outdated technology for non-volatile primary storage. _____ is a promising new technology for implementing non-volatile primary storage.

11. The CPU can incur one or more _____ when it accesses slower storage devices.

12. Under relative addressing, the content of a(n) _____ is added to calculate the corresponding physical memory address.

13. The number of bits used to represent a memory address determines the amount of a CPU's _____ .

Data Storage Technology

14. _____ typically is stated in milliseconds for secondary storage devices and nanoseconds for primary storage devices.

15. Average access time for a disk drive is determined by its _____, _____, and _____.

16. _____ storage is slower, cheaper, and less volatile than _____ storage.

17. In a magnetic disk drive, a read/write head is mounted on the end of a(n) _____.

18. The access method for RAM is _____ or _____ if words are considered the unit of data access. The access method is _____ if bits are considered the unit of data access.

19. DDS and AIT both use _____ to record bits and tracks on a magnetic tape.

20. _____ and _____ are two modern standards/formats for linear recording on magnetic tapes.

21. Under _____, program memory references correspond to physical memory locations. Under _____, the CPU must calculate the physical memory location that corresponds to a program memory reference.

22. A(n) _____ uses a combination of optical and magnetic storage technologies.

23. A(n) _____ is a series of sectors stored along one concentric circle on a disk _____.

24. The mechanical factors that limit disk access time are _____ and _____.

25. The data transfer rate of a secondary storage device can be calculated by multiplying its access time by its _____.

26. _____, _____, and _____ are storage formats, originally designed for music or video recording, that have been applied to computer data storage.

27. Tape drives are _____ devices. _____ are random or direct access devices.

28. Average access time usually can be improved by _____ the data content of a disk.

29. Modern personal computers generally use memory packaged on small standardized circuit boards called _____ or _____.

30. The _____ of a magnetic or optical storage medium is the ratio of bits stored to a unit of medium surface area.

31.   For most disk drives the unit of data access and transfer is a(n) _____ or _____ .

32.   Software programs permanently stored in a(n) ROM are called _____ .

33.   Many open standards for cartridge tape storage have been defined by the _____ .

Review Questions

1.   What factor(s) limit the speed of an electrically based processing device?

2.   What is/are the difference(s) between static and dynamic RAM?

3.   What improvements are offered by synchronous DRAM as compared to ordinary DRAM?

4.   Why isn't flash RAM commonly used to implement primary storage?

5.   Describe serial, random, and parallel access. What types of storage devices use each method?

6.   What is direct addressing? What is relative addressing?

7.   What are the costs and benefits of indirect addressing?

8.   How is data stored and retrieved on a magnetic mass storage device?

9.   Describe the factors that contribute to a disk drive's average access time. Which of these factors are improved if spin rate is increased? Which are improved if recording density is increased?

10.   What problems contribute to read/write errors on magnetic tapes? Are these problems also present with other magnetic storage media/devices?

11.   What is/are the advantage(s) and disadvantage(s) of helical scanning as compared to linear recording?

12.   Why is the recording density of optical discs higher than the recording density of magnetic disks? What factor(s) limit this recording density?

13.   Describe the processes of reading from and writing to a phase change optical disc? How does the performance and storage density of these discs compare to magnetic disks?

Problems and Exercises

1. Assume that a magnetic disk drive has the following characteristics:

   ► 10,000 RPM spin rate

   ► 2 nanosecond head-to-head switching time

   ► 3 microsecond average track-to-track seek time

   ► 5 platters, 1,024 tracks/platter side recorded on both sides, 50 sectors/track

   Questions:

   A. What is the storage capacity of the drive?

   B. How long will it take to read the entire contents of the drive sequentially?

   C. What is the serial access time for the drive?

   D. What is the average (random) access time for the drive (assume movement over half the tracks and half the sectors within a track)?

2. Assume that a CPU has a clock rating of 1.2 GHz. Assume further that half of each clock cycle is used for fetching and the other half for execution.

   Questions:

   A. What access time is required of primary storage for the CPU to execute with zero wait states?

   B. How many wait states per fetch will the CPU incur if all of primary storage is implemented with 5 nanosecond static RAM?

   C. How many wait states per fetch will the CPU incur if all of primary storage is implemented with 10 nanosecond SDRAM?

3. Try the following experiment. Find an analog cassette tape that was recorded at least a year ago. Start the playback in the blank section preceding a musical selection with the volume turned up fairly high. You will hear the song start at a very low volume a second or two before it actually starts at normal volume. What causes this phenomenon? What are its implications for digital data stored on magnetic tapes?

Research Problems

1. Investigate a current magneto-optical disk drive product such as those produced by Sony (www.sony.com). How does the cost per unit of storage compare to magnetic hard disk drives and phase change optical drives? How does the access time compare for both reading and writing? How does the recording density compare? Has this technology been eclipsed by CD-RW or rewritable DVD?

2. IBM has invested heavily in research and development for holographic storage systems. Investigate this research to determine the nature of holographic storage, its applicability to primary and secondary storage, and the likelihood of seeing products based on holographic technology in the near future.

3. Assume that you are the chief information officer of a rapidly growing dot-com. You manage several dozen small- and medium-size servers and are about to acquire your first mainframe to implement a data warehouse. You have a variety of DDS and AIT tape drives. You are concerned about tape drive capacity, reliability, and cost, and are considering switching to another tape format/technology. Which of the modern tape formats provides the greatest capacity at the lowest cost? Which is most reliable? Which is least likely to become obsolete five years from now? Which should you buy?

Chapter 6

System Integration and Performance

Chapter Goals

- ► Describe the implementation of the system bus and bus protocol

- ► Describe how the CPU and bus interact with peripheral devices

- ► Describe the purpose and function of device controllers

- ► Describe how interrupts coordinate actions of the CPU with secondary storage and I/O devices

- ► Describe how buffers, caches, and data compression improve computer system performance

In earlier chapters we discussed the processing and storage components of a computer system. Now it's time to describe how those components communicate with one another. This chapter explores hardware such as the system bus and device controllers that are used to connect other hardware components. The chapter also describes how the CPU communicates with components to coordinate their activities, and how bus protocols, interrupt processing, buffering, caching, and compression impact system performance (see Figure 6-1).

Figure 6-1 ►

Topics covered in this chapter

SYSTEM BUS

A **bus** is a set of parallel communication lines that connect two or more devices. The **system bus** connects the CPU with main memory and other system components, as shown in Figure 6-2. Devices other than the CPU and primary storage usually are called **peripheral devices**. Each bus line can carry a single bit value during any bus transfer operation. Computer systems have subsets of bus lines dedicated to carrying specific types of information. Three common subsets of the system bus are the data bus, the address bus, and the control bus.

Figure 6-2 ▶

Components of
the system bus
and attached
devices

System bus

As its name implies, the **data bus** moves data among computer system components. The number of lines typically is the same as, or a multiple of, the CPU word size. For example, a computer system with a 64-bit CPU would have 64 or 128 bus data lines.

The **address bus** carries the bits of a memory address. Modern address buses typically have 32 lines. When a peripheral device transfers data to main memory, it transmits the data item on the data bus while simultaneously transmitting the address of that data item on the address bus. The address bus is used only for bus transfers in which main memory is the target, or destination, device.

The **control bus** carries commands, command responses, status codes, and similar messages. Computer system components coordinate their activities by sending appropriate signals over the control bus.

Bus Clock and Data Transfer Rate

One or more control bus lines carry the **bus clock** pulse—a common timing reference for all attached devices. The frequency of bus clock pulses is measured in megahertz (MHz). It may be equivalent to the CPU clock rate or some well-defined fraction of the clock rate. Each clock pulse marks the start of a new opportunity to transmit data or a control message.

The time interval from one clock pulse to the next is called a **bus cycle**. Bus cycle time is the inverse of the bus clock rate. For example, if the bus clock rate is 133 MHz then the duration of each bus cycle is:

$$\text{bus cycle time} = \frac{1}{\text{bus clock rate}}$$

$$= \frac{1}{133{,}000{,}000 \text{ MHz}}$$

$$\approx 0.752 \text{ nanoseconds}$$

By its very nature, the bus must be relatively long because it connects many different computer system components. Miniaturization of all computer system components has reduced the length of a typical system bus but it still is 10 centimeters or more in most computer systems. A bus cycle cannot be any shorter than the time required for an electrical signal to traverse the bus from end to end. Bus length imposes a theoretical maximum on bus clock rate and a theoretical minimum on bus cycle time.

In practice, bus clock rate generally is set far below its theoretical maximum to allow for noise, interference, and the time required to operate interface circuitry in peripheral devices. Slower bus clock rates also give computer component manufacturers the ability to increase peripheral device reliability while holding cost to acceptable levels. As an analogy, consider the speed limit on a road. Even though the road itself might be able to support a maximum speed of 150 miles per hour, the speed limit is not set that high for several reasons. Vehicles that could meet the minimum speed requirement would be expensive, and reliable and safe transport would require high standards of road and vehicle maintenance, skilled drivers, and perfect weather and visibility.

The maximum capacity of a bus or any other communication channel is the product of clock rate and data transfer unit. For example, the theoretical capacity of a bus with 64 data lines and a 100 MHz clock rate is:

$$
\begin{aligned}
\text{bus capacity} &= \text{data transfer unit} \times \text{clock rate} \\
&= 64 \text{ bits} \times 100 \text{ MHz} \\
&= 8 \text{ bytes} \times 100{,}000{,}000 \text{ Hz} \\
&= 800{,}000{,}000 \text{ bytes per second}
\end{aligned}
$$

Such a measure of communication capacity is called a **data transfer rate**. Data transfer rates are expressed in bits or bytes per second, for example, 10 MB per second. See the Appendix for further discussion of data transfer rate units of measure.

There are only two ways to increase the maximum bus data transfer rate: increase clock rate or increase the data transfer unit size (the number of data lines). As discussed, bus length, conservative engineering, and peripheral device costs limit maximum clock rates. Cost and engineering considerations also limit the number of data bus lines. But the number of data bus lines always should be at least equal to CPU word size. This helps to avoid CPU wait states for bus data transfers. Data bus width larger than CPU word size can provide performance improvements if the CPU can effectively use multiple-word inputs.

Bus Protocol

The **bus protocol** governs the format, content, and timing of data, memory addresses, and control messages sent across the bus. Every peripheral device must follow the protocol rules. In the simplest sense, a bus is just a set of communication lines. In a larger sense, a bus is a combination of communication lines, a bus protocol, and devices that implement the bus protocol.

The bus protocol has two important impacts on maximum data transfer rate. First, control signals sent across the bus consume bus cycles, reducing the cycles available to transmit data. For example, a disk drive transfers data to main memory as a result of an explicit CPU command. Sending the command requires a bus cycle. In some bus protocols the command is followed by an acknowledgment from the disk drive and later by a confirmation that the command was carried out. Each signal (command, acknowledgment, and confirmation) consumes a separate bus cycle.

An efficient bus protocol consumes a minimal number of bus cycles, maximizing bus availability for data transfers. Unfortunately, efficient bus protocols tend to be complex, increasing the complexity and cost of the bus and all peripheral devices. The SCSI bus protocol, which is covered in the Technology Focus section later in this chapter, is an example of an efficient but complex bus protocol.

The bus protocol regulates bus access to prevent devices from interfering with one another. If two peripheral devices attempt to send a message at the same time, the messages collide and produce electrical noise. A bus protocol avoids collisions by one of two access control approaches—a master-slave approach and a multiple master, or peer-to-peer, approach.

In traditional computer architecture, the CPU is the focus of all computer activity. As part of this role, the CPU is also the **bus master** and all other devices are **bus slaves**. With the master-slave relationship a relatively simple bus protocol can be used. No device other than the CPU can access the bus except in response to an explicit instruction from the CPU. There are no collisions as long as the CPU waits for a response from one device before issuing a command to another device. In essence, the CPU plays the role of bus traffic cop.

When the CPU is the only bus master, overall system performance is significantly reduced. If devices can communicate only with the CPU, then transfers among devices, such as from disk to memory, must pass through the CPU. All transfers consume at least two bus cycles—one to transfer data from the source device to the CPU and another to transfer data from the CPU to the destination device. The CPU cannot execute computation and logic instructions while transferring data among other devices.

Performance is improved if storage and I/O devices can transmit data among themselves without explicit CPU involvement. There are two commonly used approaches to implementing such transfers: direct memory access and multiple master buses. Under **direct memory access (DMA)**, a device called a **DMA controller** is attached to the bus and to main memory. The DMA controller assumes the role of bus master for all transfers between memory and other storage or I/O devices. While the DMA controller manages bus traffic, the CPU is free to execute computation and data movement instructions.

In a **multiple master bus**, any device can assume control of the bus, or act as a bus master for transfers to any other device, not just memory. A single master must be chosen when multiple devices want to become a bus master at the same time. A **bus arbitration unit** is a simple processor attached to a multiple master bus that decides which device(s) must wait when multiple devices want to become a bus master. Multiple master bus protocols are substantially more complex and expensive than master-slave bus protocols.

LOGICAL AND PHYSICAL ACCESS

An **I/O port** is a communication pathway from the CPU to a peripheral device. In most computers, an I/O port is a memory address that can be read or written by the CPU and a single peripheral device. Each peripheral device may have multiple I/O ports and use them for different purposes, for example, transmitting data, commands, and error codes (see Figure 6-3). Dedicated bus hardware controls data movement between I/O ports and peripheral devices. The CPU reads and writes I/O ports using ordinary data movement instructions or dedicated I/O instructions.

Figure 6-3 ▶

Motherboard with eight I/O ports (slots) and two devices or device controllers attached to the leftmost I/O port

An I/O port is more than a memory address or data conduit, it also is a logical abstraction used by the CPU and bus to interact with each peripheral device in a similar way. I/O ports are used by the CPU and bus to interact with a keyboard in exactly the same way they interact with a disk drive or video display. The CPU instruction set and the bus are much simpler when I/O ports are used because special instructions and bus circuitry aren't required for each different peripheral device. The computer system also is more flexible because new types of peripheral devices can be incorporated into an older computer system design simply by allocating new I/O ports.

Despite similar treatment by the CPU and bus, peripheral devices differ in significant ways including storage capacity (if any), data transfer rate, internal data coding methods, and whether the device is a storage or I/O device. Physical device details, such as how a disk read/write head is positioned, how a certain color is displayed on a video display, or the font and position of a printed character, are not directly dealt with by the simple interface between peripheral device and CPU described thus far.

In essence, the CPU and bus interact with each peripheral device as if it were a storage device containing one or more bytes stored in sequentially numbered addresses. A read or write operation from this hypothetical storage device is called a **logical access**. The set of sequentially numbered storage locations is called a **linear address space**. Logical accesses to a peripheral device are similar to access to memory. One or more bytes are "read" or "written" as each I/O instruction is executed. The bus address lines carry the position within the linear address space being read or written, and the bus data lines carry data. Complex commands and status signals also can be encoded and sent via the data lines.

As an example of a logical access, consider a disk drive physically organized into sectors, tracks, and platters. When the CPU executes an instruction to read a specific sector, it does not transmit the platter, track, and sector number as parameters of the read command. Instead, it "thinks" of the disk as a linear sequence of storage locations, each holding one sector of data, and sends a single number to identify the location it wants to read.

To access the appropriate sector physically, the location within the assumed linear address space must be converted into the corresponding platter, sector, and track and the disk hardware must be instructed to access that specific location. Linear addresses can be assigned to physical sectors in any number of ways, one of which is shown in Figure 6-4. The disk drive itself or, as described later, its device controller, translates the linear sector address into the corresponding physical sector location on a specific track on a specific platter.

With tape drives, translating logical addresses to physical addresses is much more straightforward because blocks on a tape are organized physically in a linear sequence. A logical access to a storage (block) location is translated into commands to position the appropriate part of the tape physically under the read/write head.

Some I/O devices such as keyboards and sound cards communicate only in terms of sequential streams of bytes that fill the same memory location. With these devices, the concept of an address or location is irrelevant. From the CPU's point of view, the I/O device is a storage device with a single location that is read or written repeatedly.

Other I/O devices do have storage locations in the traditional sense. For example, the individual character or pixel positions of a video display or printed page are logical storage locations. Each position is assigned an address within a linear address space and those addresses are used to manipulate individual pixels or character positions. The device or device controller translates logical write operations into the physical actions necessary to illuminate the corresponding video display pixel or place ink at the corresponding position on the page.

Figure 6-4 ►

Assignment of logical sector numbers to minimize head-to-head switching and track-to-track seeks

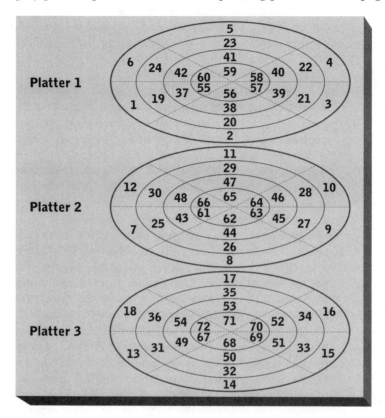

DEVICE CONTROLLERS

Storage and I/O devices normally are connected to the system bus through a **device controller** as shown in Figure 6-5. Device controllers perform the following functions:

► Implement the bus interface and access protocols
► Translate logical accesses into physical accesses
► Permit several devices to share access to a bus connection

Device controllers monitor the bus control lines for signals to peripheral devices and translate those signals into appropriate commands to the storage or I/O device. Similarly, data and status signals from the device are translated into appropriate control and data signals. A controller implements all bus interface functions for its attached peripheral device(s).

Figure 6-5 ▶

Secondary
storage and I/O
device
connections
using device
controllers

Device controllers perform some or all of the conversion between logical and physical access commands. Device controllers "know" the physical details of the attached devices and issue specific instructions to the device based on that knowledge. For example, the device controller converts a logical access to a specific disk sector within a linear address space into a command to read from a specific head, track, and sector.

Device controllers allow multiple peripheral devices to share a bus connection. This function is important because the number of physical I/O ports on the system bus usually is limited to 16 or less. Computer systems, especially minicomputers and mainframes, may have several dozen storage and I/O devices. Device controllers allow those devices to share a smaller number of bus connections. Most storage and I/O devices can't sustain the system bus data transfer rate for extended periods. Sharing a single bus connection among multiple slower devices efficiently allocates the relatively large communication capacity of a bus connection to many lower-capacity devices.

Mainframe Channels

In many mainframe computer systems, a device controller can be a dedicated special-purpose computer called an **I/O channel**,[1] or simply a **channel**. The difference

---

[1] The term channel originally was coined by IBM to describe a specific component of its 300 series mainframe computers. Due to the predominance of IBM mainframe computers, it has since gained the generic meaning described in this section. Vendors of other mainframe computer systems often use different terms, for example, **peripheral processing unit** and **front end processor**, to describe functionally similar components.

between an I/O channel and a device controller is not clear cut. It is a function of power and capability in several key areas, including:

► Number of devices that can be controlled
► Variability in type and capability of attached devices
► Maximum communication capacity

Typical I/O and storage device controllers can control only a few devices of similar type. For example, a disk controller in an IBM-compatible PC can control up to four disk drives. In contrast, a secondary storage channel in a mainframe computer can control a few dozen secondary storage devices including a mixture of device types, such as magnetic disks, optical discs, and magnetic tape drives. Another channel in the same mainframe might control up to 100 video display terminals or point-of-sale I/O devices.

Technology
Focus

SCSI

The acronym **SCSI** (pronounced "skuzzy") stands for **Small Computer System Interface**. It is a family of standard buses designed primarily for secondary storage devices. Although its name implies that it is used mainly in small computer systems, it is common in computer systems ranging from personal computers to mainframes. There are seven different SCSI standards representing seven generations of technology: SCSI-1, SCSI-2, SCSI-3, Ultra SCSI, Ultra2 SCSI, Ultra3 SCSI, and Ultra320 SCSI. There also are multiple variations of several SCSI standards resulting in a blizzard of confusing terminology, such as fast, wide, differential, high voltage differential, and low voltage differential. To keep things simple, this section will concentrate primarily on SCSI-1 and SCSI-2.

A SCSI bus implements both a low-level physical I/O protocol and a high-level logical device control protocol. The bus can connect up to 16 devices, each of which is assigned a unique device identification number of 0 through 15. The bus can be as long as 25 meters. Logically, the bus can be divided into two subsets—the control and data buses. The control bus is used to transmit control and status signals. The data bus primarily is used to transmit data and also can be used to send device identification numbers. In SCSI-1, the data bus is 8 bits wide. Later SCSI standards use a 16-bit data bus.

A SCSI bus allows multiple devices to be bus masters, thus requiring a mechanism for arbitration when more than one device wants control of the bus. In theory, a SCSI bus could be used as the system bus for an entire computer system. All computer devices, including one or more CPUs, would be attached to the bus. In practice, the SCSI bus generally is used to connect and control mass storage devices. A SCSI controller is used to translate signals between the system bus or a CPU, and the SCSI bus. This architecture is shown in Figure 6-6.

System Integration and Performance

Figure 6-6 ►

SCSI controller, bus, and secondary storage devices

A partial listing of the SCSI-1 control signals is shown in Table 6-1. When a device needs to initiate a transaction, it asks for control of the bus by sending a busy signal and placing its identification number on the data bus. If multiple devices send the busy signal simultaneously, the device with the highest identification number becomes the bus master and is designated as the initiator. The initiator then selects a target device by sending a select signal and placing the identification number of the target on the data bus.

Table 6-1 ►

A subset of the SCSI-1 control signals

Name	Description
Acknowledge	Controls asynchronous data transfer between devices in combination with the Request signal
Attention	Instructs a target to place a message to the initiator on the bus
Busy	Sent continuously by the initiator and/or target while a bus transaction is in progress
Control/Data	Sent by the target to indicate whether the data bus contains data or a control signal
Input/Output	Indicates the direction of data transfer on the bus (target to initiator or initiator to target)
Message	Sent by the target to inform the initiator that a message is being sent on the data bus
Request	Controls asynchronous data transfer between devices in combination with the Acknowledge signal
Select	Selects a particular device as the target; the identification number of the target is simultaneously placed on the data bus

Once the target device has been selected, the initiator sends it a data structure containing one or more commands. The ability to transmit multiple commands in one bus cycle is an important feature of the SCSI bus. A series of commands can be sent to a target with control of the bus released during the time the target processes those commands. The target queues the commands and executes them in sequence. This process avoids much of the control overhead associated with multiple commands. Other devices can use the bus while the target is processing commands.

A target can send several different status signals to an initiator, including:

▶ OK

▶ Busy

▶ Check condition

▶ Condition met

▶ Intermediate

The OK signal indicates successful receipt and intent to satisfy a command or command sequence. A busy signal is sent if the target is unable to process a request at the current time, for example, because it is busy processing previously received commands. A check condition signal is used to tell the initiator that an error has occurred, such as an attempt to access a nonexistent storage location. The initiator explicitly must ask the target to send a message indicating the nature of the error. The condition met status signal is sent when a search command is executed successfully. SCSI defines several different search commands including searches based on data content. An intermediate signal is sent to indicate the successful completion of one command in a command sequence.

Data transfer between target and initiator can be implemented either synchronously or asynchronously. In a synchronous transfer, the target sends a Request signal simultaneously with the data. The initiator simultaneously sends an Acknowledge signal, but the target does not wait for that signal to begin sending the next data item. Synchronous data transfer uses the full capacity of the bus, which can be up to 20 MB per second with SCSI-2, for data transfer. In asynchronous data transfer, the target sends a Request signal and data simultaneously. The target doesn't send the next data item until the initiator explicitly acknowledges the receipt of the previous data item. Asynchronous data transfer can proceed no faster than half the data transfer rate of the bus because every other cycle is used by the target to send an acknowledge signal.

A SCSI bus has several desirable characteristics including:

▶ Non-proprietary standard

▶ High data transfer rate

▶ Multiple master capability

▶ High-level (logical) data access commands

▶ Multiple command execution

▶ Interleaved command execution

Any company can freely manufacture and distribute SCSI-compatible devices. Data transfer rates can be as high as 320 MB per second with Ultra320, which is comparable to a mainframe channel. Multiple master capability attains efficient bus control with minimal control overhead. Target devices can become initiators (bus masters) in order to transfer data to a requestor. The ability to combine multiple commands and to interleave command sequences increases the likelihood that most of the bus capacity will be used to transfer data rather than transfer control and status signals.

The SCSI command set is rich and high-level. A SCSI device hides its physical organization from other SCSI devices and other computer system components. Disk devices appear as a linear address space to the CPU and/or operating system. Defective storage blocks are automatically detected by SCSI disk drives and eliminated from the linear address space. A wide variety of devices can be connected to a SCSI bus, including magnetic and optical disk drives, tape drives, and data communication devices.

Secondary storage and I/O device data transfer rates are much slower than the CPU. Table 6-2 lists access times and data transfer rates for various types of secondary storage devices. Table 6-3 shows data transfer rates of various I/O devices. Slow access times and data transfer rates primarily are due to mechanical limitations, for example, disk rotation speed. The time interval between a request by the CPU for input and the moment that input is received can span thousands, millions, or billions of CPU cycles.

Table 6-2 ►

Performance characteristics of typical memory and storage devices

Storage Device	Maximum Data Transfer Rate	Access Time
SDRAM	66 MB/second	10 nanoseconds
Hard disk	5 MB/second	8 milliseconds
Floppy disk	500 KB/second	200 milliseconds
CD-ROM	250 KB/second	100 milliseconds
DAT tape	1 MB/second	–

Table 6-3 ►

Typical speeds for various I/O devices

Device	Speed
Video display	25 to 100 MB/second
Ink-jet printer	30 to 120 lines per minute
Laser printer	4 to 120 pages per minute
Network interface	10 to 1000 Mbits/second

If the CPU waits for a device to complete an access request, the CPU cycles that could have been (but weren't) devoted to instruction execution are called **I/O wait states**. To prevent such inefficient use of the CPU, peripheral devices communicate with the CPU using interrupt signals. In a logical sense, an **interrupt** is a signal to the CPU that some event has occurred that requires the CPU to execute a specific program or process. In a physical sense, an interrupt is an electrical signal generated by a peripheral device and then sent over the control bus.

A portion of the CPU, separate from the components that fetch and execute instructions, continuously monitors the bus for interrupt signals and copies them to an **interrupt register**. The interrupt signal is a numeric value called an **interrupt code**, usually equivalent to the bus port number of the peripheral device sending the interrupt. At the conclusion of each execution cycle, the control unit checks the interrupt register for a nonzero value. If one is present, the CPU suspends execution of the current process, resets the interrupt register to zero, and proceeds to process the interrupt. When the interrupt has been processed, the CPU resumes executing the suspended process.

Communication with peripheral devices via interrupts allows the CPU to do something useful while it is waiting for an interrupt. If the CPU is executing only a single process or program then there is no performance gain. If the CPU is sharing its processing cycles among many processes then the performance improvement can be substantial. When one process requests data from a peripheral device the CPU suspends it and starts executing another process's instructions. When an interrupt is received, indicating that the access is complete, the CPU retrieves the data, suspends the process it currently is executing, and returns to executing the process that requested the data.

Interrupt Handlers

Interrupt handling is more than just a hardware feature; it also is a method of calling system software programs and processes. The operating system provides a set of processing routines, or service calls, to perform low-level processing operations such as reading data from a keyboard or writing data to a file stored on disk. An interrupt is a request by a peripheral device for operating system assistance in transferring data to or from a program. For example, an interrupt signal that indicates requested input is ready actually is a request to the operating system to retrieve the data and place it where the program that requested it can access it, such as in a register or in memory.

There is one operating system service routine, called an **interrupt handler**, to process each possible interrupt. Each interrupt handler is a separate program stored in a separate part of primary storage. In order to process an interrupt, the CPU must load the first instruction of the correct interrupt handler for execution.

An interrupt table stores the memory address of the first instruction of its interrupt handler. When the CPU detects an interrupt, it executes a master interrupt handler program called the supervisor. The **supervisor** examines the interrupt code stored in the interrupt register and uses it as an index to the interrupt table. The supervisor extracts the corresponding memory address and transfers control to the interrupt handler at that address.

Multiple Interrupts

The interrupt handling mechanism just described seems adequate until one considers the possibility of an interrupt arriving while the CPU is busy processing a previous interrupt. Which interrupt has priority? What is done with the interrupt that doesn't have priority?

Interrupts can be classified roughly into the following categories:

- I/O event
- Error condition
- Service request

The interrupt examples discussed thus far have been I/O events used to notify the operating system that an access request has been processed and that data is ready for transfer. Error condition interrupts are used to indicate errors that occur during normal processing. Error interrupts can be generated by software (for example, when attempting to open a non-existent file) or by hardware (for example, an attempt to divide by zero or to signal a dropped modem connection).

In most operating systems, the interrupt processing mechanism also is used by application programs to request operating system services. An interrupt code is assigned to each service program, and an application program requests a service by placing the corresponding interrupt number in the interrupt register. The interrupt code is detected at the conclusion of the execution cycle, the requesting process is suspended, and the service program is executed.

An operating system must have a hierarchy of interrupts in terms of their importance or priority. For example, error conditions normally are given higher precedence than other interrupts and critical hardware errors such as a power failure are given the highest priority. Priorities are used to determine whether an interrupt that arrives while another interrupt is being processed will be handled immediately or suspended until current processing is finished. For example, if a hardware error interrupt code is detected while an I/O interrupt is being processed, the I/O processing is suspended and the hardware error is processed immediately.

Stack Processing

Consider the following situation. While executing instructions in an application program, an interrupt is received from a pending I/O request. The interrupt is detected and the appropriate interrupt handler is called. The interrupt handler overwrites several general-purpose registers. When the interrupt handler terminates, processing of the application program resumes, but an error occurs because a value in a general-purpose register was altered by the interrupt handler.

How could this error have been prevented? How did the CPU know which instruction from the application program to load after the interrupt handler terminated? The operating system needs to be able to restart a suspended program at exactly the position it was interrupted. When the program resumes execution, it needs all register values to be in the same state as when it was interrupted. It would be impractical, due to wasted resources, to restart the suspended process from the beginning. The mechanism that allows a program to resume execution in exactly the same state as before interruption is called a stack.

A **stack** is an area of storage that is accessed in a last-in, first-out (LIFO) basis. Access is similar to a stack of plates or trays in a cafeteria. Items can be added to or removed only from the top of the stack. Accessing the item at the bottom of the stack requires removing all items above it.

In a computer system, the stack is a primary storage area that holds register values of interrupted processes or programs. When a process is interrupted, values in CPU registers are added to the stack in an operation called a **push**. The saved register values sometimes are called the **machine state**. When an interrupt handler finishes executing, values on the top of the stack are removed and loaded back into the appropriate registers. This operation is called a **pop**.

It is not always necessary to push the values of all registers onto the stack. At minimum, the current value of the instruction pointer must be pushed because the instruction at that address will be needed to restart the interrupted process where it left off. Think of the instruction pointer as a bookmark in an executing program. If indirect addressing is in use then the offset register also must be pushed. The general-purpose registers are pushed because they may contain intermediate results that are needed for further processing. They also might contain values awaiting output to a storage or I/O device.

Multiple machine states can be pushed onto the stack as interrupts of high precedence occur while processing interrupts of lower precedence. It is possible for the stack to fill to capacity, in which case further attempts to push values onto the stack result in a **stack overflow** error. The size of the stack limits the number of processes that can be interrupted or suspended.

The stack is a reserved area of main memory. A special-purpose register called the **stack pointer** always points to the next empty address in the stack. The stack pointer is automatically incremented or decremented each time the stack is pushed or popped. Most CPUs provide one or more instructions to push and pop the stack.

Performance Effects

The sequence of events that occurs when processing an interrupt is summarized in Figure 6-7. The sequence is complex and consumes many CPU cycles. The most complex parts are the push and pop operations, which copy many values between registers and main memory. Wait states may be incurred if the stack is not implemented within a cache. (Caching is described in the next section).

Supervisor execution also consumes CPU cycles. The table lookup procedure is relatively fast but still consumes at least 10 CPU cycles. There are other steps in the process that consume further CPU cycles; these will be discussed in a later chapter. In total, processing an interrupt typically consumes at least 100 CPU cycles in addition to the cycles consumed by the interrupt handler itself.

Figure 6-7 ▶

Interrupt
processing

BUFFERS AND CACHES

Access to I/O and storage devices is inherently slow. Mismatches in data transfer rate and data transfer unit size are cumbersome to overcome, causing interrupt processing. RAM can be used to overcome the mismatches in two distinct ways— buffering and caching. In both buffering and caching, improving overall computer system performance is the primary goal.

Buffers

A **buffer** is a small storage area (usually DRAM or SRAM) used to hold data in transit from one device to another. Buffers resolve differences in data transfer rate or data transfer unit size. A buffer is required when there is a difference in data transfer unit size. A buffer isn't required when data transfer rates differ, but using one generally improves performance.

Consider communication between a personal computer and a laser printer as shown in Figure 6-8. A personal computer usually sends data to a printer via a parallel communication port that transmits 8 bits, or 1 byte, per message. The laser printer prints an entire page at once. The data content of a printed page can range up to several million bytes. The input data transfer unit from the personal computer is a single byte and the output data transfer unit from the laser printer is a full page containing multiple bytes. A buffer is required to resolve this difference.

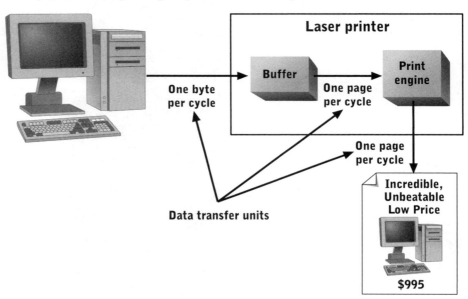

Figure 6-8 ►

A buffer is used to resolve differences in data transfer unit size between a PC and a laser printer

As single bytes are received from the parallel port they are added to the buffer in sequential order. When all bytes of a page have been received, they are removed from the buffer and transferred to the print engine as a single unit. Buffer size must be at least as large as the data output unit. If the buffer is not large enough to hold a full page, an error called a **buffer overflow** occurs. A buffer for an I/O device is typically implemented within the device, as it is in a laser printer.

A buffer also can be used to improve system performance when two devices have different data transfer rates. Consider, for example, communication between a personal computer and an Internet service provider (ISP) via an analog modem. The modem transmits data to and from the ISP at a relatively slow rate, such as 56 Kbps. Data is sent to the modem from primary storage or the CPU at the data transfer rate of the system bus, such as 133 Mbps. There is a data transfer rate difference of approximately 10^4.

Assume for the moment that the modem has a four-character buffer and that data is being transmitted from the computer to the ISP. The four-character buffer limits data transfer to the modem to four characters at a time. Once the modem places four characters in the buffer, it sends an interrupt to indicate that no further data should be sent. This message prevents more characters from arriving and causing buffer overflow before the modem can transmit the current buffer content to the ISP. Once the buffer content has been sent to the ISP, the modem transmits another interrupt to indicate that it is again ready to receive input.

How much system bus data transfer capacity is used to transmit 64 KB (65,536 bytes) via the modem? The answer is 49,152 bus cycles. Each four bytes transmitted result in two interrupts being sent across the bus—one to stop further transmission and another to restart it. The two interrupts are not data, but they do consume bus cycles. Further, those interrupts must be processed by the CPU. If we assume there are 100 CPU cycles per interrupt, then 3,276,800 CPU cycles are consumed to process interrupts while transferring 64 KB to the modem.

Now consider the same scenario with a 1,024-byte buffer in the modem. A single transfer from main memory to the modem would send as many bytes per bus cycle as could be carried on the data bus, for example, 32 bits or 4 bytes. After 256 bus transfers (1,024 bytes at 4 bytes per bus cycle) the modem would send an interrupt to stop transmission. After the buffer content was transmitted to the ISP, the modem would transmit another interrupt to restart data transfer from memory.

Computer system performance dramatically improves with the larger buffer (see Table 6-4). Two interrupts are generated for each block of 1,024 bytes transferred. The total number of bus cycles consumed is:

$$16{,}384 \text{ data transfers} + \left(\frac{65{,}536 \text{ bytes}}{1024 \text{ bytes per transfer}} \times 2 \text{ interrupts} \right) = 16{,}512 \text{ cycles}$$

The number of CPU cycles falls to 12,800 ($65{,}536 \div 1{,}024 \times 2 \times 100$). Bus and CPU cycles are thus freed for other purposes such as executing programs.

Table 6-4 ►

Bus and CPU
cycles con-
sumed for a
64 KB data
transfer with
various
buffer sizes

Buffer Size	Bus Data Transfers	Interrupts	Total Bus Transfers	Improvement	CPU Interrupt Processing Cycles	Improvement
4	16,384	32,768	49,152	n/a	3,276,800	n/a
8	16,384	16,384	32,768	33.33%	1,638,400	50%
16	16,384	8,192	24,576	25%	819,200	50%
32	16,384	4,096	20,480	16.67%	409,600	50%
64	16,384	2,048	18,432	10%	204,800	50%
128	16,384	1,024	17,408	5.56%	102,400	50%
256	16,384	512	16,896	2.94%	51,200	50%
512	16,384	256	16,640	1.52%	25,600	50%
1,024	16,384	128	16,512	0.77%	12,800	50%

Diminishing Returns

Note the rates of change in performance improvement for total bus transfers and CPU cycles shown in Table 6-4. As buffer size increases, CPU cycle consumption decreases at a nearly linear rate, but total bus cycles decrease at a diminishing rate. If we assume that there are no excess CPU cycles, then all buffer size increases shown in the table provide significant benefit. In other words, CPU cycles not used for I/O interrupts always are applied to other useful tasks.

But if we assume that extra CPU cycles are available, in other words, that the CPU is not being used at full capacity, then the only significant benefit of larger buffer size is reduced bus cycles. Because bus cycle improvement drops off rapidly, there is a point at which further buffer size increases provide no significant gain. If we also assume that buffer cost increases with buffer size (a reasonable assumption) then there also is a point at which the cost of additional RAM is greater than the monetary value of more efficient bus utilization.

If the discussion in the previous two paragraphs sounds vaguely familiar to you, then you've probably studied economics. The underlying economic principle is called the law of diminishing returns. Simply stated, the law says that when multiple resources are required to produce something useful, adding more and more of a single resource produces fewer and fewer benefits. The law is just as applicable to buffer size in a computer system as it is to labor or raw material inputs in a factory. The law has many other applications within computer systems, some of which will be described later in this book.

Table 6-5 shows bus and CPU performance improvements for a 64-byte transfer to our hypothetical modem. Note that the law of diminishing returns affects both bus and CPU performance. Both types of performance improve at the same rate as in Table 6-4 until buffer size exceeds the amount of data being transferred, after which there is no further improvement. Why do buffer size increases beyond 64 bytes provide no improvement? Because in this example there's nothing to put into the extra buffer space.

Table 6-5 ▶

Bus and CPU cycles consumed for a 64-byte data transfer with various buffer sizes

Buffer Size	Bus Data Transfers	Interrupts	Total Bus Transfers	Improvement	CPU Interrupt Processing Cycles	Improvement
4	16	32	48	n/a	3200	n/a
8	16	16	32	33.33%	1600	50%
16	16	8	24	25%	800	50%
32	16	4	20	16.67%	400	50%
64	16	2	18	10%	200	50%
128	16	2	18	0%	200	0%
256	16	2	18	0%	200	0%
512	16	2	18	0%	200	0%
1024	16	2	18	0%	200	0%

The results in Tables 6-4 and 6-5 do more than show the law of diminishing returns in action. They also show the importance of clearly defined assumptions and a clear understanding of the work that a computer system will be asked to do. Accurate computation of improvements requires a full understanding of all affected components, in this case, the CPU, bus, modem, and buffer; clearly stated operational assumptions, for example, whether the CPU is being used at full capacity; and well-understood workload characteristics, in this case, size and frequency of data transfers to/from the modem.

Cache

Like a buffer, a **cache** is a storage area (usually RAM) used to improve system performance. However, a cache differs from a buffer in several important ways including:

▶ Data content is not automatically removed as it is used

▶ Cache is used for bidirectional data transfer

► Cache is used only for storage devices accesses
► Caches are usually much larger than buffers
► Cache content must be managed intelligently

The basic idea behind caching is simple. Access to data contained within a high-speed cache can occur much more quickly than access to slower storage devices such as magnetic disk. The speed difference is due entirely to the faster access speed of the RAM used to implement the cache. Performance improvements require a sufficiently large cache and the "intelligence" to use it effectively.

Performance improvements are achieved differently for read and write accesses. During a write operation, a cache behaves as a buffer. Data first is stored in the cache and then transferred to the storage device. Performance improvements are the same as those of a buffer: reduced bus and CPU overhead due to larger continuous data transfers. Because caches usually are much larger than buffers, the performance improvement is increased. But the law of diminishing returns usually results in only slightly better performance than would be achieved with a typical buffer.

Write caching can result in more significant performance improvement when one write access must be confirmed before another can begin (see Figure 6-9). Some programs, such as transaction updates for banks and retailers, require confirmed writes. In the figure, when data is written to a cache (1), the confirmation signal is sent immediately (2), before the data is written to the storage device (3).

Figure 6-9 ►

A storage write operation using a cache

Sending a confirmation before data is written to the secondary storage device can improve program performance. Performance improves because the program immediately can proceed with other processing tasks. If the program is performing a series of write operations, performance improvements will cease as soon as the cache is full. As with a buffer, interrupts are used to coordinate the data transfer activity.

Immediate write confirmations are risky because an error might occur while copying data from the cache to the storage device. The most significant danger is total cache failure due to a power failure or other hardware error. In that case all data, possibly representing many cached write operations, is lost permanently and there is no way to inform the program(s) that wrote the data of the loss.

Data written to a cache during a write operation is not removed automatically from the cache as it is written to the underlying storage device. This can provide a performance advantage in situations where data is reread shortly after it is written. A subsequent read of the same data item will be much faster unless the data item has been purged from the cache for some other reason, for example, to make room for other data items.

Most of the performance benefits of a cache occur during read operations (see Figure 6-10). In the figure, read accesses first are routed to the cache (1). If the data already is in the cache then it is accessed from there (2). Performance is improved because access to the cache is much faster than access to the storage device. If the requested data is not in the cache, then it must be read from the storage device (3). A performance improvement is realized only if requested data already is waiting in the cache.

Figure 6-10 ▶

A storage read operation using a cache

A **cache controller** is a processor that manages cache content. A cache controller can be implemented in:

▶ A storage device storage controller or communication channel, as a special-purpose processor controlling RAM installed within the controller or channel

▶ The operating system, as a program that uses part of primary storage to implement the cache

The former approach is more common with primary storage caches, and the latter with secondary storage caches.

A cache controller guesses what data will be requested in the near future and loads that data from the storage device into the cache before it actually is requested. Guessing methods can be simple or complex. One simple method is to assume linear access to storage locations. For example, while servicing a read request for location n, the cache controller prefetches location $n+1$, and possibly several subsequent storage locations, into the cache. More complex guessing methods require more complex processing circuitry or software. The extra complexity and expense is justified only if the guessing accuracy improves and if those improvements lead to significant overall performance improvement.

When a read operation accesses data already contained within the cache, the access is called a **cache hit**. When the data needed is not in the cache, the access is called a **cache miss**. The ratio of cache hits to read accesses is called the cache's **hit ratio**. A cache miss requires that a **cache swap** to or from the storage device be performed. The cache controller guesses which data items are least likely to be needed in the near future, writes them to the storage device, and purges them from the cache. The requested data then is read from the storage device and placed in the cache.

Caches can be implemented within main memory if an operating system program is used as the cache controller. However, this approach can reduce overall system performance by reducing free memory and CPU cycles. Modern device controllers can implement a cache with RAM and the cache controller installed on the device controller. In that case, the CPU and system software are unaware that a cache is in use.

A surprisingly small cache can provide significant performance gains. Typical ratios of cache size to storage device capacity range from 10,000:1 to 1,000,000:1. Primary storage caches with a 10,000:1 size ratio typically achieve cache hits more than 90 percent of the time. Actual performance gain depends on cache size and the nature of storage device accesses. Frequent sequential read accesses tend to increase performance. Frequent random or scattered read accesses tend to reduce the hit ratio.

Primary Storage Caches   Current processor speeds generally meet or exceed the ability of dynamic RAM (DRAM) to keep them supplied with data and instructions. Although static RAM (SRAM) more closely matches the processor speed, it generally is too expensive to use for all of primary storage. When the CPU accesses DRAM, it incurs waits states. For example, a 1 GHz processor incurs 50 wait states each time it reads from 50 nanosecond DRAM.

One way to limit wait states is to use an SRAM cache between the CPU and DRAM primary storage. The SRAM cache can be integrated into the CPU, located on the same chip as the CPU, or located between the CPU and DRAM. Modern computers use a combination of approaches—for example, a relatively small (64 KB) cache within the CPU, a larger (512 KB) cache on the CPU chip, and an even larger (2 MB) cache between the microprocessor and DRAM. When all three cache types are used, the within-CPU cache is called a **level one (L1) cache**, the on-chip cache is called a **level two (L2) cache**, and the off-chip cache is called a **level three (L3) cache**.

Primary storage cache control can be quite sophisticated. The default sequential flow of instructions allows simple guesses based on linear storage access to work some of the time. But accesses to operands and conditional branch instructions are not accounted for by this simple strategy. Modern CPUs use sophisticated methods to improve the hit ratio. For example, instead of guessing whether a conditional branch will be taken, the cache controller may prefetch both subsequent instructions. This consumes more cache but guarantees that the next instruction will be in the cache. Another strategy is to examine operands as instructions are loaded into the cache. If an operand is a memory address, then the cache controller prefetches the data at that location into the cache.

Technology
Focus

Itanium Memory Cache

The Intel Itanium® microprocessor uses three levels of primary storage caching. The CPU, L1 cache, and L2 cache are implemented on a single chip. The 4 MB L3 cache is implemented on a set of separate chips. All of the chips are combined into a single integrated package that plugs into a computer system motherboard. The integrated package reduces distance between components, so the L3 cache can communicate with the CPU at the CPU's clock rate. Figure 6-11 shows the relationship among the caches, other CPU components, and the system bus.

Figure 6-11 ▶

Memory caches in the Intel Itanium® processor

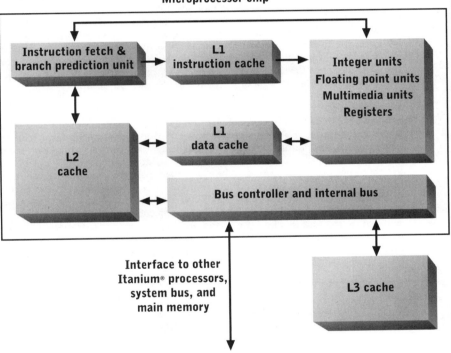

Microprocessor chip

The L1 cache is divided into two 16 KB segments—one for data and the other for instructions. The L2 cache is 96 KB and holds both instructions and data. L1 data cache contents are swapped automatically with the L2 cache as needed. The instruction fetch and branch prediction unit examines incoming instructions for conditional branches and attempts to prefetch instructions from both branch paths into the L1 instruction cache. The branch prediction unit also interacts with the CPU execution units to determine the likelihood of branch conditions being true or false.

Using separate L1 caches for instructions and data allows some streamlining of processor functions. For the instruction execution units, the L1 instruction cache is a read-only cache. The L1 data cache can be read or written as operands are moved to and from registers. Cache controller functions are optimized for each type of access. However, it is possible for a program to modify its own instructions. In this case, the code section being modified is stored in both L1 caches at the same time. The processor automatically detects this condition and maintains consistency between the two caches.

The bus controller acts as a cache controller, bus controller, and as a bus connecting the CPU chip, including its embedded L1 and L2 caches, the L3 cache, the system bus, and other Itanium® processors. The bus controller coordinates data movement between the L2 and L3 caches and between the L3 cache and main memory. It also coordinates data movement between the caches, main memory, and peripheral devices via the system bus.

The Itanium® can assign various caching modes to 2 KB, 4 KB, or 4 MB regions of main memory. Write-back caching enables full caching of both read and write operations. Main memory is updated only when necessary, such as when a modified cache line must be removed to make room for new content. Write-through caching causes all write operations to update the caches and main memory at the same time. This mode is useful for regions of memory that can be read directly without knowledge of the CPU by other devices on the system bus, for example, a disk or video controller. All caching functions can be disabled for any memory region.

The Itanium® was designed to participate in multiprocessor computer systems. Multiprocessor systems add complexity to cache management because multiple processors can cache the same memory region(s) simultaneously. Program execution errors are possible if a program executing on one processor updates a cached memory location and another program executing on another processor subsequently reads that same memory location from an outdated cache copy.

Itanium® processors use a technique called memory snooping to maintain consistency among their caches and main memory. Each processor's bus controller monitors access to main memory by up to three other processors. An access to memory that currently is cached by another processor causes the corresponding cache contents to be marked invalid. The modified cache contents automatically are reloaded from main memory the next time they are accessed. If a memory region cached by multiple processors is written, all processors directly exchange updated cache contents to ensure consistency.

Secondary Storage Caches   Disk caching is common in modern computer systems, particularly in file and database servers. In a file server, disk caching performance can be improved if information about file access is tracked and used to guide the cache controller. Specific strategies include:

► Give frequently accessed files higher priority for cache retention

► Use read-ahead caching for files that are read sequentially

► Give files opened for random access lower priority for cache retention

The operating system is the best source of file access information because it updates the information dynamically as it services file access requests. Because the operating system executes on the CPU, it is difficult to implement access-based cache control if the cache controller is a special-purpose processor within a disk controller.

Many computer system designers now rely on the operating system to implement disk caching instead of using specialized disk controller hardware. They feel that the extra cost of hardware-based disk caching solutions is better spent on more primary storage and faster or additional CPUs. The extra memory and CPU cycles can be used by the operating system to implement more intelligent access-based cache control. This approach makes sense only if the computer system has sufficient CPU capacity to devote to cache management.

COMPRESSION

Compression is a technique that reduces the number of bits used to encode a set of related data items, for example, a file or a stream of motion video images. A **compression algorithm** is a specific mathematical compression technique implemented as a program. Most compression algorithms have a corresponding **decompression algorithm** that is used to restore compressed data to its original or nearly original state. In common use, the term compression algorithm refers simultaneously to both the compression and decompression algorithm.

There are many compression algorithms, which vary in:

► Type(s) of data for which they are best suited
► Whether information is lost during compression
► Amount by which data is compressed
► Computational complexity (CPU cycles required to execute the compression program)

A compression algorithm can be lossless or lossy. With **lossless compression**, the result of compressing and then decompressing any data input is exactly the same as the original input. Lossless compression is required in many applications, for example, accounting records, executable programs, and most stored documents.

With **lossy compression**, the result of compressing and then decompressing a data input is different than, but still similar to, the original input. Lossy compression usually is applied only to audio and video data. The human brain tolerates missing audio and video data and usually can "fill in the blanks." Lossy compression commonly is used to send audio or video streams through low-capacity transmission lines or networks, for example, video conferencing.

The term **compression ratio** describes the ratio of data size in bits or bytes before and after compression. For example, if a 100 MB file is compressed to 25 MB, then the compression ratio is 4:1. Some types of data are easier to compress than others.

For example, lossless compression ratios up to 10:1 are achieved easily with word processing documents and ASCII or Unicode text files. Lossless compression ratios of greater than 3:1 are difficult or impossible to achieve with audio and video data. Lossy compression of audio and video can achieve compression ratios of up to 50:1, though at that ratio the loss of information is discerned easily by listeners and viewers.

Compression commonly is used to reduce secondary storage requirements, as shown in Figure 6-12(a). Data sent to the storage device is compressed before it is written. Data read from storage is decompressed before it is sent to the requester. Compression also can be used to increase communication channel capacity, as shown in Figure 6-12(b). Data is compressed as it enters the channel and then decompressed as it leaves the channel. Hardware-based compression is included in all modern standards for analog modems.

Figure 6-12 ►

Data compression with a secondary storage device (a) and a communication channel (b)

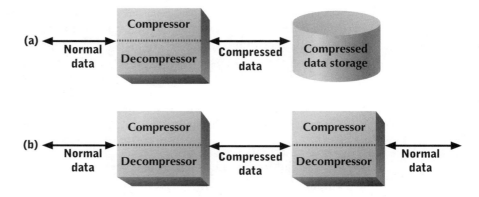

Using data compression alters the balance of processor resources and communication or storage resources in a computer system. Implementing a compression algorithm consumes processor cycles. Algorithms with high compression ratios frequently consume more processing resources than those with low compression ratios. The proper tradeoff between these resources depends on their relative cost, availability, and degree of use. Implementing data compression with CPU processing resources often is not cost effective. Implementing data compression through special-purpose processors often is advantageous. Such processors are now commonly available and are frequently used in disk controllers, video display controllers, and high-speed modems.

Technology
Focus

MPEG and MP3

The **Motion Picture Experts Group (MPEG)** is an organization that creates and evaluates standards for motion picture recording and encoding technology. The group has created three different standards: MPEG-1, MPEG-2, and MPEG-4. Motion pictures usually contain images and sound, so MPEG standards address recording and encoding formats for both data types. Each standard is divided into layers numbered 1 (systems), 2 (video), and 3 (audio). The audio encoding standard commonly known as **MP3** is actually layer 3 of the MPEG-1 standard. It is a useful encoding method for many types of audio, not just audio tracks within motion pictures.

Analog audio data is converted to digital form by sampling the audio waveform thousands of times per second and then constructing a numerical picture of each sample. On an audio CD, there are 44,100 samples per second per stereo channel. Each sample is represented by a single digital number. Sample quality is improved (due to greater precision and accuracy) if many bits are used to encode each sample. Each sample on an audio CD is encoded in 16 bits. Each second of audio data on a stereo CD requires $44,100 \times 16 \times 2 = 1,411,200$ bits.

A typical motion picture is two hours long and includes far more video than audio data. To encode both data sets within a single storage medium, for example, on a DVD-ROM, all data must be compressed. Normal CD quality music can be compressed by a ratio of 6:1 and most listeners cannot distinguish the compressed audio data from the original. MP3 is a lossy compression method. It discards significant portions of the original audio data and does not attempt to reconstruct that data for playback. Thus, MP3 has a relatively simple decompression algorithm.

MP3 takes advantage of a few well-known characteristics of human audio perception including:

▶ Sensitivity that varies with audio frequency (pitch)

▶ Inability to recognize faint tones of one frequency simultaneously with much louder tones in nearby frequencies

▶ Inability to recognize soft sounds that occur shortly after louder sounds

These three characteristics interact in complex ways. For example, loud sounds at the extremes of the human hearing range, such as 20 Hz or 20 KHz, more effectively mask nearby soft frequencies than do loud sounds in the middle of the human hearing range. MP3 analyzes digital audio data to determine which sound components are masked by others. For example, a bass drum may be masked for a moment by a bass guitar note. It then compresses the audio data stream by either discarding information about masked sounds or representing them with a reduced number of bits.

Figure 6-13 shows the conceptual components of an MP3 encoder. A frequency separator divides raw audio data into 32 narrow-frequency audio streams. In parallel, a perceptual modeler examines the entire audio signal to identify loudness peaks. The modeler then determines by how much and for how long those loudness peaks can mask nearby audio data. The masking information and the 32 data streams then are sent to an encoder that represents data within each frequency. The encoder uses the masking information from the perceptual modeler to determine what data to discard or to represent with less than 16-bit precision.

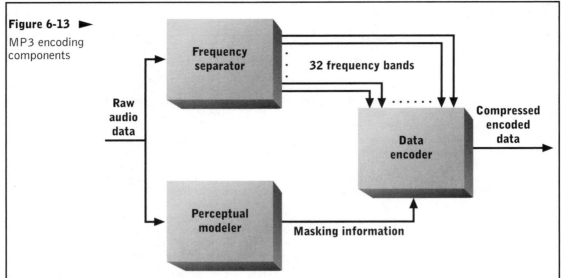

Figure 6-13 ►

MP3 encoding
components

The MP3 standard is implemented widely in MP3 player software, Internet broadcasts, MP3 hardware, and DVD playback devices. There has been much controversy over its widespread use and its effect(s) on music distribution and piracy. Attempts may be made to modify the standard, or legally mandate a variant, that makes copyright infringement more difficult, though there are doubts about whether that is even possible. The standard definitely will need enhancement to address more advanced forms of audio data such as theater-quality surround sound, but the coding method and data format have been proven effective by researchers and the marketplace. That success means MP3 will be with us in some form for many years to come.

SUMMARY

► The system bus is the communication pathway that connects the CPU with memory and other devices. The bus is implemented as a set of data lines, control lines, and status lines. The number and use of these lines, as well as the procedures for controlling access to the bus, are called the bus protocol. The CPU generally is assumed to control bus access by other devices. A multiple master bus or direct memory access gives devices other than the CPU temporary control of the bus, resulting in improved computer system performance.

► The CPU communicates with peripheral devices through I/O ports. To simplify the bus and the CPU instruction set, interaction with all peripheral devices is implemented with simple data movement instructions. The CPU treats each peripheral device as if it were a storage device with a linear address space. The device, or its controller, must translate access commands to the linear address space into whatever physical actions actually are necessary to perform the access. A peripheral device controller performs a number of important functions including logical to physical translation, physical bus interface, sharing a single bus connection among multiple devices, and (optionally) caching. A channel is an advanced type of device controller used in mainframe computers. Channels have greater data transfer capacity, a larger maximum number of attached peripheral devices, and greater variability in the types of devices that can be controlled compared with device controllers.

► Application programs use interrupt processing to coordinate data transfers to or from peripheral devices, notify the CPU of errors, and call operating system service programs. An interrupt is a signal to the CPU that some condition requires its attention. When an interrupt is detected, the currently executing program is suspended by pushing current register values onto the stack and transferring control to the appropriate interrupt handler. When the interrupt handler finishes executing the stack is popped and the suspended process resumes execution from the point of interruption.

► A buffer is a region of memory that holds a single unit of data for transfer to or from a device. Buffers allow devices with different data transfer rates and unit sizes to efficiently coordinate data transfer. A cache is a large buffer implemented within a device controller or primary storage. When used for input, a cache gives more rapid access if the data being requested already is in the cache. The cache controller guesses what data the CPU will request next and loads that data into the cache before it actually are requested.

► Compression reduces the number of bits required to encode a data set or stream, effectively increasing the capacity of a communication channel or storage device. A compression algorithm may be lossless or lossy. With lossless compression, data are the exactly the same before and after compression and decompression. Lossy compression loses some information during compression and decompression. Data compression requires increased processing resources to implement the compression and decompression algorithms while reducing resources needed for data storage and/or communication.

In Chapter 3 we described various ways in which data is represented. In Chapters 4 and 5 we described CPU, primary storage, and secondary storage hardware. In this chapter we've described how data are transmitted among all computer system hardware devices. Various hardware and software techniques for improving data transfer efficiency also were described and, thus, overall computer system performance. In the next chapter, we'll look at the technologies used to implement input/output devices.

Key Terms

address bus
buffer
buffer overflow
bus
bus arbitration unit
bus clock
bus cycle
bus master
bus protocol
bus slave
cache
cache controller
cache hit
cache miss
cache swap
channel
compression
compression algorithm
compression ratio

control bus
data bus
data transfer rate
decompression algorithm
device controller
direct memory access (DMA)
DMA controller
front end processor
hit ratio
I/O channel
I/O port
I/O wait state
interrupt
interrupt code
interrupt handler
interrupt register
level one (L1) cache
level two (L2) cache
level three (L3) cache
linear address space

logical access
lossless compression
lossy compression
machine state
Motion Picture Experts Group
    (MPEG)
MP3
multiple master bus
peripheral devices
peripheral processing unit
pop
push
Small Computer System
    Interface (SCSI)
stack
stack overflow
stack pointer
supervisor
system bus

Vocabulary Exercises

1. The _____ is the communication channel that connects all computer system components.

2. A(n) _____ cache generally is implemented on the same chip as the CPU.

3. The CPU always is capable of being a(n) _____, thus controlling access to the bus by all other devices in the computer system.

4. A(n) _____ is a small area of memory used to resolve differences in data transfer rate or data transfer unit size.

5. A(n) _____ is an area of fast memory where data held on a storage device are prefetched in anticipation of future requests for that data.

6. A cache controller is a hardware device that initiates a(n) _____ when it detects a cache miss.

7. The _____ is used to transmit command, timing, and status signals between devices in a computer system.

8. If possible, the system bus _____ rate should equal the speed of the CPU.

9. The _____ is a register that always contains a pointer to the top of the _____.

10. Interrupt handlers are called by the _____ after it looks up a memory address in the _____.

11. The set of register values placed on the stack while processing an interrupt also is called the _____.

12. A(n) _____ is a software program that is executed in response to a specific interrupt.

13. During interrupt processing, register values of a suspended process are held on the _____.

14. A(n) _____ is a signal to the CPU or operating system that some device or program requires processing services.

15. A(n) _____ is a hardware device that intervenes when two potential bus masters want control of the bus at the same time.

16. The initiator of a bus transfer assumes the role of a(n) _____. The recipient assumes the role of a(n) _____.

17. The CPU incurs one or more _____ if it is idle pending the completion of an I/O operation.

18. The system bus can be decomposed logically into three sets of transmission lines: the _____ bus, the _____ bus, and the _____ bus.

19. During a(n) _____, one or more register values are copied to the top of the stack. During a(n) _____, one or more values are copied from the top of the stack to registers.

20. The relative size of a data set before and after data compression is described by the compression _____.

21. If the result of compressing and decompressing a file is not exactly the same as the original file, then the compression algorithm is said to be _____. If the result is the same as the original, the compression algorithm is said to be _____.

22. A(n) _____ is a special-purpose processor dedicated to managing the contents of a cache.

23. A(n) _____ is a memory address to which the CPU copies data when writing to an I/O or storage device.

24. The _____ carries interrupts, read commands, and other control signals.

25. Primary storage receives a read command on the _____, the address to be accessed on the _____, and returns the accessed data on the _____.

26. The operating system normally views any storage device as a(n) _____, ignoring the device's physical storage organization.

27. Part of the function of a storage secondary device controller is to translate _____ into physical accesses.

28. _____ is used to transfer data directly between memory and secondary storage devices, without the assistance of the CPU.

29. A(n) _____ is a high capacity device controller used in mainframe computers.

30. An access to primary storage that is found within a cache is called a(n) _____.

31. The _____ defines the control, signals, and other communication parameters of a bus.

Review Questions

1. What is the system bus? What are its primary components?

2. What is a bus master? What is the advantage of having devices other than the CPU be a bus master?

3. What characteristics of the CPU and of the system bus should be balanced to obtain maximum system performance?

4. What is an interrupt? How is an interrupt generated? How is it processed?

5. What is a stack? Why is it needed?

6. Describe the execution of the push and pop operations.

7. What is the difference between a physical access and a logical access?

8. What functions does a device controller perform?

9. What is a buffer? Why might one be used?

10. How may a cache be used to improve performance when reading data from a storage device? How may a cache be used to improve performance when writing data to a storage device?

11. What is the difference between lossy and lossless compression? For what types of data is lossy compression normally used?

Problems and Exercises

1. Assume that a PC uses a 1 GHz processor and a system bus clocked at 100 MHz. An internal modem is attached to the system bus. Communication between the modem and telephone line occurs at continuous rate of 57,600 bits per second. No parity or other error-checking mechanisms are in use. The modem uses an 8250 UART chip containing a one-byte buffer. After a character is received by the modem, it immediately sends an interrupt to the CPU.

   When the interrupt is received, the following actions are performed:

   a. The supervisor is called.

   b. The modem interrupt handler is called by the supervisor.

   c. The modem interrupt handler copies the character from the UART buffer to a memory buffer.

   d. The modem interrupt handler returns control to the supervisor.

   e. The supervisor returns control to the process that originally was interrupted, such as the program that is managing the file transfer.

   Sending the interrupt requires one bus cycle. A push or pop operation consumes 30 CPU cycles. The supervisor consumes 8 CPU cycles before starting to call the interrupt handler. The serial port interrupt handler consumes 10 CPU cycles before starting to return to the supervisor, including the time needed to transfer the character from the UART buffer to memory.

   Question 1: How long does it take to process one character, that is, perform all of the steps outlined above? State your answer in both CPU cycles and elapsed time.

   Question 2: Assume that the computer is executing a program that is downloading a large file using the modem. The file is held in memory until the transfer is completed. What percentage of the CPU capacity is being used to process input from the serial port UART during the transfer? Assume that the bus uses a simple request/response protocol for peripheral to memory transfers; no acknowledgment is sent by the peripheral, just the requested data. What percentage of available bus capacity is being used to process input from the serial port UART during the transfer?

   Question 3: Assume the same facts presented in Question 2. Assume further that the 8250 UART has been replaced by a 16550A UART. The new UART is identical to the old UART except that it contains a 16-byte buffer. An interrupt is generated for each group of 16 incoming data bytes. This increases the execution time of the interrupt handler from 14 to 50 CPU cycles. What percentage of the CPU capacity is being used to process input from the new serial port UART? Assume that 4 bytes at a time can be transferred from the UART buffer to memory over the system bus. What percentage of available bus capacity is being used to process input from the serial port UART during the transfer?

2. A video frame displayed on a computer screen consists of many pixels. Each pixel, or cell, represents one unit of video output. The resolution of a video display typically is specified in horizontal and vertical pixels (such as 800×600); and the number of pixels on the screen is simply the product of these numbers ($800 \times 600 = 480{,}000$ pixels). The data content of a pixel is one or more unsigned integers. For a black and white display, each pixel is a single number typically between 0 and 255 that represents the intensity of the color white. Color pixel data typically is represented as either one or three unsigned integers. When three numbers are used, the numbers usually are between 0 and 255, and each number represents the intensity of a primary color (red, green, or blue). When a single number is used, it represents a predefined color selected from a table or palette of colors.

Motion video is displayed on a computer screen by rapidly copying frames to the video display controller. The video controller continuously converts frame data to an analog RGB signal that is output to the display device. Because video images or frames require many bytes of storage, they usually are copied to the display controller directly from secondary storage. Each video frame is an entire picture, and its data content, as measured in bytes, depends on the resolution at which the image is displayed and the maximum number of simultaneous colors that can be contained within the sequence of frames. For example, a single frame at 800×600 resolution with 256 simultaneous colors contains $800 \times 600 \times 1 = 480{,}000$ bytes of data. Realistic motion video requires a minimum of 20 frames per second to be copied and displayed, 24 or 30 frames per second are common professional standards. Using fewer frames per second results in perceived "jerky" motion because the frames are not being displayed quickly enough to fool the eye and the brain into thinking that they are one continuously changing image.

Assume that the computer system being studied contains a bus mastering disk controller and a video controller that copies data to the video display at least as fast as it can be delivered over the bus. Assume that the system bus can transfer data at a sustained rate of 100 MBps, as can both of the controllers' bus interfaces. This system will be used to display motion video on a monitor capable of resolutions as low as 640×400 and as high as 1024×768.

Assume that a single disk drive is attached to the disk controller and that it has a sustained data transfer rate of 20 MB per second when reading sequentially stored data. Next, assume that the channel connecting the disk drive to the disk controller has a data transfer rate of 80 Mbps. Finally, assume that the files containing the video frames are stored sequentially on the disk and that copying the content of these files from disk to the display controller is the only activity that the system will perform, no external interrupts, no multitasking, etc.

Assume that the video display controller contains 2 MB of 50 nanosecond, 8-bit buffer RAM—the video image arriving from the bus can be written to the buffer at a rate of 8 bits per 50 nanoseconds. The RAM buffer of the video display can be written from the bus while it simultaneously is being read by the display device, sometimes called dual porting. Finally, data can be received and displayed by the display device as fast as they can be read from the RAM buffer by the video controller.

Question 1: What is the maximum number of frames per second (round down to a whole number) that can be displayed by this system in 256 simultaneous colors at a resolution of 640 × 400?

Question 2: What is the maximum number of frames per second (round down to a whole number) that can be displayed by this system in 65,536 simultaneous colors at a resolution of 800 × 600?

Question 3: What is the maximum number of frames per second (round down to a whole number) that can be displayed by this system in 16,777,216 simultaneous colors at a resolution of 1024 × 768?

3. An employee listens to an Internet radio station while using his or her office computer. The Internet radio station sends MP3 encoded stereo music through the Internet and company network to the employee's computer system. The music source is compact disc and the music is compressed at an average ratio of 8:1.

The employee's computer is an IBM-compatible PC with a 100 Mbps Ethernet interface and a 100 MHz 32-bit PCI bus. The employee uses Microsoft Media Player to listen to the music.

Question 1: How much company network capacity is consumed by the employee while listening to music?

Question 2: What percentage of available bus cycles on the employee's computer are consumed when listening to the Internet broadcast?

Question 3: Does your answer to either Question 1 or Question 2 support forbidding this employee to listen to Internet broadcast music? What if there are 500 employees on the company network?

Research Problems

1. Investigate the implementation of modern PC buses such as the Peripheral Component Interconnect (PCI) bus. How wide are the data and address buses? What is the bus clock rate? What is the maximum data transfer rate? What is the set of commands and command responses? How is multiple master capability implemented? What is the bus arbitration strategy?

2. Fibre Channel (http://www.fibrechannel.org) is a relatively recent bus standard for high-speed connection to storage devices. It currently is used to implement high-speed storage arrays where storage devices are separated by large distances. It also can be used so multiple computer systems can share access to a single storage device or array. Investigate the implementation details of the Fibre Channel standard. What are its strengths and weaknesses as compared to Ultra320-SCSI? Which would be used in a typical LAN environment? Which would be used in a distributed transaction-processing environment?

3. Most IBM-compatible PCs use secondary storage devices based on one of the IDE/ATA standards. Investigate those standards and compare them to lower-cost SCSI standards such as SCSI-3. Which standard is the best choice for an average home or office workstation? Which standard is the best choice for high-end workstations such as those used by engineers and digital music producers? Which standard is the best choice for a PC-based server used to share files and databases on an office network?

Chapter 7

Input/Output Technology

Chapter Goals

► Describe manual input devices and how they are implemented

► Explain the characteristics and implementation technology of video display devices

► Understand printer characteristics and technology

► Identify the characteristics of audio I/O devices, and explain how they operate

People communicate in many different ways and those differences are reflected in the variety of methods for interacting with computer systems. This chapter describes the concepts, technology, and hardware used in communication between people and computers (see Figure 7-1). Understanding I/O technology is important because that technology helps or limits the ability of computers to assist people in problem-solving activities.

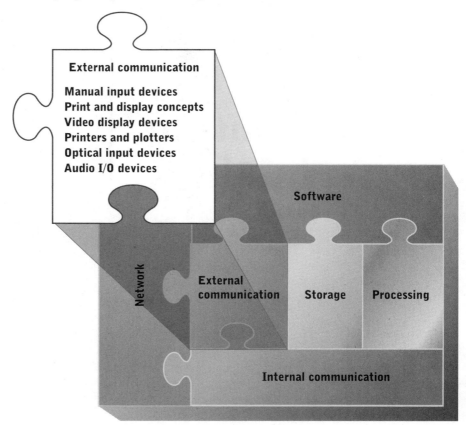

Figure 7-1 ▶

Concepts and topics covered in this chapter

External communication

Manual input devices
Print and display concepts
Video display devices
Printers and plotters
Optical input devices
Audio I/O devices

Software

Network

External communication

Storage

Processing

Internal communication

MANUAL INPUT DEVICES

Manual input devices include keyboards, mice, and other pointing devices. Keyboards have been used for computer input almost since computers were invented. Pointing devices have only recently come of age. Speech recognition is expected to supplant keyboards as the primary means of human input to computer systems, but no one is certain when this will happen. However, even when speech recognition hardware and software are reliable and affordable, there still will be some need for keyboards, for example, for people with speech impairments. Pointing devices also will be required for certain types of processing such as image manipulation.

Keyboard Input

The predominant device for human input to computer systems is a character keyboard. The exact mechanisms by which character data from a keyboard are captured for processing and storage have changed substantially over time.

Early computer systems accepted input via punched cards or tape. Keypunch machines converted manual keystrokes into punched holes in a cardboard card or in paper tape. Another device called a card reader converted punched cards into electrical inputs that could be recognized by the CPU. Punched cards were passed over a light source within the card reader. Light shining, or not shining, through the punched holes was detected and interpreted as input characters.

Modern keyboard devices translate keystrokes directly into electrical signals, eliminating the need for intermediate storage media such as punched cards or paper tape. Modern keyboards use an integrated microprocessor, also called a keyboard controller, to generate bit stream outputs. Pressing a key sends a coded signal to the controller, and the controller generates a bit stream output according to an internal program or lookup table.

A modern keyboard contains a large number of special-purpose keys. These include function keys such as F1; display control keys such as ↑, Page Down, and Scroll Lock; program control keys such as Print Screen and Esc; and modifier keys such as Shift, Caps Lock, Control, and Alt. There are many valid combinations of two or more keys, such as Control+Alt+Delete. The large number of keys and key combinations makes coding keyboard outputs via ASCII or EBCDIC impractical. There simply are not enough unused codes in those coding tables to accommodate all keys and key combinations.

A keyboard controller generates an output called a **scan code**. A scan code is a one- or two-byte data element that represents a specific keyboard event. For most keys, the event is simply a key press. For some keys, pressing and releasing the key may represent two different events, and generate two different scan codes. There are few standards for coding keyboard events into scan codes. The widespread use of IBM-compatible personal computers has created a few de facto standards such as those based on keyboard controllers used in the original IBM PC, IBM AT, and IBM PS/2.

Pointing Devices

Pointing devices include the mouse, trackball, joystick, and digitizer tablet. The devices all perform a similar function—translating the spatial position of a pointer, stylus, or other selection device into numeric values within a system of two-dimensional coordinates. Pointing devices can be used to enter drawings into the system or to control the position of a **cursor**, or pointer, on a display device.

A **mouse** is a pointing device that is moved on a flat surface such as a table, desk, or rubber pad. The position of the mouse on the surface corresponds to the position of a cursor or pointer on a video display. As the mouse is moved left or right, the pointer on the screen is moved left or right. As the mouse is moved toward or away from the user, the pointer is moved toward the bottom or top of the display.

Mouse position is translated into electrical signals by mechanical components or an optical sensor. Mechanical mice use a roller ball in contact with two wheels mounted on pins. As the mouse moves, the roller ball moves one or both wheels. Wheel movement generates electrical signals, which are sent to a device controller and then converted to numeric data describing the direction and distance the mouse is moved. An optical mouse uses a photosensor that tracks movement by detecting minute color or reflectivity changes under the mouse. Optical mice have no moving or exposed parts that can become contaminated by dust or dirt.

A mouse typically has one to three buttons mounted on the top. These buttons are clicked to set the position of the cursor in the display, make selections of text or menu items, open programs, or display pop-up or context menus. Some have a scroll wheel on or near the buttons the user can use to scroll up or down the display. Mouse buttons are implemented with simple electromechanical switches. User commands can be indicated by clicking or double-clicking a single button, or by pressing two buttons at once. Each action is interpreted as a different command by software.

A trackball is essentially an upside-down mouse. The roller ball typically is much larger than in a mechanical mouse and the selection buttons are mounted nearby. The roller is moved by the fingertips, thumb, or palm of the hand. Trackballs have the advantages of being stationary and not requiring as much space as a mouse. Because they can be used on just about any surface, including your lap, they are popular devices for portable computers. Trackballs also are easier to use for persons such as young children and persons with certain physical disabilities, who lack fine motor control.

Digitizers use a pen, or stylus, and a digitizing tablet. The tablet is sensitive to the placement of the stylus at any point on its surface. Stylus position can be recognized by pressure-sensitive contacts beneath the surface or by disturbance of a magnetic field in the tablet. Digitizers are commonly used to trace the outline of an object or drawing. The drawing is placed over the tablet and the stylus is moved manually around objects or lines on the drawing.

Digitizers also can be used to enter commands or make menu selections. For example, in some computer-aided design programs, a menu or command template can be placed over the digitizing tablet. While a drawing is on the video display device, the user selects commands by pointing the stylus to one of the command positions on the template.

BASIC CONCEPTS OF PRINT AND DISPLAY

Printing technology has a several thousand year history. Communication via video display devices is less than a century old. However, both share many features, including character representation methods, measurement systems, and methods of generating color. These topics are covered here because they are common to many I/O technologies and devices.

Matrix-Oriented Image Composition

Display surfaces vary widely in size and composition. The most commonly used display surfaces are paper and video, or television, screens. Display surfaces have an inherent background color, usually white for paper and black for video. Display surfaces can be divided into rows and columns similar to a large table or matrix. Each cell in that table represents one component of an image. A generic term for one of these cells is a **pixel** (a shortened form of "picture element"). For printed output, a pixel can be empty or contain one or more inks or dyes. For video display a pixel can be empty or display light of varying color and intensity.

The number of pixels in a display surface depends on the size (height and width) of the surface and the size of individual pixels. Pixel size for a modern video display typically is between 0.2 and 0.3 millimeters. For paper, pixel size corresponds to the smallest drop of ink that can be accurately placed on the page. Years ago, printers adopted one seventy-second of an inch as a standard pixel size, called a **point**. The term and its use as a printer's measuring system continues even though modern printing techniques can apply ink pixels smaller than one-thousandth of an inch.

Images, printed characters, and other display elements can be described by their height and width in pixels or points. For example, image size might be stated as 300 × 200, which means the image fits within a rectangular area consisting of 300 rows and 200 columns of pixels or points, though pixels usually are implied. If the measurement unit is pixels then the displayed image size depends on pixel size. The image will be 90 millimeters wide and 60 millimeters high if the dot pitch of the display is 0.3 millimeters, assuming that there are no gaps or spaces between the pixels.

The **resolution** of a display is the number of pixels displayed per linear measurement unit. In the United States, resolution generally is stated in **dots per inch (dpi)**, where a dot is equivalent to a pixel. For example, the resolution of a laser printer might be described as 600 × 600 dpi. This describes a device with 600 pixels per linear inch both horizontally and vertically. Higher resolutions correspond to smaller pixel sizes. To an observer, the quality of a printed or displayed image is directly related to the pixel size. Smaller pixel sizes correspond to higher quality print and display images because fine details can be incorporated into the image.

Fonts Written Western languages are based on systems of symbols called characters. Individual characters can be represented as a matrix of pixels as shown in Figure 7-2. Users do not require an exact match to a specific pixel map to recognize a specific printed character. For example, the symbols E, **E**, &, E, and **E** are all easily interpreted by people as the letter E even their pixel composition varies.

A collection of characters of similar style and appearance is called a **font**. Figure 7-3 shows sample characters from several different fonts. Font characters can be large or small, or any size between. The most common unit for measuring font size is points. Figure 7-4 shows a single font printed in several point sizes. Note that the measurement refers to the height of the characters but not their width, although width normally is scaled to match height. Because character size varies, for example, T is larger than a, font point size is the distance between the top of the highest and bottom of the lowest character in the font.

Figure 7-3 ▶

Sample characters printed in various fonts

AaBbCc XxYyZz 123 !@#　　　News Gothic
AaBbCc XxYyZz 123 !@#　　　Braggadocio
AaBbCc XxYyZz 123 !@#　　　Century Schoolbook
AaBbCc XxYyZz 123 !@#　　　Garamond Condensed
AaBbCc XxYyZz 123 !@#　　　Tahoma

Figure 7-4 ▶

A sample font
(Times Roman)
printed in various
point sizes

8 point Times Roman
10 point Times Roman
12 point Times Roman
16 point Times Roman
24 point Times Roman

Color   The human eye interprets different light frequencies as different colors. For video displays, color is generated directly by the display device. For printed display, color is the light frequency reflected from the page. Any color in the visible spectrum can be represented as a mixture of varying intensities of colors such as red, green, and blue. For example, mixing equal intensities of red and blue produces the color magenta, or purple. The absence of all three colors is black because no light is generated or reflected. Full intensity of all three is white.

Different sets of primary colors are used for different purposes. In paint, red, green, and yellow are used as primary pigments. The video display industry has used red, green, and blue as primary colors since the earliest color televisions. The primary colors for video display are sometimes referred to by their first letters, RGB. A video display that generates color using mixtures of these colors sometimes is called an RGB display.

The printing industry generates color using the inverse of the primary video display colors. In this context, video display colors are called **additive colors** and their inverse colors are called **subtractive colors**. The subtractive colors are cyan (absence of red), magenta (absence of green), and yellow (absence of blue) and they often are referred to by the acronym **CMY**. Black can be generated by combining all three, but a separate black dye, identified by the letter K, is generally used. Colored dyes are expensive to produce, but pure black dye is relatively inexpensive. The four-dye scheme usually is called **CMYK** color.

Pixel Content   Because computers are digital devices, pixel content must be described numerically. A stored set of numeric pixel descriptions is called a **bit map**. The number of bits required to represent an image depends on whether the display is monochrome (black and white), grayscale, or color. Monochrome display uses a single bit to indicate whether each pixel is on or off. For print display, "on" generally means black. For video display, "on" means whatever foreground color is generated by the device.

Grayscale displays can generate shades of gray between pure black and pure white. The number of shades that can be generated depends on the number of bits used to represent a pixel. Assume, for example, that four bits are used. Interpreting four bits as an unsigned integer provides numeric values between 0 and 15. Zero can be interpreted as no light; the black background is displayed. Fifteen can be interpreted as maximal light intensity, or bright white. Values between 0 and 15 can represent successively higher levels of generated light. A very dark gray could be represented by 3, while 12 could represent a very bright gray.

The number of gray shades that can be displayed increases with the number of bits used to represent a pixel. If eight bits are used per pixel, then 254 shades of gray are available in addition to pure black (0) and pure white (255). The number of distinct colors or gray shades that can be displayed is sometimes referred to as **chromatic depth,** or **chromatic resolution.**

For color display, three separate numbers are required to represent a pixel. Each number represents the intensity of an additive or subtractive color. Chromatic depth depends on the total number of bits used to represent each color intensity. For example, if eight bits are used to represent each color then the chromatic depth is $2^{(3\times8)}$ or approximately 16 million; this sometimes is called 24-bit color. Each group of eight bits represents the intensity (0 = none, 255 = maximum) of one color. For example, with RGB color the numbers 255:0:0 represent bright red, and the numbers 255:255:0 represent bright magenta.

Color display using less than 24 bits presents a problem. Computers deal with data in byte-size increments so multiples or even fractions of eight bits, such as 4, 8, or 16 bits, are most efficient. Because these numbers aren't evenly divisible by three, there is no equal division of the available bits among the primary video display colors.

The normal approach to this problem is to define a color palette. A **palette** is simply a table of colors. The number of bits used to represent each pixel determines the table size. If eight bits are used, then the table has 2^8 or 256 entries. Each table entry can contain any RGB color value such as 200:176:40. But the number of different colors that can be represented is limited by the size of the table. Table 7-1 shows the 4-bit color palette used for the Computer Graphics Adapter (CGA) video driver of the original IBM PC/XT.

Color also can be produced by a process called dithering. **Dithering** places small dots of different colors in an interlocking pattern. If the dots are small enough, the human eye interprets them as being a uniform color representing a mixture of the dot colors. A tight pattern of 50 percent red and 50 percent blue dots is interpreted as magenta. A smaller percentage of red and higher percentage of blue would be interpreted as violet. A tight pattern of black and white dots is interpreted as gray—examine a gray area of one of the figures in this book with a magnifying glass.

Dithering commonly is used when a device is not capable of applying fine differences in color to an area. For example, if an ink-jet printer can print only four different amounts of a single dye in one pixel area, then only 64 (4^3) different colors can be generated. With dithering, the same printer can simulate several shades between each of those 64 colors.

Table Entry Number	RGB Value	Color Name
0	0:0:0	black
1	0:0:191	blue
2	0:191:0	green
3	0:191:191	cyan
4	191:0:0	red
5	191:0:191	magenta
6	127:63:0	brown
7	223:223:223	white
8	191:191:191	gray
9	127:127:255	light blue
10	127:255:127	light green
11	127:255:255	light cyan
12	255:127:127	light red
13	255:127:255	light magenta
14	255:255:127	yellow
15	255:255:255	bright white

Image Storage Requirements

The number of bits used to represent a pixel, and the image height and width in pixels, determine the required amount of storage. For example, an 800×600 image has 480,000 pixels. For a monochrome image, each pixel is represented by a single bit and the image requires 60,000 bytes of storage (480,000 pixels $\times$ 1 bit/pixel $\div$ 8 bits/byte). Storing an 800×600 24-bit color image requires 1.44 MB of storage (480,000 pixels $\times$ 3 bytes/pixel).

Image storage requirements apply to images stored in primary and secondary storage, and to buffers used in I/O devices. Image storage requirements can be reduced with image compression techniques. Commonly used compression formats include **Graphics Interchange Format (GIF)** and **Joint Photographic Experts Group (JPEG)** for single images and **Moving Pictures Experts Group (MPEG)** for moving images.

Image Description Languages

There are two significant drawbacks to using bit maps to represent high-quality images. The first is their size. An $8\frac{1}{2} \times 11$ inch, 24-bit color image with 300 dpi resolution requires approximately 25 MB of uncompressed storage. A good compression algorithm might achieve compression ratios of 10:1 or 20:1, reducing the storage requirements to a few megabytes. Files of this size take a great deal of time to move over slow communication links such as the printer port of a personal computer.

The second drawback is that there is no standard storage format for raw bit maps, making them device-dependent. Bit maps vary among software packages and output device types, manufacturers, and models. Image files stored by a program usually must be converted into a different format before they are printed or displayed. The conversion process is computationally and memory intensive, and multiple conversion utilities or device drivers may be required.

Image description languages address both drawbacks of bit maps. Such languages are device-independent and compact. They use symbolic language to describe primitive image components, which reduces storage space requirements. A symbolic description of an image component often is much smaller than a bit map of that same component. The simplest image description languages use vector lists and are limited to describing image components that are straight-line segments or can be built from such segments. In graphics, a **vector** is a line segment that has a specific angle and length with respect to a point of origin, as shown in Figure 7-5(a). Line drawings can be described as **vector lists**, a series of concatenated or linked vectors. Such an image resembles a "connect the dots" drawing, as shown in Figure 7-5(b). Images and image components containing curved lines and irregular patterns, however, must be communicated as bit maps to be represented accurately.

More advanced image description languages have the ability to describe complex images such as shapes and fonts. A group of high-level commands for generating graphic output is called a display list. A **display list** describes an image as a collection of **display objects** such as circles, triangles, rectangles, and other shapes. **Display attributes** are characteristics such as position or origin, size, line thickness, line color, and fill pattern. Examples of advanced image description languages include Adobe's PostScript and Hewlett-Packard's Printer Control Language (PCL).

Although display list representation of an image generally is much more compact than a vector list or bit map of the same image, the display device must interpret vector or display list elements and convert them into whatever commands or signals are needed to display the image. An output device that accepts display lists as input must contain an embedded processor and software for processing the display list to compose a bit map within its internal buffer. Processing display or vector lists within an output device tends to increase device cost, reduce output/display speed, or both.

Figure 7-5 ►

Vector list image descriptions: the elements of a single vector (a) and a more complex image built from concatenated vectors (b)

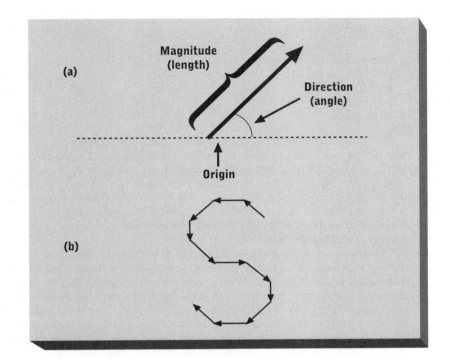

Technology
Focus

PostScript

PostScript is an image description language designed primarily for printed documents, although it also can be used to generate video display outputs. PostScript also is a programming language. A PostScript display list is a program or procedure for producing a specific image. Images may include various display objects such as characters, lines, and shapes. Display objects can be manipulated in a variety of ways including rotation, skewing, filling, and coloring. The language is sufficiently robust to represent virtually any imaginable graphic image.

A PostScript display list contains a set of commands composed of normal ASCII characters. Transfer of a display list between devices, such as a computer and a printer, can be performed by the same means as normal character-based communication. Commands may include numeric or textual data, primitive graphic operations, and procedure definitions. Data is held within a stack. Primitive operations and/or procedures remove data from the stack and use that data as control information or procedure parameters.

Consider the following short program:

```
newpath
400 400 36 0 360
arc
stroke
showpage
```

The first line declares the start of a new path—straight line, curve, or complex line. The second line contains a list of numeric data items to be pushed onto the stack. The control parameters are removed from the stack by the predefined procedure named arc (the third line) as they are used. The left two parameters specify the origin (center) of the arc as row and column coordinates. The center parameter specifies the radius of the arc in points. The remaining parameters specify the starting and ending points of the arc in degrees. In this case, the starting and ending points represent a circle.

Primitive graphic operations are specified in an imaginary drawing space containing a grid (rows and columns) of points. The first three commands specify a tracing within this space. The fourth command states that this tracing should be stroked (drawn). Because no width is specified, the circle is drawn using the current default line width. The final command instructs the output processor to display the contents of the page as defined so far. A PostScript printer receiving the above program as input would generate a single page containing one circle.

PostScript programs can define new procedures. The following program defines a procedure named *circle* composed of the primitive commands and parameters used to draw a circle:

```
/circle % stack = x-center, y-center, radius
{ newpath
0 360
arc
stroke } def
```

The first line defines the name of the procedure and includes a comment (the text following the % character) that explains the parameters used by the procedure. The procedure definition appears between the curly brackets and can be arbitrarily long. Note that the procedure defines only the final two parameters needed to draw a circle. The first three must be placed on the stack before the procedure is called. Figure 7-6 shows a program that draws three concentric circles using this procedure.

Figure 7-6 ►

A PostScript
program that
draws concentric
circles

```
400 400 36 circle
400 400 72 circle
400 400 108 circle
showpage
```

Figure 7-7 shows a PostScript program that combines text and graphics. PostScript is a subset of the Adobe Portable Document Format (PDF), which is used widely to distribute printable content over the Web for viewing by the Adobe Acrobat viewer.

Figure 7-7 ►

PostScript
program that
generates the
pop-up menu
shown on the
right

```
% outline box
newpath
245 365 moveto
340 365 lineto
340 265 lineto
245 265 lineto
closepath
stroke

% horizontal line
newpath
245 345 moveto
340 345 lineto
stroke

% text items
/Times Roman findfont 12 scalefont setfont
250 350 moveto (Applications) show
250 330 moveto (Word Processing) show
250 315 moveto (Spreadsheet) show
250 300 moveto (Database) show
250 285 moveto (Graphics) show
250 270 moveto (Games) show

% arrow
newpath
240 305 moveto
-6 6 rlineto
0 -3 rlineto
-20 0 rlineto
0 -6 rlineto
20 0 rlineto
0 -3 rlineto
closepath
fill

showpage
```

Applications
Word Processing
Spreadsheet
Database
Graphics
Games

VIDEO DISPLAY

Video display devices have changed greatly since their first widespread use in the mid-to late-1960s. Early devices were little more than modified television sets and could display only monochrome characters. Modern video displays are much more sophisticated devices. They are larger, have higher resolution, display color, and display graphic images as well as characters.

Character-Oriented Video Display Terminals

The first video display devices used widely for computer I/O consisted of an integrated keyboard and television screen. Such a device normally is called a **video display terminal (VDT)** or simply a **terminal**. VDTs were the most common form of video display during the 1970s and much of the 1980s. Their use faded after personal computers came into common use. Today, they are used primarily in systems such as retail checkout counters and factory floor environments.

A video display terminal consists of five functional components, as shown in Figure 7-8. The physical keyboard is attached to a keyboard driver that generates output signals according to scan codes received from the keyboard. The display generator generates the image actually viewed by a user. Display generators are available in a variety of types and capabilities, which will be described later in this chapter.

Figure 7-8 ▶

Functional components of a video display terminal

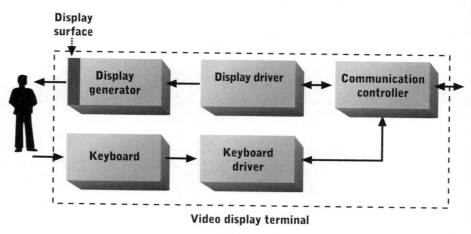

In early video displays, relatively simple electronic circuitry translated coded character input, usually ASCII, into a corresponding pixel matrix, which then was input to the display generator. In modern video display devices, the translation between coded character input and pixel patterns is controlled by a microprocessor. This allows for a great deal of flexibility in character display.

Microprocessor-based video displays are able to alter fonts, sizes, intensity, and other display features.

Graphic Video Display Devices

A few VDTs were endowed with substantial graphic display capabilities. Manufacturers of these devices included Tektronix and Digital Equipment Corporation. Graphic VDTs often were used for displaying diagrams such as construction blueprints, wireframe models, and wiring diagrams, and to produce mathematical graphs. Graphic VDTs faded quickly with the advent of personal computers. By the mid-1980s, it generally was cheaper to manufacture a general-purpose microcomputer with a high-quality video display than to manufacture a graphics terminal, due to higher production volume and the low cost of micro-computer components.

Modern video displays generally are separated from the keyboard and are called **monitors**. They operate as independent devices under control of a video controller attached to the system bus. Input is no longer in the form of coded characters transmitted via digital signals. Instead, analog signals similar to those generated by video cassette recorders are generated by the video controller. These signals are transmitted continuously at a high frequency so the display can be updated quickly. Communication of video display data is depicted in Figure 7-9.

Figure 7-9 ►

Transfer of video data from primary storage to the monitor

CRTs   Most current monitors are implemented using **cathode ray tubes (CRTs)**. CRTs are based on the same display technology as conventional television sets. The technology is highly refined and relatively inexpensive. A CRT is an enclosed vacuum tube. An electron gun in the rear of the tube generates a stream of electrons that are focused in a narrow beam toward the front surface of the tube. The interior of the display surface is coated with a phosphor. When struck by a sufficient number of electrons, the phosphor reaches an excited state and emits light. Pixel illumination is controlled by pulsing (turning on and off) the electron beam. The color of the display is determined by the chemical composition of the phosphor coating. The intensity of individual pixels or the entire display can be varied by changing the intensity of the electron beam.

The electron beam is directed toward individual pixels by magnetic deflection, which constantly is shifted so that the beam moves continually between pixel locations in a left-to-right, top-to-bottom motion. Color images are generated using three different electron beams directed toward a grid of three phosphor coatings (red, green, and blue). Different levels of intensity for each color combine to produce a continuous or nearly continuous range of color.

The phosphor atoms remain in an excited state for only a fraction of a second. To display light continously, a pixel area must be bombarded by the electron gun many times per second. The number of times per second that the entire surface is scanned by the electron gun is called the **refresh rate** of the monitor. Refresh rates are measured in hertz (Hz). Typical refresh rates range from 50 to 100 Hz. Images displayed with higher refresh rates minimize eyestrain and are regarded as having higher quality.

CRT technology is relatively old and has some disadvantages. Physical size and weight increase as the size of the display area increases. The vacuum tube is made of heavy and breakable glass. Larger screen sizes require larger and deeper vacuum tubes. A typical CRT monitor occupies 0.5 cubic meters or more—a substantial portion of the available space on an average desk. Another drawback to CRT monitors is their power consumption. CRTs consume up to several hundred watts of power and dissipate much of that as heat. Power requirements and heat generation increase with screen size. The power consumption of a typical workstation monitor is at least half that of the entire computer system.

Flat Panel Displays   Many alternative video display technologies are under development or in current production. As a group, they are called **flat panel displays** due to their characteristic shape. None has yet proven superior to CRT technology at equal or less cost. But extensive research and development activities promise that the cost/performance ratio of CRTs will be surpassed within the next decade.

A **liquid crystal display (LCD)** is a flat panel matrix that consists of encapsulated liquid crystals sandwiched between two polarizing panels. Exterior light passes through the top polarizing layer and into the liquid crystal layer. If no power is applied to the liquid crystal, the polarized light reflects back toward the front panel, appearing as a white or light gray spot. If power is applied to the liquid crystal, light passes through the crystal and is absorbed in the rear polarizing panel. The spot then appears dark, usually blue or black.

Modern LCDs use a technique called backlighting for improved contrast and brightness. In this approach, a fluorescent light source is placed behind the display panel. Liquid crystals can permit or block the passage of this light through the front of the display, depending on their electrical state. Color output is generated by using three LCD matrices with a different color filter for each.

Switching individual pixels on and off requires transistors and wiring. An **active matrix display** uses one or more transistors for every pixel. A **passive** matrix display shares transistors among rows and columns of pixels. Additional transistors increase the speed at which pixel elements can be changed. With transistors, a continuous charge can be applied to each pixel, thus generating brighter light. Transistors also increase the complexity and cost of the display panel.

LCD displays have less contrast than CRT displays because color filters reduce the total amount of light that passes through the front of the panel. LCDs also have a much more limited viewing angle than CRTs. Best contrast and brightness is achieved within a 30-to-60 degree angle. Outside that angle, contrast drops off sharply and colors shift. In contrast, CRT viewing angles usually approach 180 degrees with little loss in contrast or brightness.

The advantages of an LCD are substantially reduced size, weight, and power consumption as compared to a CRT. Power consumption generally is less than 100 watts for an active matrix display, and much less for a passive matrix display. Weight is relatively low due to small size and the use of only two flat glass panels. The cost of an LCD panel is considerably higher than a CRT of similar size and capability. LCD manufacturing methods have been slow to evolve and production yields are often poor.

Electroluminescent (EL) displays are similar in construction to LCDs. Color is generated with three matrices of different colored phosphors. There are no liquids or gases, as well as very little glass, so the devices are stable, rugged, and lightweight. The primary drawbacks of EL displays are color quality and cost. Color quality suffers due to the difficulty in developing "true blue" phosphors that operate at low voltages and have long life.

The EL manufacturing process is similar to that for microchips. The matrices of pixel activation transistors and related electrodes are built on a layer of clear silicon. Phosphors are deposited directly on the matrices and another layer of clear silicon is overlaid. Problems with material purity and low production yield are similar to other semiconducting fabrication processes. These problems are exacerbated by the relatively large size of the product.

PRINTERS AND PLOTTERS

Early printers were little more than electric typewriters with a communication port substituted for a keyboard. Printed output was generated by moving paper past a ball or wheel on which all characters of a font were embedded as raised metal images. The ball or wheel was rotated to place the desired character in a "ready to print" position and then the character was forced against the paper at high speed. A cloth ribbon soaked with ink was positioned between the print head and paper to generate a character image.

The functional components of a character printer are shown in Figure 7-10. Note the similarity to the components of a video display terminal shown in Figure 7-8. The communication controller and the print driver perform essentially the same functions as their counterparts in a VDT. The exact nature of the translation process performed by the printer driver depends on the type and capabilities of the print generator. Commonly used types of print generators include dot matrix, ink-jet, and laser print generators.

Figure 7-10 ▶

Functional components of a typical character printer

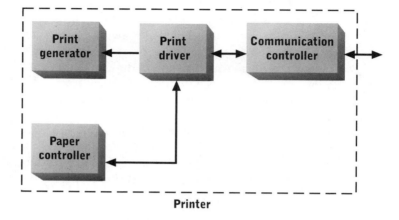

Printer

Dot Matrix Printers

Dot matrix printers are the last generation of printers to generate images through mechanical contact with ink and paper. Character codes received by the print generator are transmitted as series of control commands to the print head. The print head consists of one or more vertical lines of pins that can be forced outward from the head a fraction of an inch. An inked ribbon is moved in front of the pins, with the paper on the opposite side of the ribbon. The print head moves from left to right on each print line and pins are pushed out from the print head, impressing a dot of ink from the ribbon onto the paper. A dot matrix printer generates individual characters one or two vertical rows at a time, from left to right.

Paper is controlled by one or more motors driving rubber rollers and/or sprockets. Paper is advanced past the print head one line at a time. The printer prints characters on the page in a left-to-right, top-to-bottom order. Advanced control capabilities can move paper in increments of less than a single line. Reverse movement of paper and/or print head also can implemented.

Dot matrix printers have all but disappeared from the marketplace due to poor output quality, noisy operation, and competition from other printer technologies. The few models still available are customized for high-speed printing of preprinted multicopy forms. Print quality is sacrificed for overall speed. Fast dot matrix printers can achieve print speeds of several hundred characters per second. Sprockets and fan folded paper with matching holes along the edges move the paper rapidly through the print mechanism. Dot matrix printers can print up to a dozen pages per minute, which is faster than alternative technologies at an equivalent price.

Ink-Jet Printers

An **ink-jet printer** uses a paper movement mechanism similar to that of a dot matrix printer. Individual sheets of paper generally are used instead of fan folded paper. Another similarity is the use of a single print head that is drawn horizontally across the width of the paper. The primary difference is the operation of the print head. There is no inked ribbon and no direct physical contact between elements of the print head and the paper.

The print head of an ink-jet printer consists of an ink cartridge, a set of ink chambers, and a set of ink nozzles. Each ink nozzle can print a single pixel and nozzles are arranged in short vertical rows similar to those of a dot matrix printer. Each nozzle has a small ink chamber behind it. At the back of the chamber is a flexible piezoelectric membrane. When an electrical charge is applied to the membrane it flexes toward the nozzle. This increases pressure within the ink chamber and forces liquid ink out through the nozzle. When the electrical charge is removed, the membrane flexes back to its original shape. The now enlarged chamber has a partial vacuum that draws fresh ink in from the cartridge.

The print head nozzles are placed in very close proximity to the paper surface. Paper is fed in from an input hopper and positioned so the print nozzles are along its top edge. As the print head is drawn across the paper width by a pulley, the print generator rapidly modulates the flow of current to the ink chamber membranes. This generates a pattern of pixels across the page. The paper then is advanced and the process is repeated until the entire sheet has been drawn past the moving print head. The ink dries very quickly so there is little or no opportunity for it to smear.

Ink-jet printers use very small chambers and nozzles capable of producing ink dots as small as one six-hundredth of an inch in diameter (resolution is up to 600 pixels per inch). Color output is generated by using CMY or CMYK inks and an equal number of nozzle/chamber rows. As each row passes a horizontal point on the page, different color ink is applied. The inks mix to form pixels of various colors. Electrical charge applied to the ink chamber membranes can be varied to place variable amounts of ink on the page. CMY colors can be mixed in different proportions, producing a wide range of possible pixel colors.

Ink-jet printer technology has advanced rapidly. Typical output speeds for inexpensive models are two pages per minute for black and white and one-half page per minute for full color. Black and white print quality is good to excellent and color print quality often is better than color laser printers. Ink-jet printers are the only type of printer capable of producing high-quality color output within the budget of a typical home computer or small office user. Clogged print heads were a problem with early models. This problem was solved by integrating the print head and the ink cartridge. Both are disposed of as a unit when the cartridge is empty.

Laser Printers

A **laser printer** operates quite differently from other types of printers. No print head is used nor is any inked ribbon required. The print driver of a laser printer does not transmit individual characters or lines of an image to the print generator. Instead, it stores an internal image of an entire printed page in an internal buffer as a bit map. Once filled, the buffer contents are sent to the print driver for page generation.

The print generator operates using electrical charge and the attraction of ink to that electrical charge, as illustrated in Figure 7-11. First, a rotating drum is lightly charged over the width of its surface. The print driver then reads rows of pixel values from the buffer and modulates a tightly focused laser over the width of a metal drum. The drum then is advanced and the process repeated with the next line of pixels. The laser removes the charge wherever it shines on the drum. The drum contains an image of the page with charged areas representing black pixels and uncharged areas representing white pixels. After charging, the drum then passes a station where fine particles of toner, which is a dry powder ink, are attracted to the charged areas.

Figure 7-11 ▶

Components of a typical laser printer engine

In synchronized motion with the drum, paper is fed through a series of rollers and given a high electric charge. As the paper passes over the drum surface, toner on the drum is attracted to the paper due to its higher charge. The paper and attached toner then are fed through heated rollers that fuse the toner to the paper surface. A light source then removes all charges from the drum and excess toner is removed from the drum by a fine blade and/or vacuum.

Color laser output uses three separate print generators (laser, drum, and laser modulators), one for each color. Paper must be aligned precisely as it passes over the three drums. Also, the complexity and expense of the printer increase substantially. Color also can be generated by three passes over the same print generator with a different color ink applied during each pass. This also reduces maximum printer speed and also reduces the complexity and cost of the printer.

Monochrome laser printers are found in nearly all modern offices. Price reductions have made them affordable for most home and office users. Print speeds range from a few pages per minute to several dozen pages per minute. The devices are well developed and reliable.

Plotters

A **plotter** is a printer that generates line drawings on wide sheets or rolls of paper. Paper is mounted within a paper control mechanism that can move the paper up or down precisely. Plotters can handle paper widths up to 60 inches, which makes them ideal for producing blueprints and other engineering drawings.

Plotters vary in drawing technology, printing speed, and paper width. Pen-based printing technology was the norm until the mid 1990s, after which ink-jet technology became the norm. Pen-based technology drew images using one or more pens, which could be raised and lowered onto paper while the paper moved. Modern plotters use ink-jet technology and are little more than large-format ink-jet printers. Comparably sized laser print engines are too expensive to compete with ink-jet plotters.

OPTICAL INPUT DEVICES

Optical sensors include such devices as digital still and video cameras, optical scanners, mark sensors, bar code readers, and optical character recognition (OCR) devices. The operating principles behind all optical sensing devices are similar. A light source, usually a laser, shines onto a printed surface. Light is reflected from the surface into a photodetector. High light intensities induce relatively large current outflows from the photodetector. Low intensities produce little or no current. High and low current outflows from the photodetector are interpreted as binary digits.

Optical scanning devices can be differentiated by the following criteria:

► Input format requirements
► Normal and maximum spatial (pixel) resolution
► Normal and maximum chromatic resolution
► Embedded processing capability

Optical scanning device resolution is currently as high as 1200 dpi. Simpler devices such as bar code scanners may require relatively large marks (no smaller than 5 millimeters) for reliable recognition. Some devices require marks that appear at specific locations on a page or within specific distances of one another. Devices that attempt any form of image recognition or classification, such as optical character recognition devices, must embed processing capabilities within the device or must interact with software on a host computer.

Mark Sensors and Bar Code Scanners

Mark sensors and **bar code scanners** are simple optical recognition devices. A mark sensor scans for light or dark marks at specific locations on a page. Mark sensor input pages are familiar to students who take standardized multiple choice tests. Input locations are marked on the page with circles or boxes, and are selected by filling them in with a dark pencil. The mark sensor uses preprinted bars on the edge of the page to establish reference points, for example, the row of boxes corresponding to the possible answers to Question #5. The mark sensor then searches for dark marks at specific distances from those reference points. Marks must be of a sufficient size and intensity to be correctly recognized.

Bar code scanners operate similarly to mark sensors but without preestablished reference marks. A bar code is a series of vertical bars of varying thickness and spacing. The order and thickness of bars is a standardized representation of numeric data. Bar code readers use one or more scanning lasers to detect the vertical bars. A **scanning laser** sweeps a narrow laser beam back and forth across the bar code. Bars must have precise width and spacing, as well as high contrast for accurate decoding. Multiple scanners at oblique angles allow for variation in the position and orientation of the bar code.

Bar code readers typically are used to track large numbers of inventory items. Common uses include grocery store inventory and checkout, tracking of packages during shipment, warehouse inventory and picking, and ZIP code routing of postal mail. The U.S. Postal Service uses a modified form of bar coding with evenly spaced bars of equal thickness but varying height. These appear along the lower edge of an envelope and are used to encode five-and nine-digit ZIP codes (see Figure 7-12).

Figure 7-12 ►

A sample
address with a
standard U.S.
Postal Service
bar code

> The White House
> 1600 Pennsylvania Avenue NW
> Washington, DC 20500
> ‖ı|ıl‖ı‖ıııılı|ıl‖ıı‖ıııı‖ı|

Advanced scanning technology can now read two-dimensional bar codes. Figure 7-13 shows a sample 2-D bar code in the PDF417 format. Data is encoded in patterns of small black and white blocks surrounded by vertical bars marking the left and right boundaries of the bar code. PDF417 bar codes can hold approximately 1 KB of data, so descriptive or shipping information can be stored directly on the shipping container or inventory item. PDF417 format is used widely by governmental and industry-specific standards organizations.

Figure 7-13 ►

A PDF417
bar code

Optical scanners generate bit map representations of printed images. A bright white light is shone on the page and reflected light is detected by an array of photodetectors. Spatial resolution is determined by the size and spacing of the photodetectors. Chromatic resolution is determined by the sensitivity of the photodetectors to different light frequencies. Typical desktop scanners have spatial resolution up to 600 dpi and chromatic resolution up to 24 bits. Manual scanners, sometimes called hand scanners, require a user to move the scanning device over a printed page at a steady rate of speed. Automatic scanners use motorized rollers to move paper past a scanning surface or to move the light source and photodetector array across the paper.

Optical character recognition (OCR) devices combine optical scanning technology with hardware or software interpretation of bit map content. Once an image has been scanned, the bit map representation is searched for patterns corresponding to printed characters. In some devices, input is restricted to characters in prepositioned blocks such as the "Amount Enclosed" blocks in Figure 7-14. In these devices, as with mark sensors, preprinted marks provide a reference point for locating individual characters. This greatly simplifies the character recognition process.

Figure 7-14 ►

Sample form with
OCR input forms

PIT-ES New Mexico Individual Estimated Tax Payment Voucher   Tax Year_____

Indicate quarter the payment is to be applied to: ☐ **1ST (APR 15)**   ☐ **2ND (JUN 15)**   ☐**3RD (SEPT 15)**   ☐ **4TH (JAN 15)**

YOUR FIRST NAME AND INITIAL	LAST NAME	YOUR SOCIAL SECURITY NUMBER
SPOUSE'S FIRST NAME AND INITIAL	LAST NAME	SPOUSE'S SOCIAL SECURITY NUMBER
ADDRESS (Number & Street)	CITY, STATE AND ZIP CODE	☐ Check this box if address is new or changed

Mail payment and voucher to:
New Mexico Taxation and Revenue Department
P.O. Box 8390, Santa Fe, NM 87504-8390   AMOUNT ENCLOSED   [|].0 0

More sophisticated OCR software and hardware place no restriction on the position and orientation of symbols, creating dual problems of finding and correctly classifying printed symbols. Accurate recognition is easiest when text is printed in a single font and style with all text oriented in the same direction on the page. If the text is handwritten, contains mixed fonts or styles, is combined with images, or has varying orientations on the page, accurate recognition becomes more difficult. Many inputs are beyond the current capabilities of OCR devices.

Error rates of 10 percent or higher still are common with mixed-font text and even higher with handwritten text. The field has experienced rapid improvement in accuracy and flexibility of input, but much more progress needs to be made before error rates are acceptable for many types of applications.

AUDIO I/O DEVICES

Sound is an analog signal that must be converted to digital form for computer processing or storage. The process of converting analog sound waves to digital representation is called **sampling**. The content of the audio sound spectrum is analyzed many times per second and converted to a numeric representation. For sound reproduction that sounds natural to people, frequencies between 20 Hz and 20 KHz must be sampled at least 40,000 times per second. The standard for compact disc recordings is 48,000 samples per second.

Sound varies by frequency (pitch) and intensity (loudness). By various mathematical transformations, a complex sound such as a human voice, consisting of many pitches and intensities, can be converted to a single numeric representation. For reproduction that sounds natural to humans, 16 bits normally must be used to represent each sample point. One second of sound sampled 48,000 times per second requires 96,000 bytes for digital representation (48,000 × 16 bits × .0125 bytes/bit). A full minute of sound requires over 5 MB.

Computer generation of sampled sound is relatively simple. The digital information is input to a **digital-to-analog converter (DAC)**, then to a conventional analog amplifier, and finally to a speaker. Sampling also is relatively simple.

However, the processing power and communication capacity needed to support sampling and playback reached specialized consumer devices such as compact disc players and digital audio tape recorders in the 1980s, and ordinary desktop computers in the early 1990s.

Within a computer system, sound generation and recognition can be used in a number of ways, including:

- General-purpose sound output, such as warnings, status indicators, and music
- General-purpose sound input, such as digital recording
- Voice command input
- Speech recognition
- Speech generation

Limited general-purpose sound output has existed in computer systems for many years. Microcomputers, workstations, and VDTs can generate a single audible frequency at a specified volume for a specified duration of time. This type of sound output is called **monophonic** output because only one frequency (note) can be generated at a time.

As computers have been adapted for multimedia presentation and as general-purpose communication devices, greater sound generation capability has been added. Most computers now contain **polyphonic,** or multifrequency, sound generation hardware with built-in amplification and high-quality speakers.

Speech Recognition

Speech recognition is the process of recognizing and appropriately responding to the meaning embedded within human speech. Human speech consists of a series of individual sounds called **phonemes,** roughly corresponding to the sounds of each letter of the alphabet. A spoken word is a series of interconnected phonemes. Continuous speech is a series of phonemes interspersed with periods of silence. Recognizing individually voiced (spoken) phonemes is not a difficult computational problem. Sound can be captured in analog form by a microphone, converted to digital form by digital sampling, and the resulting digital pattern compared to a library of patterns corresponding to known phonemes, as shown in Figure 7-15.

Figure 7-15 ▶

The process of
speech
recognition

Although the process of speech recognition is simple conceptually, a number of factors complicate it, including:

▶ Speaker variability

▶ Phoneme transitions and combinations

▶ Real-time processing

Speakers vary widely in the exact characteristics of voiced phonemes. Variations arise from physical, geographic, and situational differences. For example, phonemes may differ depending on whether the speaker is male or female, is from Texas or New England, or is making polite conversation as opposed to giving a command. Phoneme matching is inexact because of these variations. Comparison to a library of known phonemes cannot be based on exact match; the closest match must be determined and decisions among multiple possible interpretations must be made. Real-time speech recognition for multiple speakers consumes much or all of the CPU power available on a typical desktop computer.

Another difficult problem in speech recognition arises from the continuous nature of speech. Individual letters sound similar when repetitively voiced by the same speaker. However, when combined with other letters in different words, their voicing varies considerably. In addition, a computer must determine where one phoneme ends and another begins, as well as where one word ends and another begins. Once again, the complexity of such processing requires complex software and powerful CPUs.

Most currently available speech recognition systems are not capable of unrestricted speech recognition for multiple speakers in real time. They must be "trained" to recognize the sounds of individual speakers. These systems are referred to as **speaker dependent**. They also are restricted to relatively limited vocabularies, perhaps only a few thousand words. The most limited of these are command recognition systems. They are designed to recognize up to a few hundred spoken words from a single speaker. Such devices are useful when a single user will use the device and when manual input is impractical. These systems have been applied in airplane cockpits, manufacturing control, and input systems for the physically disabled.

A **digital signal processor (DSP)** is a CPU specialized to the task of processing continuous streams of audio or graphical data. DSPs commonly are embedded within audio and video hardware such as personal computer sound cards and dedicated audio/video workstations. In some cases, the DSP preprocesses audio input before data is transferred to the CPU. In other cases, the CPU and DSP share processing duties, with each performing the tasks to which it is best suited. Some general-purpose CPUs, such as later generations of the Intel Pentium, now include capabilities that used to be available only in DSPs.

Speech Generation

A device that generates spoken messages based on textual input is called an **audio response unit**. Typical applications include delivering limited amounts of information over conventional telephones, such as automated telephone bank tellers and automated call routing. Audio response units allow a telephone to act as an information system's input/output device.

Simple audio response units store and play back words or word sequences. Words or messages are recorded digitally and stored. To output a message, the stored message is retrieved and sent to a device that converts the digitized voice into analog audio signals. All possible outputs must be recorded in advance. Phrases and sentences can be generated from individual stored words though pauses and transitions among the words poorly mimic human speech.

A more general and complex approach to speech generation is called **speech synthesis**. In speech synthesis, individual vocal sounds, or phonemes, are stored within the system. Character outputs are sent to a processor within the output unit, which assembles corresponding groups of phonemes to generate synthetic speech. The quality of speech output from such units varies considerably. Problems are encountered when combining groups of phonemes to form words. Transitional sounds must be generated between the phonemes and composite phonemes must be accounted for. The difficulties of speech generation are far less formidable than those of speech recognition.

General-purpose audio hardware also can be used for speech generation. Digital representations of phoneme waveforms can be combined mathematically to produce a continuous stream of digitized speech, which then is sent to the digital-to-analog converter of the audio hardware. Once converted to an analog waveform, the speech is amplified and routed to a speaker or telephone.

General-Purpose Audio Hardware

General-purpose audio hardware typically is packaged as an expansion card that connects to the system bus of a workstation. Common names for such hardware include **sound card**, **sound board**, and **multimedia controller**. The latter term often is used when a compact disc controller and drive are part of the purchased package.

At minimum, sound cards include an **analog-to-digital converter (ADC)**, a digital-to-analog converter, a low power amplifier, and connectors (jacks) for a microphone and a speaker or headphones (see Figure 7-16). More elaborate cards may provide:

► Two pairs of converters for stereo input and output

► A general-purpose Musical Instrument Digital Interface (MIDI) synthesizer

► MIDI input and output jacks

► A more powerful amplifier to drive larger speakers and generate more volume

Other differences among sound cards include degree of polyphony (the number of simultaneous sounds that can be generated), supported sampling rates, and the power and speed of embedded DSPs.

As stated earlier, sampling rates of 40,000 Hz or greater are required to capture and reproduce waveforms that sound natural to the human ear. Reduced sampling rates may be adequate when full range sound isn't required, such as in speech generation, or when processing power or disk space is limited. Halving the sampling rate also halves processing and storage requirements.

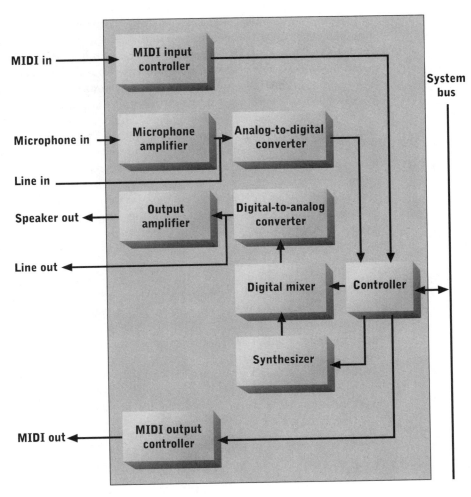

Figure 7-16 ►

Components and connections of a typical sound card

Musical Instrument Digital Interface (MIDI) is a standard for the storage and transport of control information among computers and electronic musical instruments. MIDI originally was designed so electronic keyboards could control remotely connected synthesizers. MIDI defines a serial communication channel and a standard set of 1- to 3-byte commands that can:

► Turn individual notes on and off

► Modulate pitch

► Mimic various musical instrument controls, such as the sustain pedal of a piano

► Select specific instruments or artificial sounds

Up to 16 channels of MIDI data can be sent over the same serial transmission line, so 16 separate instruments or instrument groups can be played at once.

Most sound cards include some MIDI capability. At minimum they provide a simple synthesizer capable of simulating the sound of a few dozen instruments. Files containing MIDI control data are transferred to the sound card over the system bus and then the synthesizer processes the commands and generates digital waveforms for each instrument. The digital waveforms are combined (mixed) and sent through the digital-to-analog converter, amplifier, and speakers.

More complex sound cards provide more synthesized instruments, better-quality waveforms, and the ability to accept or transmit MIDI data through external MIDI connections. The quality of synthesized sound varies widely from one synthesizer and/or sound card to another. The poorest-quality sounds are generated by inexpensive DSPs executing simple instrument simulation programs and operating at low sampling rates. The most realistic sounds are achieved with sample-based synthesis, which uses stored digitized recordings of actual instruments.

The primary advantage of MIDI is its compact storage format. A command to sound a single note on a piano for one second requires 6 bytes: 3 bytes for the note on and 3 bytes for the note off control messages. A 16-bit digital recording of that same sound sampled at 48,000 Hz requires 96,000 bytes of storage. The disadvantage of MIDI is a lack of control over the exact nature of the generated sound. A MIDI command selects instruments and notes, but doesn't control how instrument sounds are generated. Sound quality can vary widely among sound cards and synthesizers depending on sample rate, synthesis technique, and hardware capability. The sound for a given instrument may not even exist on the synthesizer receiving the MIDI commands.

SUMMARY

► Manual input devices include keyboards, mice, and other pointing devices. The predominant mechanism for human input to computer systems is a character keyboard. Devices used to capture graphic and image data include the mouse, trackball, joystick, and digitizer tablet. The position of the mouse, trackball, or digitizing pen on a surface is an analog to the position of a cursor or pointer on a video display.

► Display surfaces can be divided into rows and columns similar to a large table or matrix. Each cell, or pixel, in that table represents one simple component, or dot, of an image. Written Western languages are based on systems of symbols called characters. A collection of characters of similar style and appearance is called a font. Any color in the visible spectrum can be represented as a mixture of varying intensities of the additive (RGB) colors red, green, and blue or the subtractive (CMY) colors cyan, magenta, and yellow.

► A stored set of numeric pixel descriptions is called a bit map. Monochrome display uses a single bit to indicate whether each pixel is on or off. Multiple bits can represent varying levels of intensity, such as a grayscale. Three separate numbers or sets of bits are required to represent a pixel in full color. Significant drawbacks to the use of bit maps include their large size and device dependence. Compression can be used to reduce size. An image description language is another method of reducing image file size. Image description languages can compactly describe images using vectors, display objects, or both.

► Video display terminals consist of an integrated keyboard and television screen. A typical character-oriented VDT can display at least 80 columns and 24 rows. Characters normally are displayed using a matrix of pixels. Modern video displays are separate from keyboards and are called monitors. Monitors generally are implemented using cathode ray tubes (CRTs). Flat-panel display technologies include liquid crystal displays (LCDs) or electroluminescent (EL) displays. LCDs have less contrast and a reduced viewing angle as compared to CRT displays. Their advantages over CRTs include reduced size, weight, and power consumption.

► Commonly used paper output devices include dot matrix printers, ink-jet printers, laser printers, and plotters. Dot matrix printers are inexpensive but also are slow, noisy, and produce relatively poor quality output. Ink-jet printers produce excellent quality output but are relatively slow. Laser printers are relatively fast and produce excellent quality output. Modern plotters are similar to ink-jet printers except that they print on wide format paper.

► Optical input devices include optical scanners, mark sensors, bar code readers, and optical character recognition (OCR) devices. All use a laser or light reflected off a printed surface into a photodetector. A mark sensor scans for light or dark marks at specific locations on a page. A bar code is a series of vertical bars of varying thickness and spacing. Optical scanners generate bit map representations of printed images. Optical character recognition (OCR) devices combine optical scanning technology with intelligent interpretation of bit map content.

► General-purpose speech recognition systems can be used for command and control, or for the input of large amounts of textual material. Sound can be captured in analog form by a microphone, converted to digital form by digital sampling, and then compared to a library of patterns corresponding to known phonemes. Recognition of continuous speech is much more difficult. Digital signal processors commonly are used to address the computational complexity of speech recognition and complex sound generation.

► Sound cards include an analog-to-digital converter (ADC), a digital-to-analog converter (DAC), a low-power amplifier, and connectors for a microphone, speaker, or headphones. MIDI inputs, MIDI outputs, and a music synthesizer also can be provided.

This chapter discussed how I/O technology supports human/computer interaction. The next two chapters focus on another important class of I/O technology—data communications and computer networks. These technologies are used so computers can interact with each other and so users can interact simultaneously with multiple computer systems. Modern computer systems rely heavily on these technologies to deliver powerful information retrieval and processing capabilities to end users.

Key Terms

active matrix display
additive colors
analog-to-digital converter (ADC)
audio response unit
bar code scanner
bit map
cathode ray tube (CRT)
chromatic depth
chromatic resolution
CMY
CMYK
cursor
digital signal processor (DSP)
digital-to-analog converter (DAC)
digitizer
display attribute
display list
display object
dithering
dot matrix printer
dots per inch (dpi)

electroluminescent (EL) display
flat panel display
font
Graphics Interchange Format (GIF)
image description language
ink-jet printer
Joint Photographic Experts Group (JPEG)
laser printer
liquid crystal display (LCD)
mark sensor
monitors
monophonic
Moving Pictures Experts Group (MPEG)
mouse
multimedia controller
Musical Instrument Digital Interface (MIDI)
optical character recognition (OCR)
optical scanner

optical sensor
palette
passive matrix display
phoneme
plotter
point
polyphonic
PostScript
refresh rate
resolution
RGB
sampling
scan code
scanning laser
sound board
sound card
speaker dependent
speech synthesis
speech recognition
subtractive colors
terminal
vector
vector list
video display terminal (VDT)

Vocabulary Exercises

1. Color display can be achieved by using separate, but closely spaced, elements colored _____, _____, and _____.

2. The display device used in most computer monitors and televisions is a(n) _____.

3. A(n) _____ is used to control the location of the _____, or input pointer, on a video display.

4. A color display system that uses 8 bits to represent each primary video display color intensity is called _____.

5. The printing industry generally uses inks based on the _____ colors, which are _____, _____, and _____. A(n) _____ ink also can be used as a fourth color.

6. A(n) _____ is another name for a wide format printer.

7. _____ and _____ are commonly used display languages.

8. The print head of a(n) _____ forces liquid ink onto a page through a small nozzle.

9. The _____ format commonly is used to compress graphic images.

10. The _____ format commonly is used to compress motion video.

11. A(n) _____ sound card can generate multiple notes simultaneously. A(n) _____ sound card can generate only one note at a time.

12. The _____ of a display device is the number of colors that can be displayed simultaneously.

13. Graphic images generally are stored as a(n) _____, where each pixel is represented by one or more numbers.

14. General purpose sound cards use a(n) _____ to generate the signal sent to speakers, headphones, or an internal or external amplifier.

15. A(n) _____ is a primitive component of voiced human speech.

16. A(n) _____ display is similar in construction to a(n) _____ display except that it employs solid state elements for light generation.

17. A(n) _____ recognizes input in the form of alternating black and white vertical lines of varying width.

18. A(n) _____ converts analog sound waves to a digital representation.

19. A(n) _____ list differs from a(n) _____ list in that it can describe graphical objects other than straight lines.

20. An individual display element of a video display surface is called a(n) _____.

21. _____ speech recognition programs must be trained to recognize one person's voice.

22. The output resolution of a graphic display device such as a laser printer normally is measured in terms of _____.

23. An electronic keyboard generates a(n) _____, which can be translated into ASCII character output by a keyboard controller.

24. _____ and bar code scanners use a(n) _____ to detect marks on a page.

25. The size of an individual pixel displayed on a monitor is called the monitor's _____.

26. The _____ of a(n) _____ describes the number of times per second that the screen contents are redrawn.

27. A(n) _____ is one seventy-second of an inch.

28. A group of symbols of a similar style is called a(n) _____.

29. Text pages may be converted to bit maps by a(n) _____. These bits maps then are input to a(n) _____ program that generates an ASCII or Unicode representation of the text.

30. _____ creates a pattern of colored pixels that fools the eye into thinking a composite color is being displayed.

31. _____ is a standardized method of encoding notes and instruments for communication with synthesizers and sound cards.

Review Questions

1. Describe the process by which keystrokes are recognized by software.

2. What is a font? What is point size?

3. What are the additive colors? What are the subtractive colors? What types of I/O devices use each kind of color?

4. What is a bit map? How does the chromatic resolution of a bit map affect the size of a bit map image?

5. What is a display list? What is/are its advantages as compared to a bit map?

6. What is JPEG encoding? What is MPEG encoding?

7. What is/are the difference(s) between a video display and a monitor?

8. Describe the various technologies used to implement flat panel displays. What are their relative advantages and disadvantages?

9. How does the operation of a laser printer differ from a dot matrix printer?

10. Describe the various types of optical input devices? For what type of input is each device intended?

11. Describe the process of automated speech recognition? What types of interpretation errors are inherent in this process?

12. Describe the components and functions of a typical sound card. How is sound input captured? How is speech output generated? How is musical output generated?

Research Problems

1. Many organizations have begun to store documents as graphic images rather than encoded textual and numeric data. Examples of applications that use such storage methods include medical records, engineering data, and patents. Identify one or more commercial software products that support this form of data storage. Examine the differences between such products and more traditional data storage approaches. Concentrate on issues of efficient use of storage space, internal data representation, and methods of data search and retrieval. What unique capabilities are provided by image-based storage and retrieval? What additional costs and hardware capabilities are required? Examine sites such as http://www.belmont.com, http://www.psrw.com and http://www.virage.com.

2. Many computer scientists are researching the problem of textual data understanding. The visual recognition of character data and its conversion to ASCII or other coded representation is largely solved. However, basic understanding and classification of the textual data still is poorly developed. Examples of applications that could be built with such capabilities include automatic indexing, abstracting, and summarization of news articles or other large textual inputs. Several products that address these needs are emerging from research labs, and can be seen at http://ciir.cs.umass.edu, or can be seen at commercial and governmental domains, such as http://www.islanddata.com and http://thomas.loc.gov. Identify one or more of these products. What limitations must be placed on text input to ensure reasonable success in automated recognition? What methods are used to perform classification and summarization? What is the relative performance, or speed and error rate, of such systems compared to humans performing the same tasks? What traditional information processing applications might use this technology?

Chapter **8**

Data and Network Communication Technology

Chapter Goals

- ► Explain communication protocols

- ► Describe signals and the media used to transmit digital signals

- ► Compare and contrast methods of encoding and transmitting data using analog or digital signals

- ► Describe methods for efficiently using communication channels

- ► Describe methods for detecting and correcting data transmission errors

Successful and efficient communication is a complex endeavor. People have long labored with the intricacies of communicating concepts and data. The method of expression (such as words, pictures, and gestures), language syntax and semantics, and communication rules and conventions vary widely across cultures and contexts. This makes successful communication among people a long learning process covering all of these areas.

Understanding computer and network communication can be daunting as well because these topics also cover a broad range of complex interdependent concepts and technologies. Some of those topics were addressed in Chapter 7. This chapter extends and deepens that coverage, as shown in Figure 8-1, and lays the foundation for a detailed discussion of computer networks in Chapter 9.

Figure 8-1 ▶

Topics covered in this chapter

COMMUNICATION PROTOCOLS

A **message** is a unit of data or information transmitted from a sender to one or more recipients. A message can consist of a single byte or character such as pressing Esc to terminate a program, or many bytes such as downloading a graphic image from an Internet server. Messages can be categorized loosely into two types—data messages and command messages. Data messages vary in format and content and can include any of the primitive CPU data types or more complex data types (see Chapter 3). For purposes of computer communication, the content of a data message is not important. That is, the message simply is moved from one place to another with no attempt to understand its content.

A command message contains instructions that control some aspect of the communication process. Examples of command messages include many of the ASCII device control characters, addressing and routing instructions, and error detection and correction information. Command messages also can be used to transmit information about data messages such as format, content, length, and other information the receiver needs to successfully interpret the data message.

At the most primitive level, any message is simply a sequence of bits. Sender and receiver must agree on a common method of encoding, transmitting, and interpreting those bits. A **communication protocol** is a set of rules and conventions for communication. Although this definition sounds simple, it covers many complex details.

Consider the communication protocol that applies during a classroom lecture or presentation. First, all participants must agree on a common language. The grammatical and semantic rules of the language are part of the classroom communication protocol. For spoken languages, the air in the room serves as the transmission medium. Large classrooms might require auxiliary means, such as sound insulation or voice amplification equipment, to support adequate transmission through the medium.

Command and response sequences ensure efficient and accurate information flow. The instructor coordinates access to the shared transmission medium. Gestures such as raising a hand and pointing are used to ask for and grant permission to speak (transmit messages). Allowable messages are limited to those on specific topics. Messages that are off-topic or too long are ignored or explicitly cut off. Some rules and conventions ensure accurate transmission and reception of messages. For example, participants must speak loudly enough to be heard by all. The instructor might pause and look at the students' faces to determine whether the message is being received and interpreted correctly. The very act of entering the classroom signals participant acceptance of the communication protocol.

Communication among computers also uses complex communication protocols. Figure 8-2 presents a hierarchical organization of the components of a computer communication protocol. Each node of this hierarchy has many possible technical implementations. A complete communication protocol is a complex combination of subsidiary protocols and the technologies by which they are implemented. You might find it helpful to return to this hierarchy often while reading the remainder of this chapter so that topics can be placed in their proper context.

Figure 8-2 ▶

Components of a communication protocol

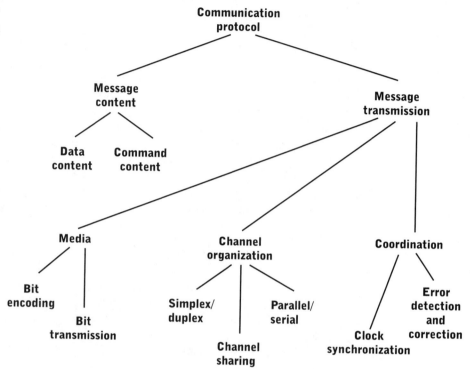

ENCODING AND TRANSMITTING BITS

Chapter 3 discussed various ways in which data items such as integers and characters are encoded within a bit string. This section describes how bits are represented and transported among computer systems and computer hardware components.

Carrier Waves

Light, radio frequencies, and electricity travel through space or cables as a **sine wave** (see Figure 8-3). The energy content of a sine wave varies continuously between positive and negative states. Three characteristics of a sine wave can be manipulated to represent data:

▶ Amplitude

▶ Phase

▶ Frequency

Figure 8-3 ▶

Characteristics of a sine wave

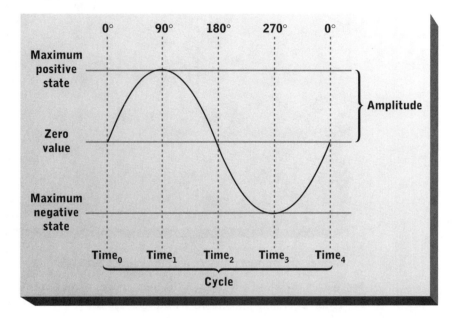

Amplitude is a measure of wave height or power—the maximum distance between a wave peak and its zero value. Note that amplitude is the same whether measured from zero to a positive peak or zero to a negative peak. A complete **cycle**, also called a period, of a sine wave follows its full travel from zero to positive peak, back to zero, to its negative peak, and then back to zero again.

Phase is a specific time point within a wave's cycle. Phase is measured in degrees, with 0° representing the beginning of the cycle and 360° representing the end. The point of positive peak is 90°, the zero point between a positive and negative peak is 180°, and the point of negative peak is 270°.

Frequency is the number of cycles that occur in one second. Frequency is measured in hertz (Hz) where hertz is interpreted as cycles per second. Figure 8-4 illustrates two sine waves with 2 Hz and 4 Hz frequencies.

Figure 8-4 ►

Sine waves with
frequencies of
2 Hz and 4 Hz

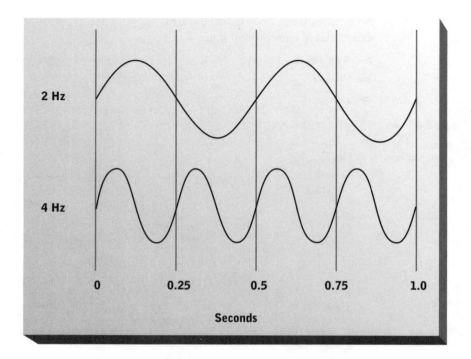

Waves are important in communications because:

► Waves travel, or propagate, through space, wires, and fibers.

► Patterns can be embedded within waves.

Bits are encoded within a wave by precisely manipulating, or **modulating**, amplitude, frequency, or phase. A wave with embedded bits is called a **carrier wave** because it transports, or carries, bits from one place to another. The receiver observes a modulated carrier wave characteristic and interprets the modulations as bit values. A **signal** is a specific data transmission event or group of events that represent a bit or group of bits.

As a simple example, consider ripples in a pond as carrier waves. Using a stick, the sender can generate waves of varying height that travel across the pond to a receiver on the other side. Strong strikes make large waves and weak strikes make smaller waves. A large wave can be interpreted as a one bit and a small wave as a zero bit. The opposite interpretation also can be used. It doesn't matter which interpretation is used as long as the sender and receiver use the same interpretation.

Data can be encoded as bits (large and small waves) by any shared coding method. For example, text messages could be encoded with Morse code (see Table 8-1) with one bits representing dashes and zero bits representing dots (see Figure 8-5). Data also could be encoded using ASCII, Unicode, or any of the other coding methods previously discussed in Chapter 3.

Table 8-1 ▶

Morse code

Letter	Code	Letter	Code	Char	Code
A	•—	N	—•	0	—————
B	—•••	O	———	1	•————
C	—•—•	P	•——•	2	••———
D	—••	Q	——•—	3	•••——
E	•	R	•—•	4	••••—
F	••—•	S	•••	5	•••••
G	——•	T	—	6	—••••
H	••••	U	••—	7	——•••
I	••	V	•••—	8	———••
J	•———	W	•——	9	————•
K	—•—	X	—••—	?	••——••
L	•—••	Y	—•——	.	•—•—•—
M	——	Z	——••	,	——••——

Figure 8-5 ▶

Wave amplitude representing Morse code dots and dashes

Dot   Dash

Modulation Methods

Amplitude modulation (AM) represents bit values as specific wave amplitudes. Amplitude modulation sometimes is called **amplitude shift keying (ASK)**. Figure 8-6 shows an amplitude modulation scheme using electricity where 1 volt represents a zero bit value and 10 volts represents a one bit value. Amplitude modulation holds frequency constant while varying amplitude to represent data. The amplitude must be maintained for at least one full wave cycle.

Figure 8-6 ▶

The bit string
11010001
encoded within a
carrier wave
using amplitude
modulation

Frequency modulation (FM) represents bit values by varying carrier wave frequency while holding amplitude constant. Frequency modulation also can be called **frequency shift keying** (FSK). Figure 8-7 shows a frequency modulation scheme where 2 Hz represents zero and 4 Hz represents one.

Figure 8-7 ▶

The bit string
11010001
encoded in a
carrier wave
using frequency
modulation

Phase is a fundamentally different wave characteristic than amplitude or frequency. The nature of a sine wave makes it impossible to hold phase to a constant value. However, phase can be used to represent data by making an instantaneous shift in the phase of a signal or by switching quickly between two signals of different phases. The sudden shift in signal phase can be detected and interpreted as data, as shown in Figure 8-8. This method of data encoding is called **phase shift modulation** or **phase shift keying (PSK)**.

Multilevel coding is a technique for embedding multiple bit values within a single wave characteristic, such as frequency or amplitude. Groups of bits are treated as a single unit for the purpose of signal encoding. For example, two bits can be combined into a single amplitude level if four different levels of the modulated wave characteristic are defined. Figure 8-9 illustrates a four-level amplitude modulation coding scheme. Bit pair values of 11, 10, 01, and 00 are represented by amplitude values of 8, 6, 4, and 2, respectively.

Figure 8-9 ►

The bit string 1101001 encoded in a carrier wave using 2-bit multilevel amplitude modulation

Analog Signals

An **analog signal** uses the full range of a carrier wave characteristic to encode continuous data values. The measured value of a wave characteristic either is equivalent to a data value or can be converted to a data value by a simple mathematical function. For example, assume that data is to be encoded as an analog signal using sound frequency. The numeric value 100 could be encoded by setting the wave frequency to 100 Hz. The number 9999.9 could be encoded by setting the wave frequency to 9999.9 Hz. Data values anywhere within the range of frequencies that can be generated and detected could be encoded within the signal.

Analog signals are continuous in nature. They can represent any data value within a continuum, or range, of values. In theory, the number of values that can be encoded is infinite. For example, even if the voltage range that can be sent through a wire is limited to 0 to 10 volts, there are an infinite number of possible voltages. Possible values within that range include 0 volts, 10 volts, and an infinite number of possibilities between 0 and 10, such as 4.1 volts, 4.001 volts, 4.00001 volts, and so forth. The number of different voltages within any range is limited only by the ability of the sender to generate them, the transport mechanism to carry them accurately, and the receiver to distinguish among them.

Digital Signals

A **digital signal** can contain one of a finite number of possible values. A more precise term is **discrete signal**, where discrete means a countable number of possible values. The modulation schemes in Figure 8-6 and Figure 8-7 can encode one of two different values in each signal event, and are called **binary signals**. Other possible digital signals include trinary (three values), quadrary (four values), and so forth. The modulation scheme in Figure 8-9 generates a quadrary signal.

Digital signals also can be generated by using a square wave instead of a sine wave as the carrier wave. A square wave contains abrupt amplitude shifts between two different values. Figure 8-10 shows data transmission using an electrical square wave with one bits represented by 5 volts and zero bits represented by no voltage. Square waves are generated by rapidly switching, or pulsing, an electrical or optical power source. Binary data transmission via square waves sometimes is called **pulse code modulation (PCM)** or **on-off keying (OOK)**.

Figure 8-10 ▶

The bit string 11010001 encoded in square waves (digital signals)

Square waves are the preferred method of digital data transmission over relatively short distances, for example, on a computer system bus. However, electrical square waves cannot be reliably transmitted over long distances, such as greater than a kilometer[1]. Power loss, electromagnetic interference, and noise generated within the wiring all combine to "round off" the normally sharp edges of a square wave. Pulses also tend to spread out due to interactions with the transmission medium. Without abrupt voltage level changes, receivers are unable to decode data values reliably.

[1] In most communication and network standards, distance is stated in metric units. One meter is approximately 39 inches, and one kilometer is approximately 0.625, or 5/8, of a mile.

Wave characteristics such as frequency and amplitude are not inherently discrete. For example, electrical voltage can be 0, 0.1, 0.001, 5, 10, 100, and many other values. To send a digital signal with electrical voltage, sender and receiver choose two different values, such as 0 and 5, to represent two different bit values. However, the voltage that reaches the receiver might not be exactly 0 or 5 for a number of reasons that will be discussed shortly. How should the receiver interpret voltage that doesn't precisely match the assigned values for zero and one bits? For example, should 3.2 volts be interpreted as a 0, 1, or nothing?

A digital signaling scheme defines a range of wave characteristic values to represent each bit value. For example, Figure 8-11 shows two ranges of voltage. Values within the lower range are interpreted as a zero bit. Values within the upper range are interpreted as a one bit. The dividing line between the two ranges is called a threshold. The sender encodes a bit value by sending a specific voltage, such as 0 volts for a zero bit and 10 volts for a one bit). The receiver interprets the voltage by comparing it to the threshold. If the received voltage is above the threshold then the signal is assumed to contain a one bit. If the voltage is below the threshold then the signal is assumed to contain a zero bit. In Figure 8-11, 0, 0.1, 1.8, and 2.4999 volts would each be interpreted as a zero bit.

Figure 8-11 ▶

A binary signaling method using voltage ranges

Signal Capacity and Errors

Two important differences between analog and digital signals are their message carrying capacity and susceptibility to error. Analog signals can carry a greater amount of information than digital signals within a fixed time interval. Higher data carrying capacity results from the large number of possible messages that can be encoded within an analog signal during one time period.

For example, assume the use of an analog signaling method using electrical voltage with a signal duration of one second. Assume further that the range of voltage that can be reliably transmitted on the wire is 0 to 127 volts, and that sender and receiver can distinguish 1 volt differences in voltage accurately. Under these assumptions, there are 128 different signals that can be transmitted during one second. This number would be much larger, theoretically approaching infinity, if sender and receiver were capable of distinguishing smaller differences in voltage.

Now consider a binary electrical signal under the previous assumptions, using 64 volts as the threshold. During any single transmission event the number of possible data values sent is only two (either a zero or a one). The message carrying capacity of the analog signal is 64 times greater, 128 possible values with analog divided by two possible values with binary, than the binary signal in this example.

To transmit more than two possible values with binary coding, sender and receiver combine adjacent binary signals to form a single data value. For example, the sender could transmit seven different binary signals in succession and the receiver could combine them to form one message. It is possible to transmit 128 different messages ($2^7 = 128$ possible combinations of 7 binary signals) during 7 consecutive time periods. But an analog signal could have communicated any of 128 different values in *each* time period.

Although analog signals have higher message carrying capacity than digital signals, they are much more susceptible to transmission error. If the mechanisms for encoding, transmitting, and decoding electrical analog signals were perfect, this would not be an issue. But errors always are possible. Electrical signals and devices are subject to noise and disruption because of electrical and magnetic disturbances. The noise that one hears on a telephone during a thunderstorm is an example of such a disturbance. Voltage, amperage, and other electrical wave characteristics can be altered easily by such interference.

Consider the following hypothetical scenario. Assume that your bank's computer communicates with its automated teller machines via analog electrical signals. You are in the process of making a $100 withdrawal. The value $100 is sent from the automated teller to the bank's computer by a signal of 100 millivolts. The computer checks your balance and decides that you have enough money so its sends an analog signal of 100 millivolts back to the automated teller as a confirmation. During the transmission of this signal, a bolt of lightning strikes, inducing a 2-volt (2,000 millivolt) surge in the wire carrying the signal. When the signal reaches the automated teller, it dutifully dispenses $2,000 dollars to you in clean crisp $10 and $20 bills.

Susceptibility of electrical signals to noise and interference never can be completely eliminated. Computer equipment is well shielded to prevent noise from interfering with internal signals but external communications are more difficult to protect. Also, errors can arise from resistance within wires or from the magnetic fields generated within a device, such as by a transformer or fan.

A digital electrical signal is not nearly as susceptible to noise and interference. Consider the previous example using digital encoding instead of analog encoding. If we use a threshold of 2.5 volts, 0 volts to represent zero, and 5 volts to represent one, almost 2.5 volts of noise can be tolerated without misinterpretation of a signal (see Figure 8-12). If a zero was being sent at the instant the lightning struck, the automated teller still would interpret the signal as a zero because the 2 volts induced by the lightning are below the threshold value of 2.5. If a one was being sent then the lightning would raise the voltage from 5 to 7 volts, still above the threshold. Resistance in the wire or other factors degrading voltage would not be a problem as long as the voltage did not drop more than 2.5 volts.

Figure 8-12 ▶

Margin of transmission error (voltage drop or surge) allowed before the data value encoded within a binary signal is altered

TRANSMISSION MEDIA

The communication path used to transport signals is called a **transmission medium**. Copper wire and optical fiber are two types of transmission media. The atmosphere or space also can be used as a transmission medium for radio or microwave transmissions. Messages must be encoded in signals that can be conducted or propagated through the transmission medium. In copper wires, signals are streams of electrons. In optical fiber, signals are pulses of light. Microwave and satellite transmissions are broadcast as **radio frequency** (**RF**) radiation through space.

A **communication channel** (see Figure 8-13) consists of a sending device, receiving device, and the transmission medium that connects them. In a less physical sense, it also includes the communication protocol(s) used in the channel. Figure 8-13 implies a relatively simple channel design, but most channels are of complex construction. A complex channel has multiple media segments using different signal types and communication protocols. As a message moves from one media segment to another, it must be altered to conform the next segment's signal characteristics and protocol.

Data and Network Communication Technology

Figure 8-13 ▶

Basic elements
of a
communication
channel

Consider an analog modem connection from a home computer to an Internet service provider (ISP). Digital electrical signals travel across the computer's system bus to an internal modem. The modem converts the signal to a limited bandwidth analog signal, which then travels through twisted pair copper wiring to the house telephone interface. From there it travels via lower gauge (thicker) copper wiring to the nearest telephone pole, and then to the local telephone switching center. The switching center converts the signal to a digital optical signal and routs it via fiber-optic cable to the ISP. Messages sent from the ISP to the home computer travel the same path in reverse.

Figure 8-14 ▶

Analog modem
connection to an
Internet service
provider

Several characteristics of transmission media affect their ability to transmit messages successfully and efficiently. These include:

▶ Speed and capacity
▶ Bandwidth
▶ Noise, distortion, and susceptibility to external interference

In general, transmission media are better when their speed, capacity, and bandwidth are high and their noise, distortion, and external interference are low. However, media with all of these desirable characteristics are costly. Reliability, efficiency, and cost also depend on the signaling method and communication protocol. Different combinations of transmission medium, signaling method, and protocol are cost-effective for different types of communication links. Those differences account for the variety of media, signaling methods, and protocols that comprise the communication channel between a modem and an ISP, or between any sender and receiver.

Speed and Capacity

The fundamental transmission speed limit for any medium is the rate at which a carrier wave propagates through the medium. Electrical signals travel through wires at close to the speed of light. Optical signals transmitted through fibers and radio frequencies transmitted through the atmosphere or space also travel at close to the speed of light. Raw speed varies little among commonly used transmission media. What does vary is length of the media, the ways in which multiple media segments are interconnected, and the rate at which bits are encoded in signals and are recognized by the receiver. It is these factors that account for transmission speed differences among different media.

Speed and capacity are interdependent. That is, a faster communication channel has a higher transmission capacity per time unit than a slower channel. Capacity is not solely a function of transmission speed. It also is affected by the relative efficiency with which the channel is used. Different communication protocols and different methods of dividing a channel to carry multiple signals alter transmission capacity.

The speed and data transmission capacity of a channel are jointly described as a data transfer rate. Several different types of data transfer rates can be stated. A **raw data transfer rate** is the maximum number of bits or bytes per second that the channel can carry. A raw data transfer rate ignores the effects of the communication protocol.

A net or **effective data transfer rate** describes the transmission capacity actually achieved with a particular communication protocol. It typically is less than the raw data transfer rate because most communication protocols use part of the channel capacity to do things other than transmit raw data. Examples include sending command messages and retransmitting data when errors are detected.

Frequency and Bandwidth

Carrier wave frequency is a fundamental measure of data carrying capacity. Carrier wave frequency limits data carrying capacity because a change in amplitude, frequency, or phase must be held constant for at least one wave cycle to be detected reliably by a receiver. If a single bit is encoded in each wave cycle, then the maximum raw data transfer rate is equivalent to the carrier wave frequency. If multiple bits are encoded in each wave cycle via multilevel coding, then the maximum raw data transfer rate is an integer multiple of the carrier wave frequency.

Figure 8-15 shows the range of electromagnetic frequencies and the terms that describe various subsets of that range. Frequencies in the **infrared** and visible light ranges are used most frequently in fiber-optic communication. Electrical signals propagated through wires are generally below 1 GHz. Fiber optics have higher potential data transmission capacity than electricity because light is a higher frequency carrier wave. However, note the word *potential* in the previous sentence. Achieving the bit transmission rates implied by carrier wave frequency requires encoding and decoding equipment that can operate at the carrier wave frequency. This is not possible with current optical switching technology.

Figure 8-15 ▶

Spectrum of electromagnetic frequency between 10^1 and 10^{19} Hz

The difference between the maximum and minimum frequencies of a signal is called the signal **bandwidth**. The difference between the maximum and minimum frequencies that can be propagated through a transmission medium is called the medium bandwidth. Maximum frequency and bandwidth limit the raw transmission capacity of a communication channel. For example, consider the frequency range of an analog voice-grade telephone channel, which is 300 Hz to 3,400 Hz.

The bandwidth of the channel is the difference between the highest and lowest frequencies that can be carried (3400 − 300 = 3100 Hz). In contrast, the frequency range of human speech is approximately 200 Hz to 5000 Hz, a band width of 4800 Hz; the frequency range of the human ear is approximately 20 Hz to 20,000 Hz, a bandwidth of 19,980 Hz. Analog telephone channels can transmit only an approximation of human speech because many high and low frequencies cannot be transmitted.

Modulator-demodulator (modem) technology is used to send digital signals over voice-grade telephone channels. Digital signals from the computer bus are encoded in an analog carrier wave. Early modulation techniques used four separate frequencies to carry bit values (see Table 8-2). The lowest of these frequencies (1070 Hz) created an effective upper limit on the speed of transmission. Assuming that a single bit is transmitted per signal cycle, the maximum transmission rate is 1070 bps.

Table 8-2 ▶

Frequency assignments for data transmission over 300 bps analog telephone lines

Mode	Signal Frequency (Hz)	Binary Value
Transmit	1070	0
Transmit	1270	1
Receive	2025	0
Receive	2225	1

Subsequent modem standards have exceeded this theoretical capacity substantially. But they have had to employ ever more complex schemes to do so, including:

▶ Hardware-based data compression

▶ Multilevel coding of up to 32 bits in a single carrier wave cycle

▶ An elaborate combination of amplitude and phase modulation

Current modem transmission rates range as high as 56,000 bps. But the 3100 Hz bandwidth of the voice-grade telephone channel is an effective barrier against higher transmission rates.

It is interesting to note that digital signaling methods using ordinary telephone wire can achieve transmission rates of up to 100 Mbps. Telephone wiring can transmit frequencies much higher than 3400 Hz. But older analog telephone switching equipment was designed to use only the 300 to 3400 Hz frequency band. Much of this equipment has been or is being replaced with modern digital switching equipment that converts the electrical voice signals to digital form. Bandwidth above 3400 Hz can be used to transmit digital signals as long as analog voice signals below 3400 Hz are filtered out. Modern digital subscriber lines (DSL) use this method to transmit analog voice and digital data signals over the same telephone line.

Signal-to-Noise Ratio

Within a communication channel, **noise** refers to any extraneous signals that might be interpreted incorrectly as data. Noise can be heard in many common household devices. Turn on a radio and tune it to a frequency (channel) on which no local broadcasting station is transmitting. The hissing sound is amplified background radio frequency noise. Noise and distortion also can be heard on any home stereo system. Set the amplifier of receiver input to a device that isn't turned on, such as a tape or CD player, and turn the volume up relatively high (be sure to turn it back down when you're finished!). The hissing sound is the amplified noise in the signal transmission and amplification circuitry. Any sound other than hissing represents distortion introduced by the amplifier, speakers, or cabling among them.

Noise can be introduced into copper, aluminum, and other wire types by **electromagnetic interference (EMI)**. EMI can be produced by a variety of sources, including electric motors, radio equipment, and nearby power transmission or communication lines. In an area that is dense with wires and cables, EMI problems are compounded because each wire can both radiate and respond to EMI.

Attenuation is a reduction in the strength (amplitude) of a signal as it passes through a transmission medium. It is caused by interactions between the signal energy and the transmission medium. Different types of signals and media cause different amounts of attenuation. Electrical resistance is the primary cause of electrical signal attenuation. Optical and radio signals also are subject to attenuation. Attenuation generally is proportional to the length of the medium. Doubling the length of the medium doubles the amount of attenuation.

Another source of errors in communication is **distortion**, or any characteristic of a communication channel that causes a data signal to be altered. The transmission medium itself is a primary source of distortion. Resonances within a transmission medium can amplify certain portions of a complex signal. Attenuation within any medium varies with signal frequency, distorting complex signals with many frequency components. Echoes and other unwanted artifacts can be introduced by the medium or by other components of the channel, such as an amplifier.

For a receiving device to interpret encoded data correctly, it must be able to distinguish the encoded bits from noise and distortion. Distinguishing valid signals from extraneous noise becomes increasingly difficult as transmission speed increases. The effective speed limit of any given channel is determined by the power of the message-carrying signal in relation to the power of the noise in the channel. This relationship is called the **signal-to-noise (S/N) ratio** of the channel. S/N ratio is measured at the receiving end of the channel.

As an example, consider listening to a speech. The difficulty or ease of understanding the speech is directly related to the speed at which it is delivered and to the volume (amplitude) of the speech in relation to background noise. Accurate reception of speech is impaired by sources of noise such as other people talking, an air conditioning fan, or a nearby construction project. The speaker can compensate for noise by increasing speech volume, thus increasing the signal-to-noise ratio. Accurate reception also is impaired if the speaker speaks too quickly. This is due to insufficient time for the listener to interpret one word or speech fragment before the next is received.

The S/N ratio incorporates the effects of signal attenuation and distortion. In Figure 8-16, the S/N ratio is positive for distances up to 5 kilometers. Successful transmission theoretically is impossible beyond 5 kilometers because the S/N ratio is negative.

Figure 8-16 ▶

S/N ratio as a function of distance for a hypothetical channel

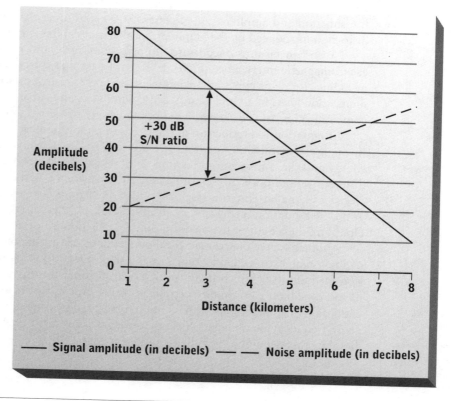

Data and Network Communication Technology

Each transmitted bit is a period of time during which a signal representing a zero or one is present on the channel. As transmission speed is increased, the duration of each bit in the signal, known as the **bit time**, decreases. If signal-generating equipment could generate a full amplitude signal instantaneously, then this would not be a problem. But no device, including the human voice, can go from zero amplitude to full power instantaneously. Short bit times don't give the signal-generating device sufficient time to "ramp up" to full power before the next bit must be transmitted. The S/N ratio decreases because the amplitude of each individual signal is decreased. Eventually, a limit is reached at which the bit time is so short that the signal is no louder than background noise. This is the point at which the S/N ratio is small or zero.

As higher speeds are attempted, the rate of errors increases. A transmission error represents a wasted opportunity to send a message, reducing the effective, or net, data transfer rate. A further difficulty is that noise usually is not constant. In electronic signals, noise usually occurs in short intermittent bursts, for example, due to a nearby lightning strike or the startup of an electric motor. As discussed later in this chapter, the receiver can request retransmission of a message if errors are detected. If noise bursts are infrequent, retransmissions might not diminish the overall data transfer rate significantly and a relatively high raw transmission speed could be used.

Electrical and Optical Cabling

Electrical signals usually are transmitted through copper wiring. Copper is used because it is relatively inexpensive, highly conductive, and easily formed into wire. Copper wiring can vary widely in construction details including gauge (diameter), insulation, purity and consistency, and number and configuration of conductors per cable. The two most common forms of copper wiring used for data communication are twisted pair and coaxial cable.

Twisted pair wire is the most common transmission medium for telephone and local area network connections. It contains two copper wires that are twisted around one another. The wires usually are encased, or shielded, by nonconductive material such as plastic. The primary advantages to twisted pair are its low cost and ease of installation. Its primary disadvantages are a high susceptibility to noise and limited transmission capacity. These are due to low bandwidth, generally less than 1 MHz, and a relatively low amplitude (voltage) limit.

Various standard forms of twisted pair wire and connectors are available. The Electronics Industries Association and the Telecommunications Industries Association have defined the most commonly used standards. The most widely implemented LAN wiring standard is called Category 5. A **Category 5** cable contains four twisted pairs and can transmit at speeds up to 1 Gbps. All four wire pairs are bundled in a single thin cable and standardized modular RJ-45 jacks similar to modular telephone connectors usually are used at both ends (see Figure 8-17).

Figure 8-17 ►

Common cable
connectors

RJ-45

Coaxial

**MT-RJ
(fiber-optic)**

Coaxial cable contains a single copper conductor surrounded by a thick plastic insulator, a metallic shield, and a tough outer plastic wrapping. Due to its heavy shielding, coaxial cable is very resistant to EMI. It has a relatively high bandwidth of up to 50 MHz and relatively high data transmission capacity. Coaxial cable is used widely for cable television because of its EMI resistance and its ability to carry many channels of analog video signals.

Its primary disadvantages are that it is more costly and harder to install than Category 5 twisted pair wire. Coaxial modular connectors (see Figure 8-17) are not as easily installed as RJ-45 connectors. Coaxial cable also is very stiff, making it more difficult to install than Category 5 wire. Coaxial cable is infrequently used in modern computer networks due to improving twisted pair quality and declining costs of both twisted pair and fiber-optic cable and devices.

Fiber-optic cable contains one or more strands of light-conducting filaments made of plastic or glass. Electrical shielding is not used because light waves neither generate nor respond to EMI. A tough, thin outer plastic coating protects the inner fibers from physical damage. Older fiber-optic cables called **multimode cables** surround the glass fibers with a reflective coating called cladding. The cladding acts as a mirror that bounces light waves back into the fiber core. Multimode cables split light waves into multiple signals, each of which travels a different path through the cable. Modern optical cables use fiber cores of varying density so that the core itself contains light waves. Maximum transmission speed is limited because light pulses spread out as they travel through the cable.

The highest quality optical cables are called **single-mode cables**. The fiber core is very narrow and constructed so all light waves travel a straight path down the center of the fiber. Single-mode cables can transmit data at much higher speeds than multimode cables, but they also are also much more expensive. Single-mode cable can transmit data at up to 2 Gbps, and higher speeds will be achieved as signal encoding and decoding devices improve.

The primary disadvantages of fiber-optic cable are its high cost and relatively difficult installation. Optical cable is much more expensive than Category 5 wiring. It is more expensive than coaxial cable, but is more often used because of its higher bandwidth and immunity to EMI. Until recently, optical cable was much more expensive to install than other cable types due to a lack of standardized connectors and lack of personnel trained in splicing techniques. Standardized connectors have been developed (see Figure 8-17) and installation cost will soon approximate that of twisted pair cable.

No cable type can carry signals more than a few dozen kilometers at high transmission rates with low error rates. Transmitting data over longer distances requires devices such as amplifiers and repeaters to increase signal strength and to remove unwanted noise and distortion.

An **amplifier** increases the strength, or amplitude, of a signal. In data communications, amplifiers can extend the range of an electrical signal by boosting signal power to overcome attenuation. However, the effective length over which the signal travels cannot be increased indefinitely. It is limited by two factors. The first is noise and interference introduced during transmission. An amplifier amplifies whatever signal is present at its input. If that signal includes noise along with the intended message, the noise is amplified as well as the message. In addition, amplifiers are never perfect. Some distortion or noise generally is introduced in the amplification process. A signal that is amplified many times will contain noise from multiple transmission line segments as well as distortion introduced by each stage of amplification.

A **repeater** fills much the same role as an amplifier, but operates on a different principle. It can be used with electrical or optical cable carrying analog or digital signals. Rather than amplifying whatever is sent to it, a repeater interprets the message it receives and retransmits it. A repeater does not retransmit noise or distortion. As long as the noise introduced since the last transmitting or repeating device does not cause a misinterpretation of the message, the retransmitted message will be the same as the original. Repeaters typically are required every 2 to 5 kilometers for coaxial cable and every 40 or 50 kilometers for single-made fiber-optic cable.

Wireless Data Transmission

Wireless transmission uses short wave radio or infrared light waves to transmit data through the atmosphere or space. The short wave radio spectrum covers frequency bands commonly used for FM radio, VHF and UHF broadcast television, cellular telephones, land-based microwave transmission, and satellite relay microwave transmission (Figure 8-15). Infrared transmissions occupy higher frequency bands that can carry more data but are more susceptible to atmospheric interference, limiting transmission distance to a few hundred meters at most.

The primary advantage of wireless transmission is its relatively high bandwidth, and the fact that it avoids costly wired infrastructure. Another advantage in some situations is broadcast transmission because many receivers can receive simultaneously. The primary weaknesses of wireless transmission are susceptibility to many forms of external interference and the cost of transmitting and receiving stations.

Wireless transmission frequencies are regulated heavily and regulations vary in different parts of the world. The United States, Europe, and Japan each have their own regulatory agencies. Different regulations apply to different transmission power levels. In general, low power transmission doesn't require a license, but the transmission equipment must be certified to meet appropriate regulations. Current wireless local area network technology uses unlicensed but regulated radio frequency (RF) bands. High power transmissions, such as long-distance microwave transmission, require certified transmitters and a license to use a specific frequency band. License costs are high because available bandwidth is a limited commodity.

Long-distance wireless networks seldom are implemented by a single user or organization because of the relatively high cost of long-distance wireless transmission equipment and licenses. Instead, companies are formed to purchase and maintain the required licenses and infrastructure and users purchase capacity from these companies on a fee-per-use or bandwidth-per-time-interval basis.

The Technology Focus on Wireless Data Communication that appears later in this chapter discusses wireless standards and encoding techniques.

CHANNEL ORGANIZATION

The configuration and organization of a communication channel has a direct impact on its cost and efficiency. Configuration and organization issues include the number of transmission wires or bandwidth assigned to each channel, the assignment of those wires to carry specific signals or signal components, and the sharing, or lack thereof, of channel components among multiple senders and receivers.

Simplex, Half Duplex, and Full Duplex

The most basic form of channel organization for electrical transmission through wires is shown in Figure 8-18. A single communication channel requires two wires—a **signal wire**, which carries data, and a **return wire**, which completes an electrical circuit between the sending and receiving devices. Optical transmission requires only a single optical fiber per channel because a complete circuit is not required. In some transmission modes, it may be necessary to provide two or more communication paths between sender and receiver. This type of channel organization is shown in Figure 8-19. If multiple electrical transmission lines are used they may share a single return wire to complete all electrical circuits.

Figure 8-18 ▶

Basic
configuration of
two-conductor
electrical
transmission line

Figure 8-19 ▶

Multichannel
communication
with two or more
signal wires and
one common
return wire

A single transmission line can be used in either simplex or half-duplex mode. In **simplex** mode, messages flow in only one direction, as shown in Figure 8-20(a). This is a useful communication mode when data flows in only one direction and when the chance of transmission error is small or irrelevant, such as in ordinary radio broadcasts. However, if transmission errors are common or if error correction is important, then simplex mode is inadequate. If the receiver detects transmission errors there is no way to notify the sender or to request retransmission. A typical use of simplex mode is to send file updates or system status messages from a host processor to distributed data storage devices within a network. In such cases, the same message is transmitted to all devices on the network simultaneously, or in **broadcast mode**.

Half-duplex mode uses a single shared channel. Each node takes turns using the transmission line to transmit and receive. The nodes must agree which node is going to transmit first. The transmitting node signals its intent to cease transmission by sending a special control message called a **line turnaround**. The receiver recognizes this message and subsequently assumes the role of transmitter. When its transmission is finished, it sends a line turnaround message and the original transmitter once again becomes the sender.

Half-duplex mode transmission allows the receiver to request retransmission of a message if it detects errors, as shown in Figure 8-20(b). If ASCII control characters are used, a **negative acknowledge (NAK)** character is sent if errors are detected and an **acknowledge (ACK)** control character is sent if no errors are detected. In half-duplex mode, receipt of NAK causes the sender to retransmit the preceding message after turning the line around again.

Figure 8-20 ▶

Configurations for simplex (a), half-duplex (b), and full-duplex (c) transmission modes

(a)

(b)

(c)

Communication line cost is the same in simplex and half-duplex modes. However, the added reliability of half-duplex mode is achieved at a sacrifice in overall data transfer rate. When the receiver detects an error, the receiver must wait until the sender transmits a line turnaround before it can send a NAK. If the error occurs near the beginning of a message, the transmission time used to send the remainder of the message is wasted. Also, because errors typically occur in bursts, many retransmissions of the same message might be required before the message is correctly received. If a channel is persistently noisy, half-duplex transmission will be relatively inefficient and slow.

The inefficiencies of half-duplex mode may be avoided by using two transmission lines, as shown in Figure 8-20(c). This two-channel organization permits **full-duplex** communication. In full-duplex mode, the receiver can communicate with the sender at any time by using the second transmission line, while the data transmission still is under way on the first line. If an error is sensed, the receiver

Data and Network Communication Technology

immediately can notify the sender, which can halt the data transmission and retransmit. The speed of full-duplex transmissions is relatively high even if the channel is noisy. Error bursts are corrected promptly, with minimal disruption to message flow. Improved speed is achieved at the cost of a second transmission line.

Parallel and Serial Transmission

Parallel transmission uses a separate transmission line for each bit position (see Figure 8-21). The width, or number of lines, of a parallel channel typically is one byte or one word, plus a common return line. Parallel communication is relatively expensive due to the cost of multiple transmission lines. Its comparative advantage is the higher data transfer rate that results from combining the capacity of multiple transmission lines.

Figure 8-21 ▶

Parallel transmission of a data byte (8 bits)

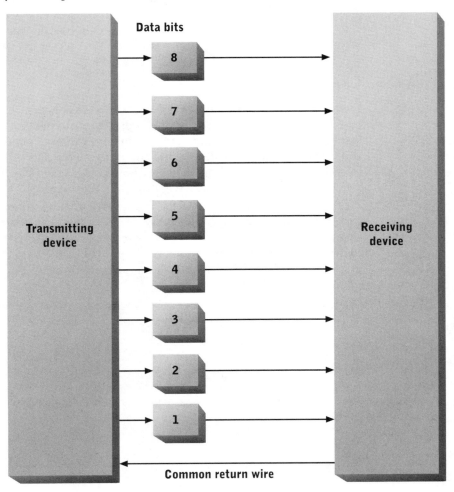

The maximum distance over which data can be sent reliably via parallel transmission is limited by an effect called **skew**. Because of slight differences among parallel transmission lines, data bits might arrive at the receiver at slightly different times. The timing difference, which is the skew, among bits increases with distance and transmission rate. At some point, usually less than 10 meters, skew is significant enough to cause errors in signal interpretation. Skew can be corrected by placing repeaters along the communication route but this adds to the expense of parallel channels. Parallel channels are used only over short distances where high data transfer rates are required, for example, a system bus or a connection between a workstation and a printer.

Serial transmission uses only a single transmission line or line pair for electrical signals. Bits are sent sequentially through the single transmission line and reassembled by the receiver into larger groups (see Figure 8-22). Digital communication over distances greater than a few meters usually uses serial transmission. Serial transmission avoids skew and minimizes the cost of wiring. LAN, WAN, and long-distance telecommunication lines all use serial transmission.

Channel Sharing

Few users require large amounts of data transmission capacity on a continuous basis. Rather, transmission capacity typically is needed for short periods, or bursts. Channel sharing techniques combine the traffic of multiple users and generally result in more efficient utilization of available data transfer capacity. As long as many users don't need large amounts of capacity at the same time, the combined load on the channel averages to an acceptable level.

Most local telephone service is based on a channel sharing strategy called **circuit switching**. When a user makes a telephone call, the full capacity of a channel between the user and the nearest telephone switching center is devoted to that call. The user has a continuous supply of data transfer capacity whether it is needed or not. The channel is unavailable to other users until the call is terminated.

Data and Network Communication Technology

Figure 8-22 ▶

Serial
transmission of a
data byte (8 bits)

Transmitting device

Receiving device

Common return wire

Signal wire

Time division multiplexing (TDM) describes any technique by which data transfer capacity is split into small time slices and allocated to multiple users and/or applications. Circuit switching is an example of TDM using this definition. **Packet switching** is the most common type of TDM. Packet switching divides messages from all users/applications into relatively small pieces called **packets** (see Figure 8-23). Each packet contains a header that identifies the sender, recipient, sequence number, and other information about its contents. Packets are routed to their destination as channel capacity becomes available.

Packets can be held temporarily in a buffer pending channel availability. If multiple channels are available, multiple packets can be sent simultaneously.

Figure 8-23 ►

Packet switching using time division multiplexing (TDM)

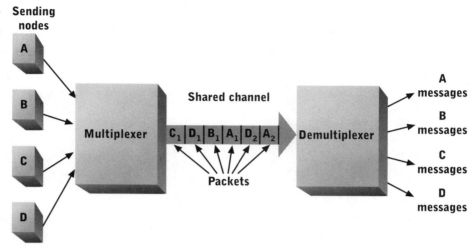

The primary advantage of packet switching is that the telecommunication service provider determines how to use available data transfer capacity and channels most effectively. In most situations, the provider can make rapid and automatic decisions that efficiently allocate available capacity among users. The result is substantially reduced total cost for telecommunication service.

The primary disadvantages of packet switching are uncertain delays in message transmission and the complexity of creating and routing packets. Because a user does not have a dedicated channel, the time required to send a message is unpredictable. It rises and falls with the total demand for the available data transfer capacity. Delays can be substantial when many users simultaneously send messages and insufficient capacity is available.

Packets add substantial complexity to a communication protocol. Messages must be broken into packet-sized chunks and appropriate headers must be created. The packets must be routed through the network to their intended destination. The receiver must reassemble packets into the original message in their intended sequence even if they arrive in a different order. Transmission errors must be localized to a specific packet and retransmission of that packet must be requested and completed. The hardware and system software required to perform these tasks are understandably complex.

Despite its complexities, packet switching is the dominant form of intercomputer communication. Its cost and complexity are more than offset by more efficient utilization of available communication channels. Circuit switching is used only in situations where data transfer delay and available data transfer capacity must be within precise and predictable limits.

Another method for sharing communication channels, as shown in Figure 8-24, is **frequency division multiplexing (FDM)**. Under FDM, a single **broadband** (high bandwidth) channel is partitioned into multiple **baseband** (low bandwidth) subchannels. Each subchannel represents a different frequency range, or **band,** within the overall broadband channel bandwidth. Signals are transmitted within each subchannel at a fixed frequency or narrow frequency range. Cable television (CATV) is the most common example of FDM. Long-distance telecommunication providers sometimes use FDM to multiplex single mode optical fibers.

Figure 8-24 ▶

Channel sharing using frequency division multiplexing (FDM)

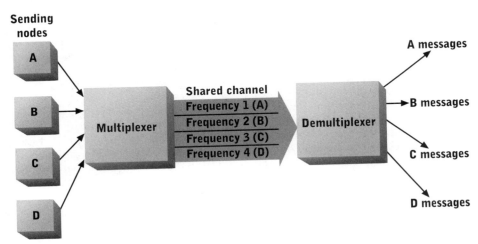

FDM might require moving signals originally intended for one frequency band into another frequency band. Cable television is an excellent example of this process. In most cable TV systems, channels 2 through 13 are carried across a coaxial cable at their original broadcast frequencies. Channels above 13 are remapped to frequencies other than their assigned broadcast frequencies so they occupy contiguous subchannels within the coaxial broadband channel.

Baseband channels within a single broadband channel can use different signaling methods, communication protocols, and transmission speeds. Baseband channels can be shared using packet switching or other time division multiplexing methods. Multiple baseband channels can be combined into a parallel transmission channel.

Technology
Focus
Asynchronous Transfer Mode

Asynchronous transfer mode (ATM) is a protocol for high-speed broadband data transmission. ATM was developed by a consortium of telecommunications companies as a standardized method for carrying any type of data including voice and video. To accommodate the real-time nature of voice and video data, ATM minimizes packet relay delays and guarantees minimum data transfer rates between senders and receivers.

There are separate ATM standards for twisted pair, coaxial, and fiber-optic cable. The twisted pair standard provides 25 Mbps over Category 5 cable with cable segments limited to 100 meters. The coaxial cable standard provides 155 Mbps and the multimode fiber-optic cable standard provides 622 Mbps. Senders, receivers, and cable segments are connected through hardware devices called ATM switches.

ATM uses time division multiplexing based on 53-byte packets called cells. Each cell carries 48 data bytes and a 5-byte cell header. Error detection information is included for the header but not the data. Data error checking was excluded to increase packet relay speed and because ATM designers assumed that highly reliable physical connections would be used. Network protocols that use ATM for physical data transport are free to add data error checking information to the normal data stream.

ATM establishes a specific path called a **virtual circuit (VC)** between sender and receiver before any data is transmitted, much like the connection established when dialing a telephone number. An ATM network allocates a guaranteed portion of its data transfer capacity to each VC and guarantees the availability of that capacity for the duration of the VC. The guarantee is called a **quality of service** level. Guaranteed data transfer capacity is well suited to continuous audio and video data transmission.

Creating a VC is a complex process. A VC creation request from a potential sender, including the intended recipient and desired quality of service level, first goes to the nearest switch. A complex series of requests then passes from switch to switch until the intended receiver is located. Each ATM switch stores information about currently defined VCs and unused transmission capacity with nearby switches. If sufficient uncommitted transmission capacity exists on all cable segments between sender and receiver, a new VC is created and each switch on the transmission path updates its internal information. If there is insufficient uncommitted transmission capacity, the VC creation request is denied or the request is granted at a lower quality of service level.

Once created, a VC is assigned a unique identifier. The identifier is added to the control message of each cell by the sender. Each switch stores VC routing information in an internal table and uses that information to send cells to their destination. Cells sent within a single VC are guaranteed to arrive in sequential order, which makes the task of reassembling them into larger data units such as video frames simpler and more efficient.

ATM is used by most long-distance telecommunication vendors. ATM has been less popular for computer networks. Standards for electrical and multimode fiber were not introduced until the mid-1990s. By that time, networks based on other standards such as Ethernet and Token Ring were well established. At present, ATM data transmission rates generally are slower than competing standards using equivalent transmission media. The ATM Forum has been slow to consider standard updates to improve transmission speed.

Cost also has been a major impediment. Until recently, ATM switches cost $10,000 or more and the cost of connecting a desktop computer typically was as much as the computer itself. Costs have dropped as more products have been brought to market and production quantities have increased. But ATM still is considerably more expensive than Ethernet at comparable speeds.

ATM may or may not become the connection of choice for local and wide area networks. Its success or failure will depend on several factors including comparative speed, cost, and service features. At this writing, ATM is the only well-established transmission standard that combines high speed, minimal propagation delay, support for multiple data types, and guaranteed quality of service levels. Will ATM improve its transmission speed and reduce its costs before competing standards add support for mixed data types and guaranteed service quality? Uncertainty over the answer to that question has caused many sleepless nights for investors, telecommunication managers, and system administrators.

CLOCK SYNCHRONIZATION

An important part of any communication protocol is a common transmission rate. Senders place bits onto a transmission line at precise intervals. Receivers examine the signal at or during specific time intervals to extract encoded bits. Sender and receiver must share a common timing reference and must use equivalent transmission rates or data cannot be extracted correctly from the signal. When sender and receiver share a common timing reference they are said to be synchronized.

There are two primary synchronization problems to address during message transmission:

► Keeping sender and receiver clocks synchronized during transmission

► Synchronizing the start of each message

Most computer and communication devices use internal clocks that generate voltage pulses at a specific rate when a specific input voltage is applied. Unfortunately, such clocks are not perfect. Timing pulse rates can vary up or down with temperature and with minor variations in the power input.

Minor timing fluctuations aren't problematic when all devices use the same clock. Most parallel transmission standards assign a separate transmission line to carry the sender's clock pulse. The receiver monitors the clock line continuously and uses it to ensure that it reads and interprets bits at the same rate that the sender encodes and transmits them.

With serial transmission, sending and receiving devices each have their own clock. Because the clocks are independent, there is no guarantee that their timing pulses are generated at exactly the same point in time. Most communication protocols transmit clock pulses from sender to receiver when a connection first is established. Even if perfect synchronization is achieved at that time, however, clocks subsequently can drift out of synchronization. As clocks drift out of synchronization, bit interpretation errors by the receiver become more likely.

An example of the errors resulting from unsynchronized clocks is illustrated in Figure 8-25. The receiving device clock pulses are being generated at a slower rate than the sender clock pulses, resulting in misalignment of bit time boundaries. In this example, the receiver receives fewer bits than were sent and incorrectly interprets several of them.

Figure 8-25 ▶

Communication errors resulting from lack of clock synchronization

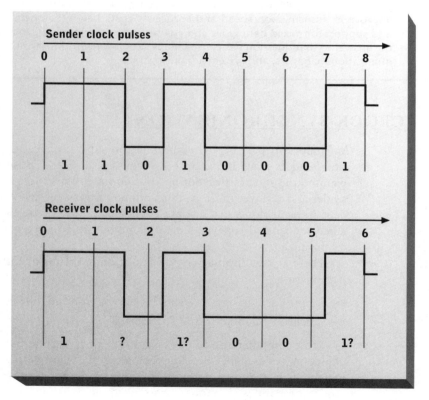

Sender and receiver occasionally must transmit timing signals over the data transmission line to ensure that their clocks are synchronized. They also must agree on the boundaries of each message. To do so they must agree on message length. Two approaches to synchronizing clocks and coordinating message boundaries are commonly used—synchronous transmission and asynchronous transmission. These methods are sometimes referred to as **character framing** methods when messages consist of ASCII or Unicode characters.

Synchronous transmission ensures that sender and receiver clocks are always synchronized by sending continuous data streams. Messages are transmitted in fixed-size byte groups called blocks (see Figure 8-26). Because block size is always the same, the receiver always knows where one block ends and another begins. If necessary, transmission blocks are separated by a continuous stream of **synchronous idle messages**. A synchronous idle message has a predetermined pattern of signal transitions designed for easy clock synchronization. The data start flag is an easily recognizable bit pattern that differs from the synchronous idle message and data stop flag bit patterns. Those differences are used by the receiver to detect the beginning of a new transmission block.

Figure 8-26 ▶

Typical format for messages to be transmitted using synchronous character framing techniques

In **asynchronous transmission**, messages are sent on an as-needed basis. Messages can be sent one after another or there can be periods of inactivity between messages. From the standpoint of the receiver, messages arrive at random or intermittent times. Clock drift is a significant problem in asynchronous transmission. During idle periods, sender and receiver clocks can drift out of synchronization because no data or timing signals are being transmitted. When the sender does transmit data, the receiver must resynchronize its clock quickly in order to interpret incoming signals correctly.

Asynchronous transmission appends one or more **start bits** to the beginning of each message. The start bit "wakes up" the receiver, informs it that a message is about to arrive, and allows the receiver to synchronize its clock before data bits arrive. Figure 8-27 shows a message containing a single byte preceded by a start bit. In network transmissions, dozens or hundreds of bytes usually are transmitted as a unit. Each byte group is preceded by one or more start bits.

Figure 8-27 ▶

Asynchronous
character
framing for serial
transmission,
including a
start bit

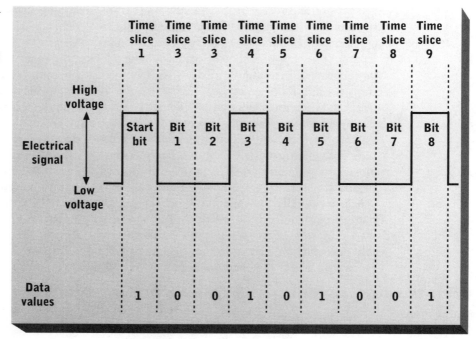

Synchronous transmission uses channel capacity more efficiently than asynchronous transmission because data is transmitted in large blocks with relatively few bits used to mark message boundaries and synchronize clocks. However, because transmission is continuous, synchronous transmission can't be used when multiple senders and receivers share the same channel. Asynchronous transmission is used in most computer networks, modem communication, and communication between computer systems and peripheral devices, such as keyboards and printers. Synchronous transmission is used only for high-speed communication between a dedicated sender and receiver, such as between two mirrored or clustered mainframe computer systems.

ERROR DETECTION AND CORRECTION

A crucial component of any communication protocol is a method or methods for detecting and correcting errors in data transmission, reception, or interpretation. All widely used error detection methods are based on some form of redundant transmission. A redundant message or message component is transmitted with or immediately after the original message. The receiver compares the original message with the redundant transmission and if the two match, the original message is assumed to have been transmitted, received, and interpreted correctly. If the two don't match, then a transmission error is assumed to have occurred and the receiver asks the sender to retransmit the message.

Error detection and correction methods vary in the:

- Size and content of the redundant transmission
- Efficient use of the communication channel
- Probability that an error will be detected
- Probability that an error-free message will be identified as an error
- Complexity of the error detection method

Size and content of the redundant transmission are inversely related to efficient use of the channel. For example, one possible error detection message is to send three copies of each message and verify that they are identical. This method is simple to implement but it uses the channel at only one-third of its capacity, or less if many errors are detected correctly or incorrectly. Changing the method to send only two copies of each method increases maximum channel utilization to 50 percent of capacity.

For most methods and channels, the probabilities of detecting errors can be mathematically or statistically computed. The probability of *not* detecting a real error is called **Type I error**. The probability of incorrectly identifying good data as an error is called **Type II error**. Type I and Type II error are inversely related. That is, a decrease in Type I error is accompanied by an increase in Type II error.

Type II errors result in needless retransmission of data that was correctly received but assumed to be in error. Increasing Type II error decreases the efficiency of channel utilization because a greater proportion of channel capacity is used to needlessly retransmit data.

In some types of communication channels, for example a system bus or a short-haul fiber-optic cable, the probability of a transmission or reception error is extremely remote. In other types of channels, for example high-speed analog modem channels, errors are common. Different methods of error detection are appropriate to different channels and different purposes. For example, error detection normally is not employed for digital voice transmissions because users are relatively insensitive to occasional bit errors. At the other extreme, communication between bank computers and automated teller machines over copper wires normally employs extensive error checking because of the importance of the data and because of the relatively high error rate in long-distance electrical transmission.

Commonly used methods of error detection include:

- Parity checking (vertical redundancy checking)
- Block checking (longitudinal redundancy checking)
- Cyclical redundancy checking

Each of these is described in detail in the sections that follow.

Parity Checking

Error detection for character data often is implemented through **parity checking** or **vertical redundancy checking**. In a character-oriented transmission, one parity bit is appended to each character. The **parity bit** value is a count of other bit values within the character. Parity checking can be based on even or odd bit counts.

With **odd parity**, the sender sets the value of the parity bit to zero if the number of data bits within the character with a value of one is odd. If the number of bits with a value of one is even, the parity bit is set to zero. Under **even parity**, the sender sets the parity bit to zero if the count of one bits is even or to one if the count of one bits is odd. The receiver counts the one bits in each character as they arrive and then compares the count against the parity bit. If the count doesn't match the parity bit, then the receiver asks the sender to retransmit the character.

Parity checking has high Type I error. For example, a transmission error that flipped the values of 2, 4, or 6 bits within an ASCII-7 character would not be detected. Parity checking is more reliable in channels with rare errors that usually are confined to widely spaced bits than in channels that are subject to error bursts affecting many adjacent bits.

Block Checking

Parity checking can be expanded to groups of characters or bytes using **block checking** or **longitudinal redundancy checking (LRC)**. To implement LRC, the sending device counts the number of one bits at each bit position within a block. The sender combines parity bits for each position into a **block check character (BCC)** and appends it to the end of the block (see Figure 8-28). The receiver counts one bits in each position and derives its own BCC to compare to the BCC transmitted by the sender. If the BCCs are not the same, the receiver requests retransmission of the entire block. In Figure 8-28, an even parity bit is computed for each bit position of a block of 8 bytes. The set of parity bits forms a block check character that is appended to the block for error detection.

LRC is vulnerable to the same types of errors as parity checking. Type I error can be reduced by combining parity checking and LRC. However, even with this approach, some compensating errors might go undetected.

Figure 8-28 ▶

Block checking
(longitudinal
redundancy
checking)

Bit positions

	0	1	2	3	4	5	6	7	
	1	0	1	0	0	0	1	0	1st byte
	1	0	0	0	1	0	1	0	2nd byte
	1	0	1	0	1	0	0	1	3rd byte
Data block	1	0	0	1	0	1	0	1	4th byte
	0	1	0	1	1	0	1	0	5th byte
	0	1	1	0	0	1	1	0	6th byte
	1	1	1	0	1	1	0	1	7th byte
	1	0	1	0	1	0	1	1	8th byte

Block check character (even parity)

0	1	1	0	1	1	1	0

Cyclic Redundancy Checking

Cyclic redundancy checking (CRC) is currently the most widely used error detection technique. Like LRC, CRC produces a block check character for a group of characters or bytes. The CRC character is generated by a complex mathematical algorithm. The CRC usually is more than 8 bits long and can be as large as 128 bits. CRC bit strings can be computed by software or by special-purpose microprocessors incorporated directly into data communication and network hardware.

CRC has much lower Type I and Type II errors than parity checking and LRC checking. Both error rates depend on the size of the transmitted data block and CRC bit string. CRC bit strings of 64 or 128 bits commonly are used within network packets and for magnetic tape data backups.

Technology
Focus

802.11 Wireless Network Standards

The Institute of Electrical and Electronics Engineers (IEEE) creates many network and telecommunication standards. (These standards will be discussed more fully in Chapter 9.) The 802.11 committee was formed in the early 1990s to develop standards for metropolitan and local area wireless networks. In

1997 the committee published the first version of the 802.11 standard. A second version (802.11b) followed in 1999 and a third (802.11g) currently is under development.

Some of the goals of the IEEE 802.11 standards are to:

► Meet the requirements of national and international regulatory bodies and pursue adoption by those bodies

► Support communication by radio waves and infrared or visible light

► Support stationary stations and mobile stations moving at vehicular speeds

To expedite its work, the committee initially focused on RF transmission in the 2.5 GHz band because RF waves can travel several miles and penetrate walls and other obstacles. The 2.5 GHz band was chosen because most of that band is regulated but unallocated in the United States, Europe, and Japan.

The first 802.11 standard defined two different RF transmission methods with a maximum raw data transfer rate of 2 Mbps. One of the transmission methods, **frequency hopping spread spectrum (FHSS)**, was abandoned in later standards because its transmission speed could not be increased within FCC mandated transmitter power limits. The other method, **direct sequence spread spectrum (DSSS)**, was enhanced in the 802.11b standard to provide raw data transmission at 5.5 and 11 Mbps.

Error detection is a major problem in any RF data transmission scheme. RF transmissions are subject to many types of interference, and mobile receivers encounter varying degrees of interference as they move. One of the most significant problems is multipath distortion, which occurs when an RF signal bounces off stationary objects (see Figure 8-29), causing the receiving antenna to receive multiple copies of the same signal at different times. Multipath distortion blurs or smears the signal content, causing bit detection errors. The problem is especially severe indoors where there are many hard flat surfaces that reflect RF waves.

The 802.11b standard divides the 2.4 GHz band into 14 channels, each with 22 MHz of bandwidth. Transmissions within each band encode bits using two-level, four-level, or eight-level phase shift modulation, which yields raw data transfer rates of 22, 44, and 88 Mbps, respectively. A substantial portion of the raw data transfer rate is used for redundant bit transmission to improve error detection. The four standard transmission speeds (1, 2, 5.5, and 11 Mbps) use different bit encoding and signal modulation methods, each representing a different trade-off among reliability, speed, and transmission distance. The fastest method uses eight-level modulation and encodes bits with minimal redundancy. Slower transmission uses two-level modulation and greater redundancy.

Transmitters continuously monitor signal quality and error rates. As noise, interference, and errors increase, or as the sender and receiver move further apart, transmitters change modulation and bit encoding methods to compensate. Transmitters step down through the 5.5, 2, and 1 Mbps transmission methods until reliable transmission is achieved and then move upward again as conditions improve.

Data and Network Communication Technology

The 802.11g standard will attempt to boost best-case throughput to at least 20 Mbps. Although the IEEE usually strives for backward compatibility, it might not be possible to achieve higher throughput by modifying current transmission methods in a compatible way. Our demands for ever faster and more reliable communication push standards toward ever more complex technologies. We reap the benefit of these improvements, but part of the price is rapid obsolescence of existing technology.

Business
Focus
Upgrading Network Capacity (Part I)

The Bradley Advertising Agency (BAA) specializes in designing and developing printed and video advertisements. Most of their work is performed under contract to other advertising agencies. BAA employs 30 people, most of whom are actively engaged in design and production work.

Printed ad copy is designed and produced almost entirely by computer. BAA uses Adobe Illustrator and other software on 15 networked IBM-compatible and Macintosh personal computers. A variety of peripheral devices are used, including several scanners, various types of color printers, and a high-speed color copier. A dedicated file server stores most data and software files on two

50 GB disks. The file server also has an 8 GB DAT tape drive, a 2 GB Jaz drive, and a 250 MB Zip drive.

Until 2000, BAA used video cameras and video tape editing equipment for video production projects. In 2000, BAA purchased two Sun Microsystems workstations, an expensive ($125,000) video editing software package, and several digital video cameras and recorders. BAA expects to add two more Sun workstations this year and two or three more the following year.

The files used to store print ads normally are 100 to 250 MB per project, but can range as high as 2 GB. BAA often has several dozen print projects in process at a time. Video production files are much larger than print files. Files for a typical 30 second television ad require 10 GB of storage.

File sharing performance already was slow when the first Sun workstations were purchased. A consultant determined that the file server was being used only at about 20 percent of its capacity and that the network itself was the performance bottleneck. The existing network is an Ethernet LAN using twisted pair wiring that was installed in 1993. The LAN originally operated at 10 Mbps but the network hub was upgraded to a 10/100 Mbps switch. The server, both Sun workstations, and eight desktop computers operate at 100 Mbps. The rest of the desktop computers operate at 10 Mbps.

Network bottlenecks have become more common since the two Sun workstations were acquired. Access to print ad files that took a few seconds a year or two ago now can take up to 5 minutes during busy periods. Recently, 36 GB disks were added to each Sun workstation so video files can be stored locally during editing, but downloading and uploading a project to or from the server can take 15 minutes or more. Project files sometimes are moved between Sun workstations several times per day if both machines are being used for the same project.

BAA is studying alternatives to upgrade its network capacity. BAA is unsure whether it should attempt to upgrade its network, relying on its existing cabling, or move to a more modern network with substantially higher performance. The two alternatives that seem most promising are upgrading to Gigabit Ethernet and a partial or total conversion to ATM. BAA is unsure whether its cabling will work with Gigabit Ethernet because it isn't certified as Category 5. An 155 Mbps ATM network also might work on the existing cable. An 622 Mbps ATM network would require fiber-optic wiring.

Questions:

1. If BAA's current cable is tested and determined to be reliable with Gigabit Ethernet, should the company choose that option?

2. If the existing cable is tested and determined to be incompatible with Gigabit Ethernet, should the company consider installing coaxial cable?

3. Assume that installing fiber-optic cable will cost three times more to purchase and install than Category 5 twisted pair cable ($60,000 vs. $20,000). Is fiber worth the extra money?

SUMMARY

► A communication protocol is a set of rules and conventions covering many communication aspects including message content and format, bit encoding, signal transmission, transmission medium, and channel organization. It also includes procedures for coordinating the flow of data, including media access, clock synchronization, and error detection and correction.

► Data bits can be encoded into analog or digital signals. Signals can be electrical, optical, or radio frequency. Bits are encoded in sine waves by modulating one or more wave parameters. Possible modulation techniques include frequency, amplitude, phase shift, and pulse code modulation.

► Important characteristics of transmission media include raw data transfer rate, bandwidth, and susceptibility to noise, distortion, external interference, and attenuation. Bandwidth is the difference between the highest and lowest frequencies that can be propagated through a transmission medium. High bandwidth channels can carry multiple messages simultaneously.

► The effective data transfer rate can be much less than the raw data transfer rate due to attenuation, distortion, and noise. Attenuation is loss of signal power as it travels through the transmission medium. Distortion is caused by interactions between the signal and the transmission medium. Noise can be generated internally or added through external interference. The signal to noise (S/N) ratio is a measure of the difference between noise power and signal power.

► Electrical cables are of two primary types—twisted pair and coaxial. Both use copper but differ in construction and shielding. Twisted pair is relatively cheap but limited in bandwidth, signal to noise ratio, and transmission speed due to weak shielding. Coaxial cable is more expensive but offers higher bandwidth, greater signal to noise ratio, and lower distortion.

► Optical cables are of two types—multimode and single mode. Single-mode cable provides higher transmission rates than multimode cable but at much greater cost. Optical cables provide very high bandwidth, little internally generated noise and distortion, and immunity to electromagnetic interference.

► Data can be transmitted without wires via radio waves and infrared light. Short-distance wireless transmission increasingly is used for local area networks. Long-distance radio frequency channels are in short supply due to their limited number and intense competition for licenses. Radio channels generally are leased from telecommunication companies on an as-needed basis. Infrared transmission has very limited transmission distance.

► Channel organization describes the number of lines dedicated to a channel and the assignment of specific signals to those channels. A simplex channel uses one optical fiber or copper wire pair to transmit data in one direction only. A half-duplex channel is identical to a simplex channel but sends a specific control signal that reverses the direction of transmission. Full duplex channels use two fibers or wire pairs to allow simultaneous transmission in both directions.

▶ Parallel transmission uses multiple lines to send several bits per signal event. Serial transmission uses a single line to send one bit at a time. Parallel transmission provides higher channel throughput but it is unreliable over distances greater than a few meters due to an effect called skew. Serial transmission is reliable over much longer distances. Serial channels are cheaper to implement because fewer wires or wireless channels are used.

▶ Channels often are shared among users and applications when no one user or application needs a continuous supply of data transfer capacity. Circuit switching is a form of time division multiplexing where an entire channel is allocated to a single user for the duration of one data transfer operation. Packet switching allocates time slices on the channel by dividing many messages into smaller units called packets and intermixing them during transmission. Frequency division multiplexing divides a broadband channel into several baseband channels.

▶ Sender and receiver must synchronize clocks to ensure that they use the same time periods and boundaries to encode and decode bit values. A single shared clock is the most reliable synchronization method but it requires that clock pulses be sent continuously from sender to receiver. Asynchronous transmission relies on specific start and stop signals (usually a single character) to indicate the beginning and end of a message unit. Synchronous transmission maintains a continuous flow of signals from sender to receiver to provide continual opportunities for clock synchronization.

▶ Error detection always is based on some form of redundant transmission. The sender compares redundant copies of messages and assumes that they are correct if they match. The receiver requests retransmission if they don't match. Increasing the level of redundancy increases the chance of detecting errors at the expense of reducing channel throughput. Common error detection schemes include parity checking (vertical redundancy checking), block checking (longitudinal redundancy checking), and cyclical redundancy checking.

At this point, you have a thorough understanding of how data is encoded and transmitted among computer systems. Data communication technology is the foundation of computer networks but other software and hardware technologies are also required. In Chapter 9, we'll discuss computer network hardware, protocols, and architecture. In Chapter 13, we'll discuss system software that supports distributed resources and applications.

Key Terms

acknowledge (ACK)
amplifier
amplitude shift keying (ASK)
amplitude
amplitude modulation (AM)
analog signal
asynchronous transfer mode
   (ATM)
asynchronous transmission
attenuation
band
bandwidth
baseband
binary signal
bit time
block checking
block check character (BCC)
broadband
broadcast mode
carrier wave
Category 5
character framing
circuit switching
coaxial cable
communication channel
communication protocol
cycle
cyclic redundancy checking
   (CRC)
digital signal
direct sequence spread
   spectrum (DSSS)

discrete signal
distortion
effective data transfer rate
electromagnetic interference
   (EMI)
even parity
fiber-optic cable
frequency
frequency division multiplexing
   (FDM)
frequency hopping spread
   spectrum (FHSS)
frequency shift keying (FSK)
frequency modulation (FM)
full-duplex
half-duplex
infrared
line turnaround
longitudinal redundancy
   checking (LRC)
message
modulating
modulator-demodulator
   (modem)
multilevel coding
multimode cable
negative acknowledge (NAK)
noise
odd parity
on-off keying (OOK)
packet switching
packet

parallel transmission
parity bit
parity checking
phase
phase shift keying (PSK)
phase shift modulation
pulse code modulation (PCM)
quality of service
radio frequency (RF)
raw data transfer rate
repeater
return wire
serial transmission
signal
signal wire
signal-to-noise ratio (S/N ratio)
simplex
sine wave
single-mode cable
skew
start bits
synchronous idle message
synchronous transmission
time division multiplexing
   (TDM)
transmission medium
twisted pair wire
Type I error
Type II error
virtual circuit (VC)

Vocabulary Exercises

1. _____ transmission sends bits one at a time using a single transmission line.

2. During half-duplex transmission, sender and receiver exchange roles after a(n) _____ is transmitted.

3. _____ encodes data by varying the distance between wave peaks within an analog signal.

4. A(n) _____ converts a digital signal to an analog signal so that digital data can be transmitted over analog telephone lines.

5. In synchronous data transmission, a(n) _____ signal is transmitted continuously during periods when no data is being transmitted.

6. _____ transmission divides the capacity of a transmission medium to implement multiple channels.

7. _____ cabling meets an industry standard that specifies four twisted wire pairs and can support data transmission rates up to 1 Gbps.

8. The _____ of a sine wave is measured in hertz.

9. A local telephone grid uses _____ switching to route messages from sender to receiver.

10. Most networks use _____ switching to route messages from sender to receiver.

11. In _____, a bit is appended to each character or byte and the bit value is determined by counting the number of 1 bits.

12. A(n) _____ signal is a discrete signal that can encode only two possible values.

13. A(n) _____ wave carries encoded data through a transmission medium.

14. With parity checking, sender and receiver must agree whether error detection will be based on _____ or _____ .

15. The _____ of a channel describes the mathematical relationship between noise power and signal power.

16. _____ is any change in a signal characteristic caused by components of the communication channel.

17. Methods of error detection and correction represent specific trade-offs among effective data transfer rate, processing complexity, _____ , and _____ .

18. Wireless transmission uses the _____ or _____ frequency bands.

19. _____ cannot affect optical signals but can affect electrical or RF signals.

20. A communication channel that uses electrical signals must have at least two wires—a(n) _____ and a(n) _____ to form a complete electrical circuit.

21. _____ measures the theoretical capacity of a channel. _____ measures the actual capacity of a channel when a specific communication protocol is employed.

22. Simultaneous or interleaved transmission of multiple messages on a single transmission line can be accomplished by _____ or _____ multiplexing.

23. A(n) _____ signal can encode an infinite number of possible numeric values.

24. _____ is a measure of peak signal strength.

25. In asynchronous transmission, at least one _____ is appended to the beginning of each character or block.

26. The term _____ describes the encoding of data as variations in one or more physical parameters of a signal.

27. In _____ transmission, blocks or characters arrive at unpredictable times, and no signal is transmitted during idle periods.

28. The _____ of a communication channel is the difference between the highest and lowest frequencies that can be transmitted.

29. _____ transmission implements two-way transmission with two separate communication channels; _____ transmission implements two-way transmission with only one communication channel.

30. _____ encodes data by varying the magnitude of wave peaks within an analog signal.

31. _____ transmission uses multiple lines to send multiple bits simultaneously.

32. A(n) _____ extends the range of data transmission by retransmitting a signal.

33. _____ generates a(n) _____ consisting of a single parity bit for each bit position within the bytes of a block.

34. In _____ transmission, signals are transmitted continuously, even when there is no data to send, to ensure clock synchronization.

35. _____ uses more than two signal parameter levels to encode multiple bits within a single signal event.

36. _____ encodes bit values in brief bursts of electrical or optical power.

37. _____ cabling is generally the least expensive alternative for network wiring.

38. A(n) _____ increases the strength of a signal, including both the encoded message and any noise that may be present.

39. The length of a parallel communication channel is limited by _____, which can cause bits to arrive at slightly different times.

40. A receiver informs a sender that data was received correctly by sending a(n) _____ message. It informs the sender of a transmission or reception error by sending a(n) _____ message.

41. _____ is used in some networks (for example, cable television networks) as a transmission medium.

42. _____ is a term that describes loss of signal strength as it travels through a transmission medium.

43. Messages to be transmitted by time division multiplexing are broken up into _____ prior to physical transmission.

44. A(n) _____ is a portion of data transmission capacity between a sender and receiver that is dedicated to guarantee a minimum quality of service level.

Review Questions

1. What are the components of a communication channel?

2. What are the components of a communication protocol?

3. Describe frequency, amplitude, phase shift, and pulse code modulation.

4. How is multilevel coding used to increase the effective data transfer rate of a communication channel?

5. Why aren't square waves usually used to transmit digital signals over long distances, for example, dozens or hundreds of kilometers?

6. Describe the relationship among bandwidth, data transfer rate, and signal frequency.

7. How can noise and distortion be introduced into a transmission medium? How does a channel's signal-to-noise ratio affect the reliability of data transmission?

8. Compare and contrast twisted pair, coaxial, multimode fiber-optic, and single mode fiber-optic cable in terms of raw data transfer rate, cost, and maximum cable segment length.

9. What are the advantages of wireless transmission using RF waves as compared to infrared waves?

10. Describe simplex, half-duplex, and full-duplex transmission. Compare and contrast them in terms of cost and effective data transfer rate.

11. Why will the actual data transfer rate of a channel usually be less than the theoretical maximum implied by the technology used to implement the channel?

12. Compare and contrast serial and parallel transmission in terms of channel cost, data transfer rate, and suitability for long-distance data communication.

13. What are the differences between synchronous and asynchronous data transmission?

14. What is character framing? Why is it generally not an issue in parallel data transmission?

15. Describe the differences between even and odd parity checking.

16. What is a block check character? How is it derived and used?

17. Compare and contrast frequency and time division multiplexing. What physical characteristics of the communication channel are required by each type? Which provides greater data transmission capacity?

18. What is the difference between an amplifier and a repeater?

Problems and Exercises

1. Assume that data is transmitted on a dedicated channel in blocks of 48 bytes at 10 Mbps. Also, assume that an 8-bit LRC check character is used for error detection and that an error is detected, on average, once every 1,000 transmitted blocks. What is the effective data transfer rate of the channel?

2. Assume that data is transmitted over twisted pair cable. The power of the signal when placed on the channel is 75 dB. Signal attenuation is 5 dB per hundred meters. Average noise power on cable is 0.1 dB per meter. What is the signal-to-noise ratio of a 50 meter channel? What is the signal-to-noise ratio of a 1 kilometer channel? What is the maximum channel length?

Research Problems

1. Parity checking often is used in computer memory to detect errors. Investigate currently available memory types, including parity and error correcting circuits (ECC) memory. How is parity checking implemented within the memory module? How are errors detected and what happens when an error is detected? What types of computer systems should use parity or ECC memory and why?

2. Many leading computer companies, including Hewlett-Packard, IBM, Intel, and Sun Microsystems, have created a proposed standard for computer peripheral interconnection called the InfiniBand® Architecture (www.infinibandta.org). Investigate the data communication aspects of the architecture. What signal encoding, cables, and connectors are used? Is the channel serial or parallel? What is the maximum channel length? Is transmission synchronous or asynchronous? How are errors detected and corrected?

Chapter 9

Computer Networks

Chapter Goals

► Compare and contrast bus, ring, and star network topologies

► Describe packet routing across local and wide area networks

► Compare and contrast token passing and CSMA/CD media access control protocols

► Describe network hardware devices, including network interface units, bridges, routers, and switches

► Describe the OSI network model, the TCP/IP protocol suite, and IEEE network standards

In the previous chapter, we discussed data communication between a single sender and a single receiver, including data transfer among hardware devices within a computer system, between a computer system and a peripheral device such as a printer, and between two directly connected computer systems. Communication among computer systems in a network is much more complex. Figure 9-1 shows the additional complexities that are addressed in this chapter.

Figure 9-1 ▶

Topics covered in this chapter

NETWORK TOPOLOGY

The term **network topology** refers to the spatial organization of network devices, physical routing of network cabling, and flow of messages from one network node to another. A network node is a device, such as a computer system, printer, or router, attached to a network that can send or receive messages. Physically linking two nearby nodes usually is straightforward. A point-to-point transmission line is laid over the shortest path and connected directly to both nodes.

Using a point-to-point network topology for more than a few nodes quickly becomes very complex. To connect four nodes, six transmission lines and three line connections per node are required. As the number of nodes increases, the number of point-to-point transmission lines rises quickly (see Figure 9-2(a)), as described by the formula:

$$(n-1)! = 1+2+3+ \ldots + (n-1)$$

(a)

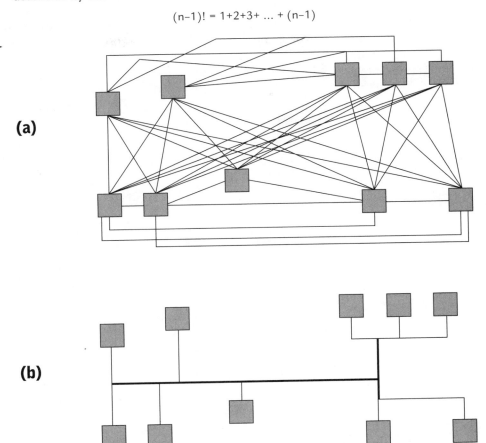

(b)

Point-to-point connections among all nodes are impractical for all but very small networks. For larger networks, it is more efficient to use a segmented approach that shares links among three or more nodes. Note the similarity of Figure 9-2(b) to a modern roadway system. Individual nodes are served by low-capacity links that connect to higher-capacity shared links in the same manner that driveways connect houses to residential streets. Smaller shared links are connected to larger shared links in the same manner that residential streets connect to higher-capacity roads and highways.

If the nodes in Figure 9-2(b) were parts factories and assembly plants, and the links among them were roads, how would you design a delivery service to move goods among them? You could buy a large number of trucks and use each truck to deliver goods between two specific locations. In the worst case of goods needing to flow between every pair of nodes, you would need (n-1)! trucks. How could you efficiently move the goods with fewer trucks?

Most large-scale delivery services use a **store and forward** system to connect source and destination nodes. Figure 9-3 shows the network in Figure 9-2(b) with three small-capacity and one large-capacity store and forward nodes. In a physical delivery system, the store and forward nodes are transfer points located at or near the junction of major roadways or air routes. Shipments move from end nodes to the nearest transfer point where they are combined into larger shipments to other transfer points.

Figure 9-3 ▶

Shared connections and store and forward nodes

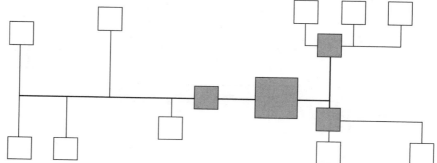

Computer networks and shipping networks have many similarities. Both connect large numbers of nodes and use a store and forward system. Efficiency is paramount for both network types to ensure reliable and rapid movement of shipments or messages among nodes with minimal investment in network infrastructure. Both network types achieve reliability and cost-efficiency by carefully organizing source and target nodes, shipping or transmission routes, and store and forward nodes.

The network topology is an essential factor in the efficiency and reliability of computer networks. Network topology can be referred to in physical or logical terms. **Physical topology** refers to the physical placement of cables and device connections to those cables. **Logical topology** refers to the path that messages traverse as they travel from node to node. Logical topology can be different from the underlying physical topology, as later examples will illustrate.

There are three common network topologies:

▶ Star

▶ Bus

▶ Ring

The characteristics that differentiate these topologies include the length and routing of network cable, type of node connections, data transfer performance, and susceptibility of the network to failure.

The **star topology** uses a central node to which all other nodes are connected (see Figure 9-4). The central node can be a transfer point, in which case it relays messages among other nodes. The central node also can be a node in its own right, such as a mainframe computer, in which case it accepts messages addressed to it and relays messages addressed to other nodes.

Figure 9-4 ▶

Star topology

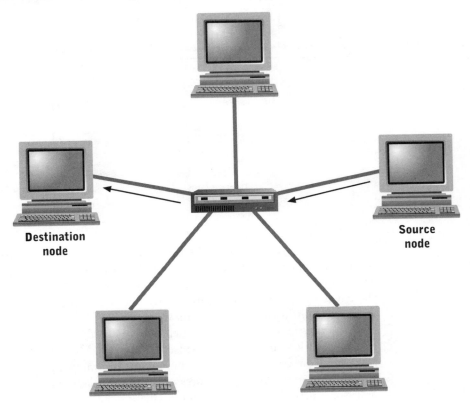

Destination node

Source node

The primary advantage of the star topology is relatively simple wiring. A transmission line is routed from each node to a hub (a network device discussed later in this chapter), which is typically located in a central location on a floor or within a building. The primary disadvantage of the star topology is that failure of the hub disables the entire network.

The **bus topology** connects each node to a common transmission line (see Figure 9-5). Transmitted messages travel from a node across the common transmission line to all other attached nodes. Each node listens for incoming messages and ignores those not addressed to itself. Terminating resistors are located at each end of the bus to absorb signals and prevent them from bouncing back down the common transmission line.

Figure 9-5 ►

Bus topology

Destination node

Source node

The primary advantages of the bus topology are relatively simple wiring and low susceptibility to failure. The bus is routed from one end of a floor or building to the other end in an easily accessible way. Each node has a transmission line routed directly to the bus. There is no central node that can fail and disable the entire network. Also, failure of any one node does not prevent other nodes from transmitting and receiving messages.

The **ring topology** connects each network node to two other nodes and the entire network forms a closed loop (see Figure 9-6). Messages are passed around the ring in one direction. Each node monitors the incoming link and removes messages addressed to itself. Messages addressed to other nodes are repeated on the outgoing link. Some networks use two rings carrying messages in opposite directions.

The primary advantages of the ring topology are long maximum network length and low susceptibility to noise and distortion. Because each node repeats outgoing messages, there is little degradation in signal quality as a message traverses the ring. The disadvantages of the ring topology are susceptibility to failure and difficulty in adding, deleting, or moving nodes. Rings are susceptible to failure because every node is an active repeater and any node failure breaks the ring. Reconfiguring the network is relatively difficult compared to other topologies because each node is connected to two others. Two links must be changed when adding, deleting, or moving a network node.

Figure 9-6 ▶

Ring topology

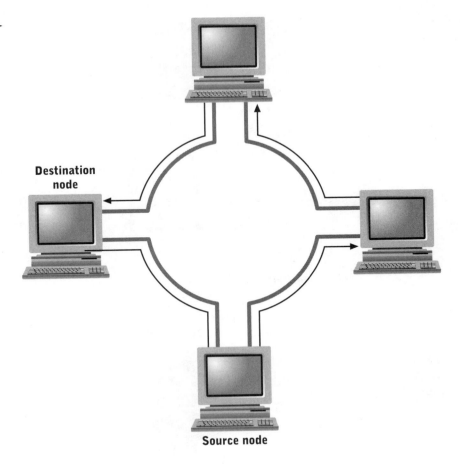

It is possible to combine the strengths of two different topologies by using one topology for physical layout and another for message routing. Star topology is the most common physical topology because it is the simplest to wire and the easiest to reconfigure. A physical star network can be a logical bus or logical ring network. The logical topology of the network is implemented within the central node. For a star topology bus network, each central node connection actually connects to an internal bus (see Figure 9-7). For a star topology ring network, each central node connector provides two connections—one for messages transmitted to the network node and another for messages transmitted by the network node. The central node internally switches the outgoing connection of one node to the incoming connection of the next node (see Figure 9-8).

Figure 9-7 ▶

Star topology
bus network

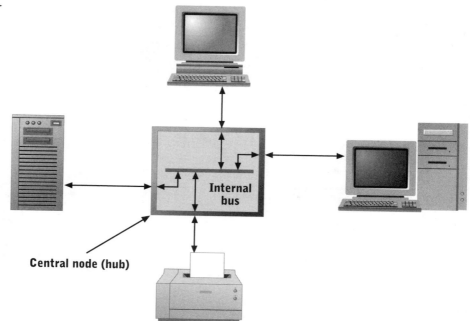

Internal bus

Central node (hub)

Figure 9-8 ▶

Star topology
ring network

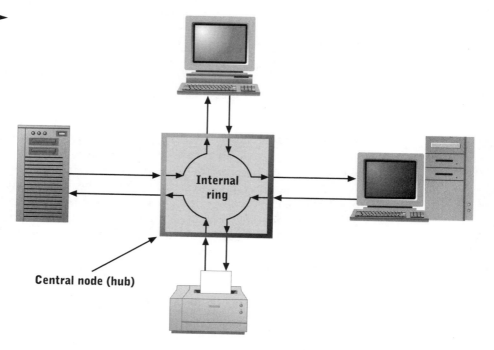

Internal ring

Central node (hub)

ADDRESSING AND ROUTING

Consider a typical college or corporate campus network (Figure 9-9). Entire buildings, such as Jones Hall, or building floors, such as Smith Hall, are organized as logical rings or buses and wired as physical stars. Each network covering a floor or building is called a **local area network (LAN)**.

All hubs in a group of buildings are connected by a zone network. Zone networks are connected by routers to the campus backbone network that is connected by a router to the Internet. Each hub and router acts as a bridge between two networks. (Hubs, routers, and bridges are discussed in the Network Hardware section later in this chapter.) The entire campus network is a **wide area network (WAN)**, including LANs, zone networks, the campus backbone network, and routers.

Messages can travel through many transmission lines and hubs before reaching their destination. Each message includes a destination address that each hub or router examines to determine where to send the message. But how does a hub or router "know" the addresses of its own nodes and how to forward a message to a distant address?

Local Area Network Routing

Each time a node is started, it sends a message announcing its presence and its address to the nearest hub. Each hub maintains a table of addresses and transmission lines or connection ports and uses that table to make routing decisions.

The exact procedure for routing a message between two local nodes depends on the logical network topology. For logical bus networks such as the floors of Smith Hall in Figure 9-9, the hub examines the destination address to see if it is local. If the address is local, the hub permits the message to propagate to all other local nodes. If the address is not local, the hub forwards the message to the next hub or router.

Routing within logical ring networks is performed in a similar manner. Messages to local addresses are repeated around the local ring and messages to distant addresses are forwarded to the next hub or router. For both bus and ring logical topologies, the hub needs to distinguish only between local and nonlocal addresses to determine whether to forward the message.

Routing of local messages within a logical star network is more complex. Because each node is attached to its own transmission line, the hub must keep track of which transmission line is connected to which address. A local message is routed from the source node's transmission line to the destination node's transmission line. The hub acts as a switch connecting local transmission lines as needed for each incoming message.

Figure 9-9 ►

A typical large
campus network

Messages coming from outside the local network to a local node are handled
in a manner similar to messages between local nodes. For logical bus and ring
networks, messages arriving from another hub or router are copied onto the

local bus or ring. For a logical star network, the hub looks up the target address and copies the message to the target node's transmission line. Table 9-1 summarizes the routing decisions made by LAN hubs.

Table 9-1

LAN hub routing decisions

Source and Destination	Routing Action
Local node to local node	For ring or bus: Propagate message through local medium
	For star: Copy message from source node's transmission line to target node's transmission line
Nonlocal node to local node	For ring or bus: Copy message to local medium
	For star: Copy message from source hub or router transmission line to target node's transmission line
Local node to nonlocal node	Forward message to the next hub or router

Wide Area Network Routing

The zone and campus routers in Figure 9-9 must make more sophisticated routing decisions than the LAN hubs. Zone and campus routers need to know the route or routes to any network address.

There are several ways to store and access information on all addresses. One way is to store a master directory of all known network addresses on one or more central servers and have routers obtain routing information by sending messages to those servers. Such an approach works in theory, but in practice it has many shortcomings, including:

► Every router would need to know the physical route to one or more directory servers. If a server were moved, then all routers would have to update their server routing information.

► The size of the directory would be very large, as would the volume of directory service requests. Many powerful computers would be required to store the directory and answer queries.

► Every directory query would travel over long distances. Wide area networks would be clogged with directory queries and responses.

► Directory updates would have to be made to each server requiring either multiple update messages (one for each server) or regular synchronization of directory content among servers. Either approach would further clog wide area networks.

Most wide area network routing protocols use a distributed approach to maintaining directories. Each router knows the addresses and physical locations of its own nodes and knows other nearby routers and the groups of addresses that they control. Each router also has a default destination for messages to addresses that it doesn't know. Routers periodically exchange information, so they all are kept abreast of changes to routers, network topology, and nodes.

For example, consider a message sent from a node in Jones Hall to a node in the south campus zone network in Figure 9-9. The Jones Hall LAN hub knows the physical location of all Jones Hall nodes and the physical location of the north campus zone router. When the hub receives a message to any node outside its LAN it forwards the message to the north campus zone router.

The north campus zone router examines the destination address and compares it to addresses in its routing directory. The LAN hubs in Jones Hall and Smith Hall periodically exchange summary information with the north campus router so it knows the range of addresses serviced by each of those hubs. Because the destination address doesn't match any of its known address ranges, the north campus router forwards the message to the campus router.

The campus router knows that the destination address falls within a range of addresses controlled by the south campus router, so it forwards the message to that router. The south campus router knows which addresses are controlled by which LAN hubs and forwards the message to the appropriate hub, which in turn forwards the message to its final destination.

This description of distributed wide area routing is generic and omits many details. There are many ways to implement distributed wide area routing, but the method just described approximates the standard Internet routing method used in most modern networks.

MEDIA ACCESS CONTROL

If multiple nodes attempt to transmit at the same time, their messages will mix, producing noise or interference, which is called a **collision**. Methods of dealing with collisions fall into two broad categories—those that allow collisions but detect and recover from them, such as CSMA/CD, and those that avoid collisions altogether, such as token passing. These methods use a **media access control** (**MAC**) protocol to control access to the shared transmission medium.

Carrier Sense Multiple Access/Collision Detection (CSMA/CD) is a MAC protocol commonly, but not exclusively, used in bus network topologies. The basic strategy is not to avoid collisions, but to detect and recover from their occurrence. The protocol details are as follows:

► A node that wants to transmit listens (carrier sense) until no traffic is detected.

► The node then transmits its message.

- The node listens during and immediately after its transmission. If abnormally high signal levels are heard, which is collision detection, then the node ceases transmission.
- If a collision is detected, the node waits for a random time interval and then retransmits its message.

The **token passing** MAC protocol commonly, but not exclusively, is used in ring network topologies. A control message called a **token** is passed from node to node and only the node that "possesses" the token is allowed to transmit messages. All other nodes can only receive and repeat messages. The token is passed from node to node in a predetermined order that includes all nodes on the network. A node must relinquish the token (transmit it to another node) after a specified time interval or as soon as it has no more messages to transmit.

The primary advantage of CSMA/CD is its simplicity. There are no control messages, or tokens, to be passed between nodes and no inherent ordering of nodes in the network. Nodes can be added or deleted without modifying the media access control protocol. The hardware and software that implement the protocol are simpler, faster, and less expensive than hardware and software that implement more complex protocols.

The primary disadvantage of CSMA/CD is potentially inefficient use of data transfer capacity. Network transmission capacity is wasted each time a collision occurs. As network traffic increases, collisions also increase. At high traffic levels, network throughput decreases due to excessive numbers of collisions and message retransmissions. Figure 9-10 shows the effect of this phenomenon on effective communication capacity. In networks that use CSMA/CD, network throughput is typically no more than one-half the theoretical data transfer rate.

Figure 9-10 ►

Effects of CSMA/CD on network throughput

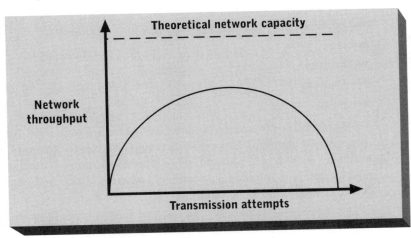

Token passing avoids the potential inefficiencies of CSMA/CD because no transmission capacity is wasted for collisions and retransmissions. A small but insignificant portion of network capacity is used to transmit the token among nodes. Thus, it is possible for token ring networks to achieve an effective data transfer rate nearly equal to the potential data transfer rate.

Other advantages of token passing include the ability to tune network performance and a superior ability to support certain types of network applications. Network performance can be tuned allowing some nodes to hold the token longer than others. Longer token hold times can be given to nodes that need to transmit large amounts of data, such as file and Web servers. Because each node has a maximum token hold time, the maximum time that a node must wait between message transmission opportunities is the sum of the maximum times that all other nodes can hold the token. Predictable maximum wait times are essential in many applications such as video conferencing and network telephony.

The disadvantage of token passing is its complexity. It is more susceptible to failure and requires special procedures when the network is first started. Some device must be responsible for creating and transmitting the token when the network is first started. There must be a strategy for generating a new token if a node "dies" while in possession of the original token. Further, each node must know the next node in the token passing sequence. The failure of a node requires the previous node to bypass the failed node.

NETWORK HARDWARE

Network topology, addressing and routing, and MAC functions are carried out by network hardware devices that include:

► Network interface units or network interface cards
► Hubs
► Bridges
► Routers
► Switches

Table 9-2 summarizes the functions of each device type. Network hardware is constantly evolving, and device functions frequently are combined or modified within a single device, particularly in later generations of hardware. When selecting devices, it is important to understand their functional capabilities, because device names might be misleading.

Table 9-2

Network
hardware
devices

Device	Function
NIU	• Connects a node to the network • Performs appropriate MAC and message forwarding functions • Acts as a bridge between computer system bus and network transmission protocols
Hub	• Acts as a central connection point for LAN wiring • Implements the logical network topology • Often bridges a LAN to a WAN
Bridge	• Connects two separate networks to form a single virtual network • Copies packets between networks as needed
Router	• Connects two or more networks • Forwards packets between networks as needed • Makes intelligent choices among alternate routes
Switch	• Connects networks or network nodes to form virtual networks on a per-packet basis • Quickly routes packets using hardware-based switching

Network Interface Units

A device that connects a network node such as a computer or a network printer to a network transmission cable is called a **network interface unit** (**NIU**) or **network interface card** (**NIC**). An NIU for a single computer is usually a printed circuit board, or card, attached directly to or inserted in a slot in the system bus. An operating system device driver controls the NIU and directs the hardware actions that move packets between the NIU and primary storage. An NIU for a peripheral device such as a printer is more complex because it can't rely on the processing and storage resources of a complete computer system.

In a bus network, the NIU scans the destination address of all packets and ignores all those not addressed to itself. In ring networks, the NIU scans the destination address of all packets and retransmits all packets not addressed to itself. When a correctly addressed packet is received, the NIU stores the packet in a buffer and generates an interrupt on the system bus. The NIU also implements the media access control protocol functions, including listening for transmission activity, detecting collisions, and retransmitting messages in CSMA/CD networks and receiving and forwarding the token in token-passing networks.

Hubs

A **hub** is usually the first connection point between a network node and a LAN. Hubs vary widely in capability and functions. The simplest hubs merely concentrate wiring connections at a single point and sometimes are called **wiring concentrators**. More sophisticated network hubs incorporate routing, bridging, or switching functions. Hubs for token-ring LANs are more sophisticated than hubs for other LAN types because they must generate a token when the network is started or if the original token is lost, and reroute transmission around disconnected or failed nodes.

Transmission lines are connected to an NIU or network hub with standard connectors. The RJ-45 connector is used for the vast majority of twisted pair cable connections. The RJ-45 connector is similar to an RJ-11 modular telephone connector except that it is larger and connects eight wires instead of four. There are several competing standards for fiber-optic connectors including ST, SC, LT, and MT-RJ. The MT-RJ standard has the support of several major network equipment vendors including Cisco and 3Com. See Figure 8-17 in the previous chapter for RJ-45 and MT-RJ connector examples.

Bridges

A **bridge**, sometimes called a **repeater**, copies or repeats packets from one network segment to another. A bridge's complexity and exact functions depend on the differences between the connected network segments. The simplest bridges connect network segments that use identical transmission speeds, packet types, and protocols. More complex bridges connect dissimilar network segments and translate packet formats and network protocols.

A bridge scans packets on each network for destination addresses of nodes on the other network and repeats those packets onto the other network. As the bridge scans network packets it also looks at source addresses and updates internal tables of node addresses on each network segment.

Bridges are most commonly used in two ways:

► To construct a virtual LAN from two separate LANs
► To segment a LAN in order to minimize network congestion

A network designer sometimes will need to build a LAN that is larger than allowed by design standards. For example, a 100 Mbps Ethernet LAN cannot be longer than 210 meters. If a 300 meter LAN is required, then two shorter LANs can be joined with a bridge. Bridged LANs sometimes are called a **virtual LAN**.

If a LAN is routinely overloaded with traffic, throughput can be improved by dividing the LAN into two or more segments and joining the segments with bridges. Nodes that have high communication volume with one another are attached to the same network segment, which minimizes the number of packets that need to cross bridges.

Routers

A **router** performs the same function as a bridge but does so in a more intelligent manner. A router constantly scans the network to monitor traffic patterns and network node additions, modifications, and deletions. Routers use this information to build an internal "map" of the network. Routers periodically exchange information in their internal tables with other routers to gain knowledge of networks beyond those to which they are directly connected. They use this information to forward messages from local nodes to distant recipients and to make intelligent decisions when there are multiple possible routes to a recipient.

A stand-alone router is essentially a special-purpose computer with its own processor and storage. Routing functions can be embedded within other devices such as LAN hubs or general-purpose computers. Any computer system with multiple NIUs connected to different segments or networks can be a router if the proper software is installed. Routing software usually is a standard component of a network operating system, and might or might not be enabled by the server administrator. Routing functions often are enabled on servers in small LANs to avoid the added expense of a dedicated router.

Routing is not a computationally complex task, but it does require fairly extensive I/O capabilities. Every network packet must be examined and forwarded. In a busy network, packet volume can consume most or all of the bus capacity of a general-purpose computer. Such a heavy load can leave insufficient bus or network I/O capacity to perform server functions such as file transfer and printer sharing.

Switches

A **switch** functions in a manner similar to a bridge and a hub. Like a hub, a switch generally has a dozen or more input connections for computers and other network nodes. Each input connection is treated as a separate LAN. A switch examines the destination address of each incoming packet and connects the transmission line of the sender to the transmission line of the receiver. The switch creates a new virtual LAN for each packet and destroys the virtual LAN as soon as the packet has reached its destination.

Switches dramatically increase network performance because:

► Switching is done in hardware
► Each virtual LAN has only one sending and one receiving node, thereby eliminating congestion

Switching is particularly advantageous for LANs that use CSMA/CD. Switches internally segment the LAN, reducing or eliminating collisions and retransmissions. Switches also can be used in place of bridges to join LANs with multiple segments. As with bridging, the network designer must group network nodes into LANs based on shared traffic in order to minimize the number of packets that must be replicated across LANs. Unlike bridges, switches can connect more than two LANs, creating larger virtual LANs.

Technology
Focus

Home Networks

Home networking has been touted as an emerging trend for at least a decade but only recently has it become a marketplace reality. Many homes now have multiple computers, several computer peripherals such as printers and scanners, and a DSL or cable modem Internet connection. Networking these devices requires networking standards, many of which already exist or are under development, including:

- Ethernet
- IrDA
- HomeRF
- IEEE 802.11
- Bluetooth

Home network users can benefit from over two decades of development in Ethernet products, including hubs, NIUs, and cabling (Ethernet is described in detail later in this chapter). Ethernet hubs with a dozen or fewer ports cost $100 or less at this writing. NIUs for computers and peripheral devices are even less expensive. The primary drawback to Ethernet for home networking is its reliance on Category 5 cabling. Few existing homes have Category 5 cabling, and adding networking cable of any type can be an expensive and messy job.

Infrared Data Association (IrDA) and **Home Radio Frequency (HomeRF)** are two existing standards used for home networking. IrDA is based on the same infrared light technology used by television remote controls. It can support megabit data transfer rates over distances less than 1 meter and kilobit rates over greater distances. The primary drawback of IrDA for home networks is the requirement for a clear line of sight between infrared sensors.

HomeRF uses 2.4 GHz radio transmission to connect up to 127 devices within a 50 meter range, and signals can be transmitted through walls and floors. HomeRF version 1.2 supports data transfer rates up to 2 Mbps and version 2.0 supports rates up to 10 Mbps. HomeRF version 2.0 supports Internet standards for data and streaming audio or video, which will expand the network to include devices such as televisions, stereos, and multifunction I/O devices (see Figure 9-11). HomeRF is supported by several major computer manufacturers including Compaq, Intel, and Motorola.

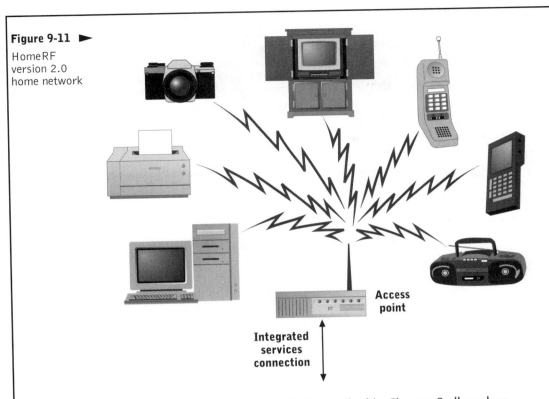

Figure 9-11 ▶

HomeRF
version 2.0
home network

Access point

Integrated services connection

The IEEE 802.11b wireless LAN standard described in Chapter 8 allows hundreds of devices to communicate via RF signals at speeds up to 11 Mbps. IEEE 802.11 devices are deployed widely in business environments and they probably will migrate into the home. Wireless 802.11 network hubs with a wired Internet connection can be purchased for a few hundred dollars and wireless NIUs rapidly are approaching the sub $100 price range.

The **Bluetooth** wireless standard, soon to become part of the IEEE 802.15 standard, forms ad-hoc networks using 2.4 GHz RF transmission over distances up to 10 meters. It is more flexible and theoretically cheaper to implement than IEEE 802.11, but the data transfer rate is currently only 1 Mbps and IEEE 802.11 has a three-year head start. Bluetooth is primarily geared to intelligent devices such as appliances and cameras. Bluetooth devices probably will coexist with HomeRF or IEEE 802.11 devices in many home networks.

Present-day home networking revolves around personal computers, computer I/O peripherals, and wired Internet connections, but three trends probably will change this constellation of devices in the near future:

▶ Voice and video communication are converting from analog to digital transmission rapidly. Most homes now have separate connections for voice, video, and data transmission but these connections probably will merge over the next decade.

OSI NETWORK LAYERS

In the late 1970s, the International Standards Organization (ISO) developed a conceptual model for network hardware and software called the **Open Systems Interconnection (OSI) model**. The OSI model is useful as a general model of networks, a framework for comparing networks, and an architectural roadmap that enhances interoperability among various network architectures and products based on the OSI model. The OSI model organizes network functions into seven layers (see Figure 9-12). Each layer uses the services of the layer immediately below it and is unaware of the internal functions of other layers.

Application Layer

The **application layer** contains programs that make and respond to high-level requests for network services. Application layer programs include end-user network utilities such as Web browsers and e-mail clients, network services embedded in the operating system such as service routines that access remote files, and network service providers such as FTP and Web server programs.

The OSI application layer of one network node generates a service request and forwards it to the OSI application layer of another network node. The application layers of the two network nodes "talk" to each other (shown by the topmost dashed arrow in Figure 9-12) using all of the lower-level OSI layers on both nodes as a communication channel (shown by solid arrows in Figure 9-12). Other layers also communicate with their counterparts in this way and use the layers below to transmit messages. The layers are cooperating processes or hardware devices.

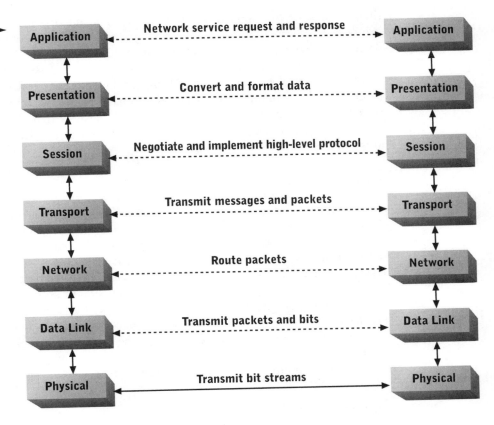

Figure 9-12 ►

Layers of the OSI network model

Application — Network service request and response — Application

Presentation — Convert and format data — Presentation

Session — Negotiate and implement high-level protocol — Session

Transport — Transmit messages and packets — Transport

Network — Route packets — Network

Data Link — Transmit packets and bits — Data Link

Physical — Transmit bit streams — Physical

Presentation Layer

The **presentation layer** ensures that data transmitted by one network node is correctly interpreted by the other network node. Specific tasks that the presentation layer might perform include encryption and decryption, compression and decompression, converting data between EBCDIC and ASCII, and font substitution. The presentation layer is primarily used by applications that format data for user display. For example, the presentation layer of a Web browser formats compressed JPEG images for visual display and decides which font(s) to use for text display.

Session Layer

The **session layer** establishes and manages communication sessions. When a session is first established, the session layers negotiate protocol parameters such as timeout, half or full duplex, synchronization, and quality of service. Once the protocol has been established, the session layers monitor communication to detect and deal with any problems that arise.

For example, when placing an order over the Web, session layers negotiate session parameters such as encrypted transmission and monitor the session for disruptions. If the communication channel is broken before the transaction is completed, the session layer on the vendor's node ensures that the transaction is completely abandoned, such as canceling an unfinished order so that charges aren't applied to the customer's credit card.

Transport Layer

The **transport layer** formats messages into packets suitable for transmission over the network. The transport layer places messages within a packet data area and adds required header and trailer information including network addresses, error detection data, and packet sequencing data. Packets are given to the network layer for delivery to the recipient. When packets are received, the transport layer examines them for errors and requests retransmission as necessary.

If a message can't fit within a single packet, the sending transport layer divides the message content among several packets, adds a sequence number to each packet, and adds special coding within the first and last packets. The receiving transport layer requests retransmission of any missing packets and reassembles data from multiple packets in the proper order.

Network Layer

The **network layer** routes packets to their proper destination. As described earlier in this chapter, the network layers of sending nodes typically route packets to the nearest LAN hub or router. Network layers within hubs and routers interact with one another to exchange routing information and update internal routing tables. Network layers within end nodes announce their presence to other network nodes, hubs, and routers when they are initialized.

Data Link Layer

The **data link layer** is the interface between network software and hardware. In an end node such as a computer system or network printer, the data link layer is implemented by the NIU and device drivers. Device drivers manage packet transfer from secondary or primary storage to the NIU, which transmits packets as bit streams across the physical network link. The media access protocol also is implemented within the data link layer.

Physical Layer

The **physical layer** is the layer at which communication between devices actually takes place. The physical layer includes hardware devices that encode and decode bit streams and the transmission lines that transport them among network nodes. The physical layer also includes other communication channel hardware such as repeaters and amplifiers.

TCP/IP

The **Transmission Control Protocol (TCP)** and **Internet Protocol (IP)**, jointly called **TCP/IP**, originally were developed for the U.S. Department of Defense (DOD) Advanced Research Projects Agency Network (ARPANET) in the late 1960s and early 1970s. Many DOD researchers worked at universities, which soon adopted ARPANET technology in their own networks. Networks based on TCP/IP eventually evolved into the Internet.

Most of the services that we normally associate with the Internet are delivered via TCP/IP. These services include file transfer via the File Transfer Protocol (FTP), remote login via the Telnet protocol, electronic mail distribution via the Simple Mail Transfer Protocol (SMTP), and access to Web pages via the Hypertext Transfer Protocol (HTTP). TCP/IP is the glue that binds private networks together to form the Internet and World Wide Web.

TCP/IP protocol layers correspond poorly to the OSI model. IP is roughly equivalent to the OSI data link, network, and transport layers. TCP is roughly equivalent to the OSI session layer. TCP/IP predates the OSI model by almost a decade so the lack of correspondence is understandable.

IP provides packet routing and forwarding services to higher network layers. IP is independent of the underlying physical network layer and effectively hides it from layers above. An IP layer is implemented for virtually every known physical network layer. IP accepts packets called **datagrams** from TCP and other higher session layer protocols.

IP translates datagrams into a format suitable for transport by the physical network. If a datagram is larger than physical layer data transfer units, such as an ATM cell, then the IP layer divides the datagram into appropriately sized units and transmits them individually. IP attaches header information to each unit which includes its sequence within the datagram. The IP layer at the receiving end reassembles units in the proper order and passes the datagram to TCP.

IP assumes that a datagram will traverse multiple networks via nodes called gateways. It determines the transmission route via a number of related protocols including the **Internet Control Message Protocol (ICMP)** and the **Routing Information Protocol (RIP)**. A **gateway** is any node that connects two or more networks or network segments. A gateway might be implemented physically as a workstation, server, hub, bridge, router, or switch. Figure 9-13 shows the TCP layers involved in a route between two nodes that includes two gateways. Note that only the IP layer is implemented within the gateways.

Figure 9-13 ▶

TCP/IP connection between sender and receiver using multiple networks and gateways

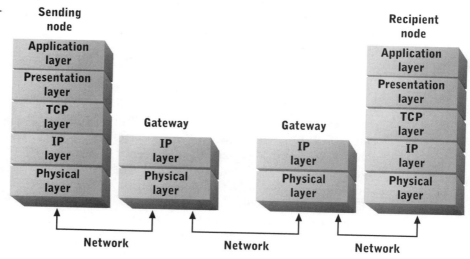

All IP nodes are identified by a unique 32-bit address of the form *nnn.nnn.nnn.nnn*, where *nnn* is a decimal number between 0 and 255. Each gateway maintains an internal table of node IP addresses and physical network addresses such as Ethernet addresses. Each node and gateway knows the physical network address of at least one other gateway. If a received message contains an IP address that doesn't appear within a gateway's internal tables, the packet is forwarded to the default gateway.

IP tables can include entries for partial IP addresses. For example, most nodes within the University of New Mexico have 129.24 as the first 16 bits of their IP address. A node or gateway might not know the physical address for a specific node, such as 129.24.8.1, but its internal tables might contain the physical address of a gateway to any node with 129.24 in the first 16 address bits. IP nodes periodically exchange routing information to keep their tables current. Nodes and gateways announce their presence to nearby devices whenever they are started.

When TCP/IP was first developed, 32 bits seemed large for a network address. Today, there aren't enough unique 32-bit IP addresses for all of the world's Internet-attached devices. In the late 1990s, a new IP standard called **IPv6** was developed with 128-bit addresses. To date, the standard hasn't been adopted widely due to the high cost of updating the IP layers of all devices currently attached to the Internet.

IP is an example of a **connectionless protocol**. In a connectionless protocol, the sender does not attempt to verify the existence of a recipient or ask permission before sending it data. IP sends a datagram to the appropriate gateway and assumes that the datagram will find its way successfully to the intended recipient. IP cannot implement some forms of error control because it is connectionless. Because datagrams are "launched" toward a recipient IP layer without telling it that data is coming, there is no way for the recipient to know if a datagram is "lost" in transit. The recipient node cannot send a negative acknowledgment or request retransmission.

TCP is a **connection-oriented protocol**. A connection-oriented protocol provides the required framework to check for lost messages by explicitly establishing a connection with an intended recipient before transmitting messages. TCP can perform several connection management functions, including verifying receipt, verifying data integrity, controlling message flow, and securing message content. The sender and recipient TCP layers maintain information about one another including message routes, errors encountered, transmission delays, and the status of ongoing data transfers.

TCP uses a positive acknowledgment protocol to ensure data delivery. A recipient TCP layer sends an ACK signal to the sender TCP layer to acknowledge data receipt. The sender TCP layer waits for ACK signals for a specific time interval. The time interval initially is set based on the time required to establish a connection. It can be extended or shortened during the life of the connection to adapt to changing network conditions. If an ACK is not received, the sending TCP layer assumes that the message has been lost and retransmits it. If several consecutive messages are lost the connection "times out" and the sender assumes that the connection itself has been lost.

TCP connections are established through a port and socket. A **port** is a TCP connection with a unique integer number. A **socket** is the combination of an IP address and a port number. Many ports are standardized to specific Internet services. For example, port 21 is the default TCP connection for FTP and port 53 is the default TCP connection for name service queries. The socket 129.24.8.1:53 is the TCP connection for Internet name services at the University of New Mexico.

NETWORK STANDARDS

The Institute of Electrical and Electronic Engineers (IEEE) has drafted a number of telecommunication and network standards, collectively referred to as the **IEEE 802 standards**. These standards describe network hardware, transmission media, transmission methods, and protocols. Current standards and standards under development are listed in Table 9-3.

Table 9-3

IEEE 802
network
standards

IEEE Standard	Functional Description
802.1	Media access control (MAC)
802.2	Logical link control (LLC)
802.3	CSMA/CD and Ethernet
802.4	Token bus
802.5	Token ring
802.6	Metropolitan area networks (MANs)
802.7	Broadband LANs
802.8	Fiber-optic LANs
802.9	Integrated data and voice networks
802.10	LAN/MAN security
802.11	Wireless LANs
802.12	Demand priority
802.14	Cable TV broadband networking
802.15	Wireless personal area networks
802.16	Broadband wireless access

The 802.1 and 802.2 standards roughly correspond to the data link layer of the OSI model. The 802.1 standard addresses a number of issues including media access and bridging. It also describes an architectural framework into which the other standards fit. The 802.2 standard addresses issues such as routing, error control, and flow control. The methods defined in these standards are incorporated into most of the other 802 standards.

The most common standards for local area networks are 802.3 (CSMA/CD in bus networks), 802.5 (token ring), and 802.11 (wireless LANs). Each standard has a number of subsidiary standards, indicated by letters added to the standard numbers, such as 802.11b, that cover additional implementation parameters.

Many commercial networking products are based on the IEEE standards. For example, Ethernet is based on the 802.3 standard. IBM's token ring products are based on a subsidiary standard of 802.5, and Switched Multimegabit Data Services (SMDS) is based on 802.6.

The IEEE standards development process is influenced heavily by companies and organizations in the telecommunications and networking industries. Standards are developed by committees whose membership is drawn from industry, government, and academia. Standards committees work closely with other standards and regulatory agencies such as the American National Standards Institute

(ANSI), the ISO, and the U.S. Federal Communications Commission and its international counterparts. The development of standards can be influenced as much by politics as by technological considerations.

The committees set criteria for evaluation and then invite proposals for adoption as standards. Proposals usually are supplied by vendors based on products and technologies already under development. The lag time between standard publication and the appearance of related commercial products is less than otherwise might be assumed. Occasionally the process backfires. Adopted standards sometimes represent significant compromises among very different proposals. Companies might decide not to adjust their developmental products to meet the compromise standard, thus generating an "orphan standard," such as 802.4. A company also might choose to release products that don't adhere to any published standard and hope that its technology becomes a de facto standard.

The marketplace ultimately decides which technologies and products will succeed. The standard-setting process usually provides an accepted target for implementation and, in the case of a widely implemented standard, a degree of compatibility among competing products.

Technology
Focus

Ethernet

Ethernet is a local area network technology closely related to the IEEE 802.3 standard. Ethernet was developed by Xerox in the early 1970s. Digital Equipment Corporation (DEC), which now is part of Compaq, was the driving force in developing commercial Ethernet products in the late 1970s. DEC, Intel, and Xerox jointly published a specification for Ethernet networks in 1980. The 802.3 standard was published a few years later and incorporated most aspects of the 1980 standard. The 802.3 standard has been extended many times and upgrades have continued.

Slower speed Ethernet standards use a bus topology, Category 5 twisted pair cable, and CSMA/CD. The 802.3 packet format is shown in Figure 9-14. There is no provision for packet priorities or quality of service guarantees, though incorporating those capabilities is a high priority for vendors and standard-setting organizations.

Figure 9-14 ▶
Ethernet packet format

Synch (56 bits)	Start flag (8 bits)	Recipient address (48 bits)	Sender address (48 bits)	Length (16 bits)	Data (12,208 bits maximum)	CRC (32 bits)

The original Ethernet standard transmits data at 10 Mbps. The standard was updated (802.3u) in the late 1990s to increase transmission speed to 100 Mbps. There were several competing and incompatible proposals for 100 Mbps Ethernet including the never-completed 802.13 and the dormant 802.12 standards.

Gigabit Ethernet is based on the 802.3z standard published in 1998 and the 802.3ab standard published in 1999. Several physical implementations are defined, each representing a trade-off between cost and maximum cable length (see Table 9-4). The 802.3ae standard currently is under development with a transmission goal of 10 Gbps or more.

Table 9-4 ►

Gigabit ethernet specifications

IEEE Standard	Configuration Name	Cable and Laser Type	Maximum Length
802.3z	1000Base-SX	Short wavelength laser over multimode fiber	550 meters
802.3z	1000Base-LX	Long wavelength laser over multimode fiber	550 meters
802.3z	1000Base-LX	Long wavelength laser over single mode fiber	5 kilometers
802.3z	1000Base-CX	Category 5 twisted pair	25 meters
802.3ab	1000Base-T	Category 5 twisted pair	100 meters

Ethernet has been a popular networking technology for over two decades. Many competitors, such as the IBM token ring, have fallen by the wayside and others, such as ATM, might soon follow. If ongoing efforts to update the technology are successful, then Ethernet should continue to dominate other network standards well into the current decade.

Business
Focus
Upgrading Network Capacity (Part II)

Note: Part I of this Business Focus was presented in Chapter 9.

The Bradley Advertising Agency (BAA) has decided to implement a 1000Base-T Ethernet network. Their reasons are:

► The existing twisted pair cable was tested and found to meet Category 5 requirements

► Gigabit Ethernet should meet current and near-future demands for video file sharing among video editing workstations

▶ Gigabit Ethernet equipment is significantly cheaper than ATM equipment

BAA is uncertain of how to integrate its microcomputers, video editing workstations, and server into the Gigabit Ethernet network. They are considering three options:

1. Connect all microcomputers to the existing 10/100 Mbps Ethernet hub. Connect the video editing workstations to a new 10/100 Mbps switching hub. Connect the server and both 10/100 hubs to a Gigabit Ethernet switch.

2. Connect all microcomputers to the existing 10/100 Ethernet hub. Connect the video editing workstations, the server, and the 10/100 hub to a Gigabit Ethernet switch.

3. Connect all computers to a single 100/1000 Mbps Ethernet switch. Upgrade existing 10 Mbps microcomputer NICs to 100 Mbps and install Gigabit NICs in the video workstations and server.

Questions

1. What are the costs of each option? Which option provides the greatest network performance?

2. What are the upgrade paths to options 2 or 3 if option 1 is implemented now, and options 2 or 3 are implemented two years from now?

3. Which option should BAA choose to implement now? Why?

4. BAA anticipates that they will need to transmit high-quality video to and from client locations within the next three years. Under this assumption, will their rejection of ATM have long-term consequences?

SUMMARY

▶ Network topology refers to the spatial organization of network devices, the physical routing of network cabling, and the flow of messages from one network node to another. Topology can be physical (arrangement of node connections) or logical (message flow). Star topology connects every node to a device, which is typically a LAN hub. Bus topology connects every node to a shared transmission medium. Ring topology connects each node to two neighbors, and all connections form a closed loop.

▶ LANs are interconnected to form WANs. Specialized network hardware devices route packets from source to destination. Packet routing within a LAN usually is handled by the LAN hub or switch. Packet routing across WANs uses a store and forward approach similar to the delivery and warehouse network of a package delivery service. Forwarding stations can be implemented using bridges, routers, or switches.

▶ A media access control (MAC) protocol specifies rules for accessing a shared transmission medium. CSMA/CD is commonly used in bus networks. Each node listens for an idle state, transmits a packet, listens for a collision, and retransmits after a random waiting period, if necessary. Token passing is commonly used in ring networks. A special packet called a token is passed around the ring and only the node that has the token can transmit.

► Network hardware devices include NIUs, hubs, bridges, routers, and switches. An NIU is the interface between a network node and the network transmission medium. A hub connects nodes in a LAN and also can perform bridging, routing, or switching functions. A bridge is a device that connects two networks or network segments and forwards packets between them. Routers are intelligent bridges that can connect to more than two networks and exchange information to improve routing decisions. Switches are high-speed devices that create virtual LANs on a per-packet basis.

► The Open Systems Interconnection (OSI) model is an ISO conceptual model that divides network architecture into seven layers—application, presentation, session, transport, network, data link, and physical. Each layer uses the services of the layer below and is unaware of other layers' implementations.

► TCP/IP is the core Internet protocol suite. IP provides connectionless packet transport across LANs and WANs. Internet addresses are 32-bit values, though a 128-bit address standard currently is approved but not implemented. TCP provides connection-oriented packet transport to higher-level Internet service protocols including HTTP, FTP, and Telnet.

► The IEEE 802 standards cover many types of networks. The 802 standards are developed by committees composed of interested parties from industry, government, and academia. IEEE standards help ensure compatibility among products from competing vendors. Some standards, such as the Ethernet standard (802.3) are very successful and others are never implemented.

In this chapter and the previous chapter we described all of the hardware and only some of the software technology that underlies modern computer networks. In Chapter 10, we'll examine software tools used to build application software. In Chapters 11 and 12, we'll look at operating system technology, including resource allocation and file management systems. Then, we'll return to computer networks in Chapter 13 and discuss distributed resources and applications.

--- **Key Terms** ---

application layer	Carrier Sense Multiple	connectionless protocol
Bluetooth	Access/Collision Detection	data link layer
bridge	(CSMA/CD)	datagram
bus topology	collision	Ethernet
	connection-oriented protocol	gateway

Gigabit Ethernet
Home Radio Frequency
  (HomeRF)
hub
IEEE 802 standards
Infrared Data Association
  (IrDA)
Internet Control Message
  Protocol (ICMP)
Internet Protocol (IP)
IPv6
local area network (LAN)
logical topology
media access control (MAC)
network interface card (NIC)

network interface unit (NIU)
network layer
network topology
Open Systems Interconnection
  (OSI) model
physical layer
physical topology
port
presentation layer
repeater
ring topology
router
Routing Information Protocol
  (RIP)
session layer

socket
star topology
store and forward
switch
TCP/IP
token
token passing
Transmission Control Protocol
  (TCP)
transport layer
virtual LAN
wide area network (WAN)
wiring concentrator

Vocabulary Exercises

1. The _____ standards define many aspects of computer networking.

2. The _____ layer establishes connections between clients and servers.

3. A(n) _____ is a device that connects two separate networks into one virtual network.

4. A(n) _____ is a special-purpose packet, representing the right to originate network messages.

5. The _____ layer determines the routing and addressing of packets.

6. A TCP/IP _____ is the combination of an Internet address and a service _____ number.

7. The _____ layer refers to programs that generate requests for network services.

8. A network using physical _____ topology connects all nodes to a central device.

9. A physical connection between two different networks is implemented using a(n) _____ , _____ , _____ , or _____ .

10. Packet loss can't always be detected by a receiver if a(n) _____ protocol is in use.

11. Under TCP, a(n) _____ is the basic data transfer unit.

12. _____ transmits at 10 Mbps over twisted pair cabling.

13. The _____ defines a generic set of software and hardware layers for computer networks.

14. In a network that uses _____, a packet is passed from one network node to the next until it reaches the destination node.

15. When two messages are transmitted at the same time, a(n) _____ has occurred.

16. Under the _____ media access strategy, collisions can occur, but they are detected and corrected.

17. A(n) _____ protocol defines the rules that govern a network node's access to a transmission medium.

18. A microcomputer or workstation hardware interface to a network transmission medium is called a(n) _____.

19. The _____ layer is responsible for data conversion functions, such as compression and encryption.

Review Questions

1. Describe the function of each layer of the OSI network model.

2. Describe the movement of packets on a bus network. Describe the movement of messages on a ring network.

3. What are the comparative advantages and disadvantages of ring and bus networks?

4. How does a packet from one LAN node find its way to a recipient on the same LAN? How does a packet find its way to a recipient on another LAN?

5. Describe the CSMA/CD media access control protocol.

6. Describe the token passing media access control protocol. Why do networks that use token passing generally achieve higher effective data transfer rates than networks that use CSMA/CD?

7. What is the function of a bridge? How does it differ from a router? How does it differ from a switch?

8. What is a LAN hub? What functions does it normally perform?

9. What is a connectionless protocol? What is a connection-oriented protocol? List one example of each protocol type.

10. How many bits are in an IP address? What is an IP port? What is an IP socket?

11. Describe various aspects of Ethernet networks including transmission speed, logical topology, supported transmission media, and relevant IEEE standards.

Research Problems

1. TCP/IP was designed as a protocol suite that could provide limited functionality in the event of massive network disruptions, such as might be experienced during war. Unfortunately, that heritage limits the ability of the TCP/IP to meet modern demands, including transmission of voice and video. Investigate the requirements of real-time voice and video data transmission to single or multiple recipients. In what ways is TCP/IP currently limited in its ability to handle such transmissions? How will those limitations be addressed when IPv6 is deployed widely?

2. Investigate the full-duplex version of Ethernet. What are the advantages and disadvantages of full-duplex Ethernet compared to half-duplex Ethernet? Is full-duplex Ethernet backward compatible with half-duplex Ethernet?

3. Investigate the Ethernet connectivity devices offered by a major vendor such as 3Com or Cisco. What is the range of features available in LAN hubs? What devices does the vendor offer that don't clearly fall into the categories described in this chapter? What is the cost per port of 100 Mbps switches, 100/1000 Mbps switches? What options are provided to link groups of hubs or switches?

Chapter 10

Application Development

Chapter Goals

- ► Describe the application development process and the role of methodologies, models, and tools

- ► Compare and contrast programming language generations

- ► Explain the function and operation of program translation software, including assemblers, compilers, and interpreters

- ► Describe link editing, and contrast static and dynamic linking

- ► Describe integrated application development software, including programmer's workbenches and CASE tools

Application software development is a complex process that follows a series of steps to translate users' needs into executable programs. There are automated tools to support each step, from user statements to system models to program source code to executable code. Understanding the role and function of application development tools will make you a more efficient and effective analyst, designer, and programmer. Figure 10-1 shows the topics covered in this chapter.

Figure 10-1 ▶

Topics covered in this chapter

Application programs

Software development process
Programming languages
Compilation and link editing
Interpretation
Symbolic debugging
Application development tools

System management

Application programs

System services

Resource allocation

Hardware interface

Hardware

Network

THE APPLICATION DEVELOPMENT PROCESS

The process of designing and constructing software translates users' information processing needs into CPU instructions that, when executed, address those needs. User needs are stated abstractly in a natural language, for example, "I need to process accounts receivable and payable." Software programs are detailed, precise statements of formal logic written as sequences of CPU instructions. Software development involves two translations—from an abstract need statement to a detailed implementation that satisfies the need, and from natural language to CPU instructions (see Figure 10-2).

Figure 10-2 ▶

Application
development
translates user
needs to CPU
instructions

Software development is a complex process that requires significant effort and resources. A user need stated in a single sentence might require millions or billions of CPU instructions to address. Developing and testing CPU instructions requires complex software development methods and tools, highly trained specialists, and months or years of effort. Software has surpassed hardware to become the most costly component of most information systems. Information system owners expect their investment in software development to be repaid with software that meets their needs.

Unfortunately, the complexity of software development creates many possibilities for error. Errors lead to software that doesn't work or fails to meet user needs. The economic cost of software development errors can be much greater than the cost of developing the software. Reduced productivity, dissatisfied customers, and poor managerial decisions are just a few of the indirect costs of software that fails or doesn't completely or correctly address user needs.

Systems Development Life Cycle

In this chapter we return to the systems development life cycle (SDLC) introduced in Chapter 1, but with a specific focus on how it is applied in developing software. Figure 10-3 shows the phases common to most SDLCs. Each phase produces outputs that are input to the next phase. The SDLC is a way of breaking the complex process of software development into smaller and more manageable pieces. In this chapter, we're primarily concerned with the analysis, design, and implementation phases.

Figure 10-3 ▶

Systems development life cycle (SDLC) phases

During the analysis phase, members of the software development team interact with users to understand and document users' needs. Much of the translation from abstract concepts to specific and detailed facts occurs during analysis. The components of a general statement such as, "I need to process accounts receivable and payable," are examined and documented in minute detail. What is an account payable or account receivable? What specific tasks are implied by the term process? What is the data content of system inflows and outflows? What data must be internally stored? What tasks are performed when, and by which people or programs? The answers to these questions are summarized in a variety of forms including text descriptions and diagrams that together compose a system requirements model.

The **system requirements model** provides the detail needed to develop a specific system to meet user needs. During the design phase, computer and related hardware, system software, and application software requirements are determined. The output of the design phase is a **design model** that provides an architectural blueprint for system implementation. The blueprint specifies required features of system software such as operating systems and network server software, and required functions of application software subroutines, procedures, methods, objects, and programs.

The design model is input to the implementation phase where each automated system component is either acquired or constructed. Typically, computer hardware and system software are purchased and some or all applications software is constructed, sometimes using purchased or previously developed software components. As acquisition and construction proceed, hardware and software are tested to ensure that they function correctly and meet user needs as described in the analysis model. The output of the implementation phase is a fully tested system that is ready to be put to productive use.

Methodologies and Models

Developers attempt to minimize errors during all SDLC phases by using proven development methodologies. A methodology is an integrated collection of models, tools, and techniques for performing all SDLC phases. Structured system development and object-oriented system development are two widely practiced methodologies.

Structured methods document requirements by modeling the flow of data (see Figure 10-4) or the relationship of various parts of the process, and use structure charts and system flowcharts to document architectural plans. Object-oriented development bases its models on objects and classes rather than on structure and process, for example, using class diagrams (see Figure 10-5) to document analysis requirements. With both methods, graphical models usually are supplemented with textual description.

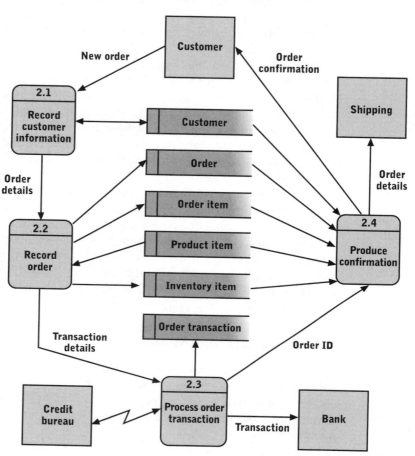

Figure 10-4 ▶

Sample data
flow diagram

Diagram 2: Create new order

Figure 10-5 ►

Sample class
diagram

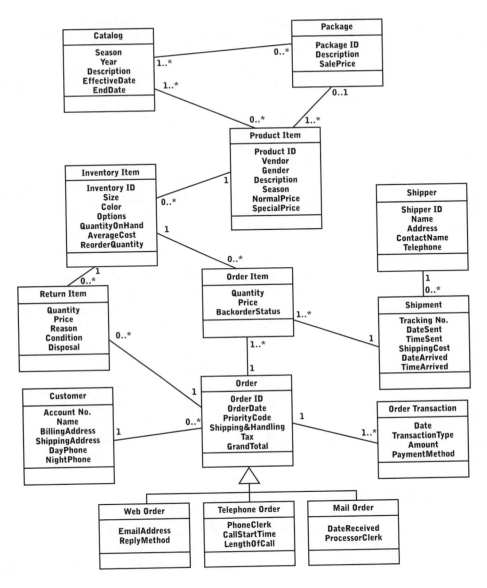

Tools

Software development is supported by many automated tools. Some tools, such as word processors and general-purpose drawing programs, can be used to build many types of models. Other tools are customized to specific models and development methodologies. Related tools often are integrated into a single suite that supports an entire system development methodology in much the same fashion that word processing, graphics, database, and spreadsheet programs are integrated into a single suite.

Software development tools vary with the target system deployment environment. Tools can be specific to a particular operating system, programming language, database management system, or type of computer hardware. Wide variation in underlying methodology, supported models and techniques, and target deployment environment makes proper tool selection a critical, but difficult, undertaking. The problem is further compounded by a dizzying array of descriptive terminology and support for different parts of the SDLC.

The declining cost of computer hardware and the increasing cost of application development labor have spurred the rapid development and widespread adoption of automated system and software development tools. Substituting automated tools for manual methods is a classic economic tradeoff between labor (such as analysts, designers, and programmers) and capital resources, (sophisticated application development tools and the computer hardware required to run them).

In the early days of computers, hardware was so expensive that its use could be justified only for high value applications. It made more economic sense to use labor-intensive processes for application development because labor was relatively cheap compared to computer hardware. As hardware costs decreased, this economic balance shifted. It was this shift that led to the introduction of automated tools to support software development.

Programming languages, program translators, and operating system service layers are the earliest examples of this shift. They are all methods by which computing hardware can support software development directly. The modern proliferation of application development tools is a continuation of this economic shift. Modern analysts, designers, and programmers use a wide array of tools to support or completely automate SDLC tasks or phases.

PROGRAMMING LANGUAGES

A **programming language** is language used to instruct a computer to perform a task. Program instructions are sometimes called **code** and the person who writes the instructions is called a **coder**, or **programmer**. Programming language development has been driven by a desire to make software easier and less labor-intensive to develop. Two ways to make programs easier to write and applications easier to develop are:

► Make the programming language easier for people to understand
► Develop languages and program development approaches that require people to write fewer instructions to accomplish a given task

Programming languages have gone through several "generations" that have each come closer to achieving these goals. Table 10-1 summarizes the generations of programming languages.

Table 10-1 ►

Programming language generations

Generation	Name or sample languages	Description or characteristics
1st	Machine language	Binary instructions for a specific CPU
2nd	Assembly language	Mnemonic instructions for a specific CPU
3rd	FORTRAN, COBOL, BASIC, PL/1, Pascal, and C	Instruction explosion, machine independence, and (usually) standardization
4th	Visual Basic, SQL	Higher instruction explosion, interactive and graphical I/O support, database support, limited nonprocedural programming, and proprietary standards
5th	LISP, Prolog	High instruction explosion, nonprocedural programming, expert systems, and artificial intelligence applications
Object-oriented	C++, Java, SmallTalk	High instruction explosion, support for modern code development, and code reuse methods such as inheritance and message passing

First Generation Languages

Binary CPU instructions, called **machine languages,** or **first generation languages** (**1GLs**) are the earliest programming languages. It is difficult for people to remember and manipulate long strings of binary digits. Yet that is what early machine language programmers did. Programmers had to remember the binary codes that represented each CPU instruction and had to specify all operands as binary numbers. Programming with binary numbers is tedious and error prone. As primary storage capacity and program size grew, it became less and less feasible for people to develop error-free programs using only binary numbers.

Second Generation Languages

A **second generation language** (**2GL**) more commonly is known as an **assembly language.** The instructions and programming examples defined in Chapter 4 used a simplified assembly language. Assembly languages use a short character sequence called a mnemonic to represent CPU instructions. Assembly language programmers can define their own mnemonic to represent the memory address of an instruction or data item. Today, we usually use the term **variable** to describe a mnemonic that represents a data item memory address and the term **label** to describe a mnemonic that represents a program instruction memory address.

2GLs made programming much simpler because for CPU instructions and program data items they used short names, which are easier for people to remember and manipulate than binary numbers. They were the most commonly used programming languages during the 1940s and early 1950s and are still used today in some types of system and application programming. For example, operating system device drivers often are written in assembly language to provide the programmer precise control over machine actions.

Any language other than a 1GL must be translated into a 1GL before it can be executed by a CPU because the CPU processes only binary digits. An assembly language program is translated into binary CPU instructions by another program called an **assembler**. An assembler reads a file containing the assembly language program one instruction at a time. It translates each mnemonic into its corresponding binary digit sequence and writes those digits to an output file ready to be loaded into memory and executed by the CPU. Assemblers are the earliest example of automated program development tools.

Although 2GLs made programming easier, they did nothing to minimize the number of instructions that a programmer had to write to implement a task. Each assembler instruction is translated into a CPU instruction, so a 2GL still directs every individual CPU action. As memory capacity and program size grew, it became more difficult for a programmer to specify every CPU action. To increase programmer productivity, the 1:1 correspondence between program instructions and CPU actions had to be broken.

Third Generation Languages

Programming languages beyond the second generation allow programmers to specify many CPU actions with a single program instruction or statement. The one-to-many (1:N) relationship between later generation programming language statements and the CPU actions that implement them is called **instruction explosion**. A **third generation language** (3GL) is a programming language that uses mnemonics to represent instructions, variables, and labels, and has a degree of instruction explosion greater than 1:1. The first 3GL was FORTRAN. Later 3GLs include COBOL, BASIC, PL/1, Pascal, and C. Some 3GLs still are used today though their popularity fades with each passing year.

Different programming languages have different degrees of instruction explosion. For example, if 1000 machine instructions result from translating a 20-statement FORTRAN program, then the degree of instruction explosion can be stated as a 50:1 ratio. Within the same programming language different types of statements have different degrees of instruction explosion. In general, statements that describe mathematical computation typically have low instruction explosion (10:1 or less) and statements that describe I/O operations typically have high instruction explosion (100:1 or greater).

Like 2GL programs, 3GL programs must be translated into binary CPU instructions before the program is executed. Compilers, link editors, and interpreters are automated tools used to translate 3GL programs; they are covered in more detail later in this chapter. 3GL program translation is much more complex than 2GL program translation, so 3GL translation programs generally are much larger and consume more computer resources than assemblers.

3GL programs are the same regardless of the CPU that executes the translated program, a characteristic that is called **machine independence**. Compilers, link editors, and interpreters translate machine independent instructions into machine language instructions for a specific CPU. Each CPU uses a different 3GL translation program that is customized to that CPU's instruction set.

Most 3GLs have significant limitations with respect to modern hardware, interactive and graphical user interfaces, and database management. Many 3GLs were developed before modern user interfaces and database management systems existed or were widely used. These limitations account for much of the decline in popularity of languages such as FORTRAN, COBOL, and C.

Fourth Generation Languages

In the 1970s and 1980s, **fourth generation languages (4GLs)** were designed to address 3GL limitations. They shared many features, including:

► Significantly higher instruction explosion than 3GLs

► Instructions or prewritten functions to implement interactive and graphical user interfaces

► Instructions to interact directly with an internal or external relational database

► Ability to describe processing requirements without precisely specifying solution procedures

The majority of 4GLs were proprietary, and many were optional components of database management systems. Few of the original 4GLs survive today though many of their modern counterparts are direct descendants. Two notable exceptions are Visual Basic and Structured Query Language (SQL). Visual Basic evolved from the 3GL BASIC and is adapted to modern software requirements. Visual Basic has a rich variety of graphical user interface capabilities and can interact directly with Microsoft Access databases. Visual Basic is used widely but it lacks some features needed by "industrial strength" software.

SQL is specialized for adding, modifying, and retrieving data from relational databases, and usually is used in combination with database management systems such as Oracle or Microsoft SQL Server. SQL is not a full-fledged programming language because it lacks many features of a general-purpose programming language, including many control structures and user interface capabilities. Figure 10-6 shows two programs that perform similar functions, one in C and the other in SQL. The number of program statements is much larger in the C program, which indicates a lower level of instruction explosion.

Figure 10-6 ►

Equivalent programs in SQL and C

```
                    Structured Query Language Example

open database banking;
select customer.acct_num, customer.name,balance+sum(transaction.
 amount)
from       customer,transaction
where      customer.acct_num=transaction.acct_num
group by   customer.acct_num,customer.name;



                         C Example

balance_report ()   {

FILE  *cust_file, *trans_file;
int   status,acct_num,a_num,balance,amount;
char name [256];

cust_file=fopen("customer","r");
trans_file=fopen("transaction","r");
status=scanf(cust_file, "%d%s%d\n", &acct_num,name, &balance);
while (status != EOF {
status=scanf (trans_file, "%d%d\n," &a_num, &amount);
while (status != EOF){
     if (acct_num == a_num) {
       balance+=amount;
     }
     status=scanf(trans_file,"%d%d\n", &a_num,&amount);
}
printf("%d  %s  %d",acct_num,name,balance);
trans_file=freopen("transaction","r");
   status=scanf(cust_file,"%d%s%d\n", &acct_num,name,&balance);
}
close(trans_file);
close(cust_file);
exit(0);
} /* end balance_report */
```

The programs also differ in their approach to solving a specific data retrieval problem. The C program describes a specific procedure for extracting and displaying data that includes control structures and many database management functions such as opening and closing files, and testing for end of file while reading records. In contrast, the SQL program contains no control structures, file manipulation, or output formatting commands. The SQL program states what fields should be extracted, what conditions extracted data should satisfy, and how data should be grouped for display. SQL is a **nonprocedural language** because it describes a processing requirement without describing a specific procedure for satisfying the requirement. The compiler or interpreter determines the most appropriate procedure for retrieving the data and generates or executes whatever CPU instructions are needed to do so.

4GLs are sometimes called nonprocedural languages, though the description is not completely accurate. Most 4GLs support a mixture of procedural and nonprocedural instructions. A 4GL typically provides nonprocedural instructions for database manipulation and report generation. Some 4GLs also provide nonprocedural instructions for interactively updating database contents. Other types of information processing such as complex computations require procedural instructions.

Fifth Generation Languages

A **fifth generation language** (5GL) is a nonprocedural language suitable for developing software that mimics human intelligence. 5GLs first appeared in the late 1960s with LISP but weren't used commonly until the 1980s. LISP and Prolog are the most common general-purpose 5GLs but there are many proprietary 5GLs used to build applications in areas such as medical diagnosis and credit application scoring.

A 5GL program contains a set of nonprocedural rules that mimic rules people use to solve problems. A rule processor accepts a starting state as input and iteratively applies rules to achieve a solution. Figure 10-7 shows a sample Prolog program for solving the problem of moving a farmer, fox, chicken, and corn across a river on a boat that only can hold two. The rules describe valid moves and unsafe states, for example, the chicken and corn can't be left without the farmer because the chicken would eat the corn. The program prints trace messages as it applies rules and prints the set of moves that solve the problem.

Figure 10-7 ▶

Sample Prolog
program

```
go (S,G) :- write ('start at state'),write (S),nl,path(S,G,[S]).

path (G,G,L) :- write ('SUCCESS, the path is:'),nl,ppt(L).

path (s(B,F,C,F,Cn),G,L) :-
move(s(B,F,C,F,Cn),s(Boat,Farmer,Chicken,Fox,Corn)),
   not(member(s(Boat,Farmer,Chicken,Fox,Corn),L)),
   not(unsafe(Boat,Farmer,Chicken,Fox,corn)),
   write('use this move to get to s('),
   write(Boat), write(','), write(Farmer), write(','), write(Chicken),
   write(','), write(Fox), write(','), write(Corn), write(')'), nl,
   path(s(Boat,Farmer,Chicken,Fox,Corn),
        G,
        [s(Boat,Farmer,Chicken,Fox,Corn) |L]), !.
/* s(Boat,Farmer,Chicken,Fox,Corn)   */

Move(s(e,e,e,Fox,corn),s(w,w,w,Fox,Corn)) :-
   Write('move farmer and chicken from east to west'), nl.

Move(s(w,w,w,Fox,Corn),s(e,e,e,Fox,Corn)) :-
   Write('move farmer and fox from west to east'), nl.

Move(s(e,e,Chicken,e,Corn),s(w,w,Chicken,e,Corn)) :-
   Write('move farmer and fox from east to west'), nl.

Move(s(w,w,Chicken,e,Corn),s(e,e,Chicken,e,Corn)) :-
   Write('move farmer and fox from west to east'), nl.

Move(s(e,e,Chicken,Fox,e),s(w,w,Chicken,Fox,w)) :-
   Write('move farmer and corn from east to west'), nl.

Move(s(w,w,Chicken,Fox,w),s(e,e,Chicken,Fox,e)) :-
   Write('move farmer and corn from west to east'), nl.

Move(s(e,e,Chicken,Fox,Corn),s(w,w,Chicken,Fox,Corn)) :-
   Write('move farmer only from east to west'), nl.

Move(s(w,w,Chicken,Fox,Corn),s(e,e,Chicken,Fox,Corn)) :-
   Write('move farmer only from west to east'), nl.

unsafe(X,X,Y,Y,_).
unsafe(X,X,Y,_,Y).

number(X,[X|T]).
number(X,[Y|T]) :- member(X,T).

ppt([]).
ppt([H|T]) :- ppt(T), write(H), nl.
```

Object-Oriented Programming Languages

3GLs and 4GLs have a similar underlying programming paradigm under which data and functions are two distinct things. Program instructions manipulate data and move it from source to destination. Data is processed by passing it among code modules, functions, or subroutines, each of which performs one well-defined task. A data flow diagram (see Figure 10-4) is a direct representation of the traditional program/data dichotomy.

Software researchers in the late 1970s and early 1980s began to question the efficacy of this program/data dichotomy. They developed a new programming paradigm called **object-oriented programming (OOP)** that better addressed program reuse and long-term software maintenance. The fundamental difference between the object-oriented and traditional programming paradigms is that OOP views data and programs as two parts of an integrated whole called an **object**. Objects contain data and programs or procedures that manipulate the data called **methods**.

Objects reside in a specific location and wait for messages to arrive. A **message** is a request to execute a specific method and return a response. The response can include data, but the data included in a response is only a copy. The original data can be accessed and manipulated only by object methods. Figure 10-8 shows the relationships among data, methods, messages, and responses.

Figure 10-8 ►

Objects, data, methods, messages, and responses in OOP

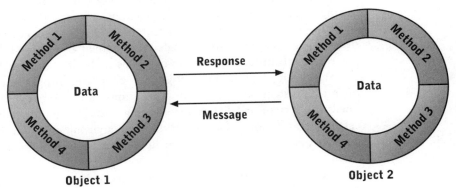

Object 1 **Object 2**

SmallTalk was the first commercial OOP language. C++ (an extension of C) and Java are more widely used OOP languages. Other languages, such as Visual Basic, have been extended with OOP concepts and many proprietary OOP languages such as Oracle Power Objects have been developed.

Object-oriented programming and design is uniquely suited to developing real-time programs such as operating systems and interactive user interfaces. Most current operating systems and windowing environments such as Windows 2000, UNIX, and X-Windows were constructed with OOP languages. OOP languages and good object-oriented software design promote reusability and portability of source code. New objects are defined that incorporate data and methods from existing objects through a feature called inheritance.

The fact that OOP languages are described by a name rather than a generation number is evidence that the traditional concept of language generations is starting to break down. OOP languages are not a clear successor to 5GLs because they are neither nonprocedural nor exclusively used for developing artificial intelligence and expert system applications. Some OOP languages such as C++ are direct extensions of 3GLs, and others such as SmallTalk and Java are more closely related to 4GLs.

Programming Language Standards

The American National Standards Institute (ANSI) and the International Standards Organization (ISO) set standards for some programming languages. Examples of ANSI standard languages include FORTRAN, COBOL, C, and C++. A programming language standard defines:

► Language syntax and grammar
► Machine behavior for each instruction or statement
► Test programs with expected warnings, errors, and execution behavior

Compilers, link editors, and interpreters are submitted to the standard-setting body for certification. Test programs are translated, and the behavior of the translator program and the executable code is examined. The behavior of the program is compared to the behavior defined by the standard to determine whether the translation program complies with the standard.

Standard programming languages guarantee program portability among operating systems and application programs. A program can be moved to a new operating system or CPU by recompiling and relinking or reinterpreting it with certified translation programs in the new environment. Portability is important because operating systems and hardware change more frequently than most program source code.

Standards specify a minimum guarantee of behavioral compatibility but they do not prohibit adding language constructs and capabilities beyond the standard. For example, the C programming language contains no primitive data type for strings. But a compiler writer can define a String data type and implement processing operations on that data type. A program translator that supports the data type still can be certified because the nonstandard data type and its associated processing operations aren't used by standard test programs.

Many compiler and interpreter vendors develop and support such language extensions to address limitations in the language or speed up program development. However, using such extensions voids the portability guarantee of the standard because there is no guarantee that the extensions will exist or result in equivalent behavior in another program translator.

Setting standards for programming languages has become much less relevant over the last two decades. There are many reasons for this, including a significant movement of basic software research from academia into industry, a reluctance by software companies to share proprietary technology, much greater variability in applications and run-time environments, and increasing interdependence between application and system software. Modern software standards frequently are more concerned with system software and interfaces between programs than with the languages used to build programs.

COMPILATION

Figure 10-9 shows the process of developing an application program using a program editor, compiler, and link editor. Input to the program editor comes from a program template, CASE tool, programmer, or any combination thereof. The output of the program editor is a partial or complete program, called **source code**, in a programming language such as C++ or Visual Basic. Source code normally is stored in a file that is named to indicate both its function and programming language; for example, ComputePayroll.cpp, where cpp is an abbreviation for C++.

Figure 10-9 ►

Program development with a program editor, compiler, and link editor

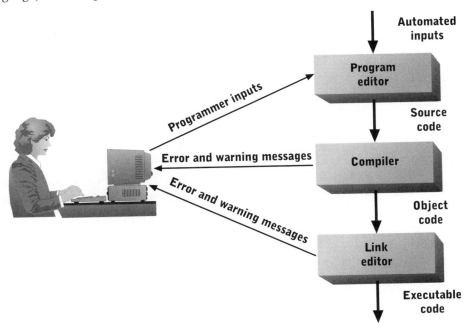

The compiler and link editor translate source code into executable code. **Executable code** is stored in a file such as an .exe file in DOS or Windows, and contains CPU instructions that are ready to be loaded and executed by an operating system. The compiler and link editor each perform a different part of the translation process. The **compiler** translates some source code instructions directly into executable code and other source code instructions into library calls, which are further processed by the link editor. Library calls and link editing are described in detail in the next section. Compiler output is called **object code**, which contains a mixture of CPU instructions, library calls, and other information needed by the link editor. Object code usually is stored in a file with an .o or .obj extension, for example, ComputePayroll.obj.

The compiler reads and translates source code instructions. While reading the source code, the compiler:

► Checks for syntax and other errors and issues appropriate warning or error messages

► Updates internal tables that store information about data items and program components such as functions and subroutines

► Generates either CPU instructions or library calls to carry out the source code instruction

The exact actions taken by the compiler vary depending on the source code instruction type. For most programming languages, there are four source code instruction types:

► Data declarations

► Data operations

► Control structures

► Function, procedure, or subroutine calls

Data Declarations

A **data declaration** defines the name and data type of one or more program variables. For example, consider the following C++ source code instructions:

```
float f_temperature;

float c_temperature;
```

These instructions declare two program variables named f_temperature and c_temperature, which both are floating point numbers. When a compiler reads a data declaration, it allocates memory to store the data item. The amount of memory the compiler allocates depends on the data type and the number of bytes used to represent that data type in the target CPU.

The compiler updates an internal table, called a **symbol table**, to keep track of the data names, types, and assigned memory addresses. Information stored in the table includes the variable name, what the name refers to, the data type, and memory location. Sample entries are shown in Table 10-2.

Table 10-2 ►

Sample symbol table entries for data declarations

Name	Type	Length	Address
f_temperature	Single precision floating point	4	1000
c_temperature	Single precision floating point	4	1004

Data Operations

A **data operation** is any instruction that updates or computes a data value. Examples of data operations include assignment statements and computations. The compiler translates data operation instructions into an equivalent sequence of data movement and data transformation instructions for the target CPU. The compiler refers to entries in the symbol table to determine source and destination memory addresses for data movement instructions.

For example, consider a source code instruction that assigns a constant value to a program variable such as:

```
f_temperature = 212;
```

The compiler looks up f_temperature in the symbol table to find its memory address (1000) and generates a CPU instruction such as:

```
MOV 1000 <212>
```

where <212> is the internal floating point representation of the value 212. An instruction that copies data from one variable to another such as:

```
f_temperature = c_temperature;
```

is translated into the CPU instruction such as:

```
MOV 1000 1004
```

Now consider a more complex example such as:

```
c_temperature = (f_temperature - 32) ÷ 1.8;
```

that combines computation and data movement. The compiler translates this statement into a sequence of CPU instructions such as:

```
MOV R1 1000     ; move f_temperature to register 1
MOV R2 <32>     ; store constant 32 to register 2
```

```
FSUB R1 R2 R3       ; subtract R2 from R1 and
                    ; store the result in register 3
MOV R2 <1.8>        ; store constant 1.8 in register 2
FDIV R3 R2 R3       ; divide register 3 by register 2 and
                    ; store the result in register 3
MOV R3 1004         ; copy register 3 to c_temperature
```

More complex formulas or larger numbers of input variables and constants would require longer CPU instruction sequences.

Control Structures

A **control structure** is a source code instruction that controls the execution of other source code instructions. Control structures include unconditional branches such as a Goto statement, conditional branches such as an If-Then-Else statement, and loops such as While-Do and Repeat-Until.

The common thread among all control structures is the transfer of control among CPU instructions. A CPU branch instruction requires a single operand containing the address of another instruction. Because all control structures require a CPU branch instruction, the compiler must keep track of where CPU instructions for each source code instruction are located in memory.

For example, consider the following source code instructions:

```
        f_temperature = 0;
loop: f_temperature = f_temperature + 1;
        goto loop;
```

and its equivalent CPU instruction sequence:

```
2000 MOV 1000 <0>     ; copy zero to f_temperature
2004 MOV R2 <1>       ; copy 1 to register 2
2008 MOV R1 1000      ; copy f_temperature to register 1
200C RADD R1 R2 R1    ; add R1 and R2 store result in R1
2010 MOV 1000 R1      ; copy result to f_temperature
2014 JMP 200C         ; loop back to add
```

The numbers to the left of each CPU instruction are the hexadecimal memory addresses of those instructions. The name *loop* in the source code is an example of a label. When the compiler reads a label, it stores the label in the symbol table along with the address of the first CPU instruction generated for the corresponding source code instruction. When the label is used as the target of a Goto instruction, the compiler retrieves the corresponding memory address from the symbol table and uses it as the operand of the corresponding CPU branch or jump instruction.

The If-Then-Else, While-Do, and Repeat-Until control structures are based on a conditional branch. For example, consider the source code instruction sequence:

```
f_temperature = 0;

while (f_temperature < 10) {

    f_temperature = f_temperature + 1;

}
```

and its equivalent CPU instruction sequence:

```
2000 MOV 1000 <0>    ; copy zero to f_temperature
2004 MOV R2 <1>      ; copy 1 to register 2
2008 MOV R3 <10>     ; copy 10 to register 3
200C MOV R1 1000     ; copy f_temperature to register 1
2010 RADD R1 R2 R1   ; add R1 and R2 store result in R1
2014 MOV 1000 R1     ; copy result to f_temperature
2018 FLT R1 R3       ; floating point less than comparison
201C CJMP 2010       ; loop back to add if less-than is true
```

As before, each CPU instruction is preceded by its memory address. When the compiler reads the While statement, it temporarily stores the contents of the condition and marks the memory address of the first instruction within the While loop (2010). The compiler then translates each statement within the loop. When there are no more statements within the loop, as indicated by the closing brace, the compiler generates a CPU instruction to evaluate the condition (2018) followed by a conditional branch statement to return to the first CPU instruction in the loop.

An If-Then-Else statement selects one of two instruction sequences using both a conditional and unconditional branch. For example, the source code instruction sequence:

```
if (c_temperature >= -273.16)

    f_temperature = (c_temperature * 1.8) + 32;

else

    f_temperature = -999;
```

causes the compiler to generate the following CPU instruction sequence:

```
2000 MOV R1 1004          ; copy c_temperature to register 1

2004 MOV R2 <-273.16>     ; store constant -273.16 to register 2

2008 FGE R1 R2            ; floating point > or =

200C CJMP 2018           ; branch to if block if true
                         ; begin else block

2010 MOV 1000 <-999>     ; copy constant -999 to f_temperature

2014 JMP 202C            ; branch past if block
                         ; begin if block (calculate f_temp)

2018 MOV R2 <1.8>        ; store constant 1.8 to register 2

201C FMUL R1 R2 R3       ; multiply registers 1 and 2 and
                         ; store the result in register 3

2020 MOV R2 <32>         ; store constant 32 in register 2

2024 FADD R2 R3 R3       ; add registers 2 and 3 and
                         ; store the result in register 3

2028 MOV 1000 R3         ; copy register 3 to f_temperature

202C                     ; next instruction after If-Then-Else
```

Function Calls

In most languages, a programmer can define a named instruction sequence called a **function, subroutine,** or **procedure**[1] that is executed by a **call** instruction. A call instruction transfers control to the first instruction in the function and the function transfers control to the instruction following the call by executing a **return** instruction.

Functions are declared in a manner similar to program variables and the compiler adds descriptive information to the symbol table when it reads the declaration. For example, consider the following function:

```
float fahrenheit_to_celsius(float F) {

/* convert F to a celsius temperature */
```

---

[1] The terms function, subroutine, and procedure all describe a named instruction sequence that receives control via a call instruction, receives and possibly modifies parameters, and returns control to the instruction after the call. For the remainder of the chapter we will use the term function exclusively although everything said about functions also applies to procedures and subroutines.

```
float C;

C = (F - 32) / 1.8;

return(C);

}
```

When the compiler reads the function declaration it adds the symbol fahrenheit_to_celsius to the symbol table and also records the memory address of the function's first CPU instruction, as well as other information such as the function type (float) and the number and type of input parameters. When the compiler reads a later source code line that refers to the function, such as:

```
c_temperature = fahrenheit_to_celsius(f_temperature);
```

it retrieves the corresponding symbol table entry. The compiler first verifies that the types of the input variable (f_temperature) and result assignment variable (c_temperature) match the types of the function and function arguments.

To implement the call and return instructions, the compiler generates CPU instructions to:

► Pass input parameters to the function

► Transfer control to the function

► Execute CPU instructions within the function

► Pass output parameters back to the calling module

► Transfer control back to the calling module at the statement immediately following the function call statement

The flow of control to and from functions is similar to the flow of control to and from an operating system interrupt handler, which was covered in Chapter 7. The calling must be suspended, the function must be executed, and the calling function then must be restored to its original state so it can resume execution. As with interrupt handlers, register values are pushed on the stack just before the function begins execution and popped from the stack just before the calling function resumes execution. The stack serves as a temporary holding area for the calling module so it can be restored to its original state after the function completes execution.

Table 10-3 shows a set of symbol table entries for the function fahrenheit_to_celsius, its internal parameters, and the variables within the function call:

```
c_temperature = fahrenheit_to_celsius(f_temperature);
```

Table 10-3 ▶

Symbol table
entries

Name	Context	Type	Address
fahrenheit_to_celsius	Global	Single precision floating point, executable	3000
F	fahrenheit_to_celsius	Single precision floating point	2000
C	fahrenheit_to_celsius	Single precision floating point	2004
f_temperature	Global	Single precision floating point	1000
c_temperature	Global	Single precision floating point	1004

The compiler reads data names used in the function call and return assignment, and looks up each name in the symbol table to extract the corresponding address. It also looks up the name of the function parameter and extracts its addresses and lengths. The compiler then creates a MOV4 (move 4 bytes) instruction using this address and places it immediately before the PUSH instruction:

```
MOV4 2000 1000    ; copy fahrenheit temperature

PUSH              ; save calling process registers

JMP 3000          ; transfer control to function
```

The return from the function call is handled in a similar manner:

```
MOV4 1004 2004    ; copy function value to caller

POP               ; restore caller register values
```

The compiler generates code to copy the function result to a variable in the calling program. Control then is returned to the caller by executing a POP instruction.

LINK EDITING

The examples in the previous section showed how a compiler directly translates assignment, computation, and control structure source code instructions into CPU instructions. The compiler translates other types of source code instructions such as file, network, and interactive I/O statements into external function calls. The compiler also generates external function calls when it encounters source call instructions without corresponding source code function declarations. An **external function call**, sometimes called an **unresolved reference**, is a placeholder for missing executable code. An external function call contains the name and type of the called function, as well as the memory addresses and types of function parameters.

A **link editor** searches an object code file for external function calls. When an external function call is found, the link editor searches other object code files or compiler libraries to find executable code that implements the function. A **compiler library** is a file that contains a related set of executable functions and an index of the library contents. When the link editor finds executable code in a library or another object code file, it inserts the executable code into the original object code file and generates instructions to copy parameters and the function return value. The result is a single file containing only executable code. If the link editor cannot find executable code for an external function call, it generates an error message and produces no executable file.

At this point, you might be wondering why the compiler couldn't have generated all of the executable code, eliminating the need for a link editor. Link editors provide two key benefits in program translation:

► A single executable program can be constructed from multiple object code files compiled at different times.

► A single compiler can generate executable programs that run under multiple operating systems.

Consider how a program to convert temperatures from fahrenheit to celsius and vice versa might be compiled and linked if each function were written by a different programmer. Each programmer would independently develop a function and store the source code in a file. Each file would be compiled separately, and none of the object code files would contain a complete program. The link editor patches together the call and return instructions in each object code file to create a single executable file (see Figure 10-10).

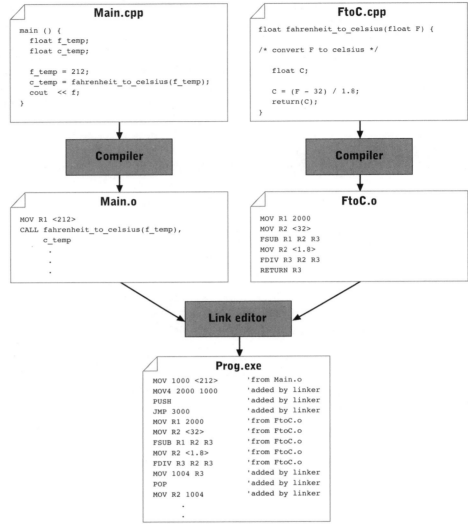

```
                    Main.cpp

main () {
  float f_temp;
  float c_temp;

  f_temp = 212;
  c_temp = fahrenheit_to_celsius(f_temp);
  cout  << f;
}
```

```
                    FtoC.cpp

float fahrenheit_to_celsius(float F) {

/* convert F to celsius */

  float C;

  C = (F - 32) / 1.8;
  return(C);
}
```

Compiler

Compiler

```
                    Main.o

MOV R1 <212>
CALL fahrenheit_to_celsius(f_temp),
     c_temp
     .
     .
     .
```

```
                    FtoC.o

MOV R1 2000
MOV R2 <32>
FSUB R1 R2 R3
MOV R2 <1.8>
FDIV R3 R2 R3
RETURN R3
```

Link editor

```
                    Prog.exe

MOV 1000 <212>        'from Main.o
MOV4 2000 1000        'added by linker
PUSH                  'added by linker
JMP 3000              'added by linker
MOV R1 2000           'from FtoC.o
MOV R2 <32>           'from FtoC.o
FSUB R1 R2 R3         'from FtoC.o
MOV R2 <1.8>          'from FtoC.o
FDIV R3 R2 R3         'from FtoC.o
MOV 1004 R3           'added by linker
POP                   'added by linker
MOV R2 1004           'added by linker
     .
     .
```

The link editor also makes it possible to use a single compiler for multiple operating systems. For example, consider the C++ temperature conversion program in Figure 10-11 to read temperature input from the keyboard (lines beginning with // are comments). Because the CPU instructions for interactive I/O vary by operating system, if the compiler were responsible for generating CPU instructions for I/O operations, then different compilers would be required to implement this program for each operating system.

Figure 10-11 ▶

A C++ program
that reads input
from the
keyboard

```
main() {
    float f;
    float c;

    // print prompt
    cout << "Enter a fahrenheit temperature: ";
    // get keyboard input
    cin >> f;
    // compute celsius equivalent
    c = fahrenheit_to_celsius (f);
    // print celsius temperature
    cout << f << " degrees fahrenheit is ";
    cout << c << " degrees celsius." << endl;

    return(0);
}
```

Using external function calls and link editing allows a single compiler to serve multiple operating systems. CPU instructions that implement I/O functions for each operating system are stored in a compiler library. When object code files are linked, the programmer instructs the link editor to use the library for the target operating system. The link editor replaces external function calls for interactive I/O with the appropriate CPU instructions for a specific operating system.

Link editing with alternate compiler libraries can provide flexibility in other ways. For example, encryption algorithms vary in complexity and maximum length of the encryption key. Different versions of encryption algorithms with the same name and parameters can be stored in different libraries. An application programmer then can choose the appropriate level of encryption to embed in an executable program by choosing which encryption library is read by the link editor.

Dynamic and Static Linking

The linking process described to this point often is called **static linking**, or **early binding**. The term static is used because library and other subroutines cannot be changed once they are inserted into the executable code. **Dynamic linking**, or **late binding**, is linking that is performed during program loading or execution.

Calls to operating system service routines sometimes are implemented with dynamic linking. For example, most Windows operating system service routines are stored in **dynamically linked libraries (DLLs)**. During program development, the link editor inserts code to call a special DLL interrupt handler and passes pointers to the service routine name and parameters in general-purpose registers. During program execution, the DLL interrupt handler accepts DLL service requests, locates the service routine by name, loads it into memory if it isn't already there, copies input parameters, and then executes the service function. When the service routine finishes, the DLL interrupt handler copies output parameters, terminates the service routine, and passes control back to the calling program.

Dynamic linking provides two key advantages as compared to static linking. The first advantage is smaller application program executable files. Static linking incorporates copies of service layer functions directly into executable programs. Depending on the nature and number of functions included, this can add dozens of megabytes to each executable program. Much of this additional storage is redundant because commonly used subroutines, such as those to open and close files or windows, are used by many application programs. Dynamic linking avoids redundant storage of service functions within application program executable files. Instead, each program contains only the code needed to call the DLL interrupt handler, which typically is only a few hundred kilobytes.

The second advantage of dynamic linking is flexibility. Portions of the operating system can be updated by installing new DLLs. If application programs use dynamic linking, then updating the operating system updates all application programs with the new service layer subroutines. The new subroutines are loaded and executed the next time an application program is executed after an upgrade. In contrast, static linking locks an application program to a particular version of the service layer functions. The functions can be updated only by providing an updated compiler library and relinking the application object code.

The primary advantage to static linking is execution speed. When a dynamically linked program calls a service layer function, the operating system must find the requested function, copy parameters, and transfer control. If the subroutine is not in memory, the application program is delayed until the service function is located and loaded into memory. Most of these steps are avoided if the application program is statically linked, resulting in significantly faster execution.

Static linking also improves the reliability and predictability of executable programs. When an application program is developed, it usually is tested with a specific set, or version, of operating system service functions. Service function updates are designed to be backward compatible with older versions, but there is always the risk that new functions dynamically linked into an older application program will result in unexpected behavior. Static linking ensures that this cannot happen. The reverse situation also can occur. That is, a new service function version might fix bugs that were present in an older version. Static linking prevents these fixes from automatically being incorporated into existing programs.

INTERPRETERS

Source and object code files are compiled and linked as a whole. In contrast, **interpretation** interleaves source code translation, link editing, and execution. An **interpreter** reads a single source code instruction, translates it into CPU instructions or a DLL call, and immediately executes the instructions or DLL call before the next program statement is read. In essence, one program (the interpreter) translates and executes another (the source code) one statement at a time.

The primary advantage of interpretation over compilation is that it gives flexibility to incorporate new or updated code into an application program. This flexibility is a direct result of dynamic linking. Dynamic linking can be used for operating system services and for other program code. Program behavior can be updated easily by installing new versions of dynamically linked code. It also is possible for programs to modify themselves directly. This flexibility is required for certain types of applications such as expert systems.

The primary disadvantage of interpretation as compared to compilation/linking is increased memory and CPU requirements during program execution (see Table 10-4). With compilation and link editing, each program used to develop executable code resides in memory for only a short time. When a program is loaded and executed, all of the software components that were used to create it (the program editor, compiler, and link editor) are no longer in memory. The memory requirements of a compiled and linked application program consist only of the memory required to store its executable code.

Table 10-4 ▶

Memory and CPU resources used during application execution

Resource	Interpretation	Compilation
Memory contents (during execution):		
Interpreter or compiler	Yes	No
Source code	Partial	No
Executable code	Yes	Yes
CPU instructions (during execution):		
Translation operations	Yes	No
Library linking	Yes	No
Application program	Yes	Yes

An interpreter is a large and complex program that must reside in memory during the entire execution of a source code program. Essentially, there are two programs in memory at run time—the interpreter itself and the application program that it is translating and executing. The memory requirements of an interpreted program are substantially greater than those of a compiled program. This is a critical performance difference for many CPU-intensive applications such as numerical modeling and data mining.

Technology
Focus

Java

Java is an object-oriented programming language and program execution environment. It was developed by Sun Microsystems during the early and mid 1990s. It first debuted in the Netscape Navigator Web browser version 3.0 and, as a result, has been described widely as a Web or network programming language. Although it does satisfy that description, Java is a full-featured programming language that supports virtually any combination of hardware platform and operating system (OS).

Java is similar in syntax and capability to C++, so programmers familiar with C++ can be trained quickly in Java. Java incorporates support for object-oriented features, including encapsulation, inheritance, and polymorphism—multiple inheritance is not supported. Java is designed to maximize reliability of applications and reusability of existing code.

What makes Java unique is a standardized target machine language for Java interpreters and compilers. Compilers and interpreters for other programming languages translate source code into executable CPU instructions and service function calls to specific operating system function calls. A Java compiler or interpreter translates Java source code into machine instructions and service function calls for a hypothetical computer system and operating system called the **Java Virtual Machine (JVM)**. Instructions and library calls to the JVM are called Java byte codes.

The term JVM also describes an interpreter that translates JVM instructions into actual CPU instructions and operating system service function calls. The JVM simulates a Java computer system with the CPU and operating system of a conventional computer system (see Figure 10-12). The JVM byte code interpreter translates JVM machine instructions into CPU instructions for a real CPU such as an Intel Pentium or Sun SPARC. Calls to JVM service functions are translated into equivalent calls to a real operating system such as Microsoft Windows or UNIX.

Java programs come in two forms: applets and stand-alone applications. Stand-alone Java applications execute within the JVM, which also handles all security considerations for stand-alone applications. A Java **applet** is designed to run within another program such as a Web browser. Applets run within a protected area called the **sandbox**. The sandbox implements extensive security controls to prevent applets from accessing unauthorized resources or damaging the hardware, operating system, or file system.

In 1997, Sun announced that it had developed a microprocessor that directly executed Java byte codes. Sun hoped that the microprocessor would be applied widely in embedded applications such as Web-enabled televisions. Sun apparently has abandoned that hardware approach because its latest chip offerings don't directly execute Java byte codes, though Sun claims that the chips are optimized to execute Java programs.

Figure 10-12 ▶

How Java and
JVM relate to
the CPU and OS

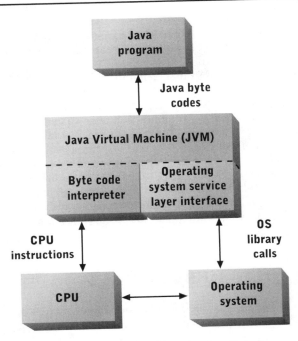

Native support for Java OS calls has been added to the Sun Solaris operating system, which is a customized version of UNIX. Other operating system vendors have not added native support for Java, though it is rare to find a general-purpose computer system without an installed JVM. Incorporating the JVM into the native operating system should reduce some of the inefficiency inherent in the current OS translation processes. It also will reduce the need for Java-enabled application software to install and use private and sometimes incompatible copies of JVM software.

The popularity of Java has far exceeded most initial expectations due to a number of factors including:

▶ Sun's strategy of providing Java compilers and virtual machines at little or no cost

▶ Incorporation of JVMs into Web browsers

▶ Increasing development of application software that uses a Web browser as the primary I/O device

▶ The ability of Java programs to execute on any combination of computer hardware and operating system

There are a few drawbacks to Java. The first is reduced execution speed due to the use of interpreted byte codes and operating system translation. Emulation processes are inherently inefficient users of hardware resources. **Native applications**, those compiled and linked for a particular CPU and operating system, typically execute

10 times faster than interpreted Java byte code programs. Some JVMs employ a just-in-time compiler to reduce the speed difference to approximately 3:1, but Java programs always will be slower than native programs when executed on anything other than a true Java machine. This is an important consideration for programs that make maximal demands of available hardware resources such as simulation applications and database management systems.

The security model of Java is also a trouble spot. Early versions of Java made several restrictions on program behavior to enhance security. These included access restrictions to the file system and memory. These restrictions enhanced security at the expense of program functionality. They led to early perceptions of Java as a programming language that was capable of implementing only cute animated images in Web pages. Current Java standards have relaxed these restrictions and, as a result, have endowed the language with the capability to implement more powerful applications. But relaxing the restrictions has created additional security problems for which complete solutions have yet to be developed.

SYMBOLIC DEBUGGING

Compilers, link editors, and interpreters produce error messages to warn programmers of actual or potential mistakes in source code. However, even if a program is compiled and linked successfully, it might still produce errors when executed. Diagnosing run-time errors based on a typical run-time error message is difficult because the error message refers to the executable code generated by the compiler and link editor instead of the source code written by the programmer. Variable names have been replaced by memory addresses and symbolic source code instructions have been replaced by CPU instructions. Determining the source code instructions and variables that correspond to these memory addresses is a difficult task.

A compiler can be instructed to incorporate its symbol table into the object code file and a link editor can be instructed to produce a **memory map**, or **link map**, based on the symbol table contents. A memory map lists the memory location of every function and program variable. A programmer can use a memory map to trace error messages containing memory addresses to corresponding program statements and variables. Tracing those addresses to particular source code statements or data items requires a very detailed memory map and a thorough understanding of machine code and the compiling and linking processes.

For example, consider the partial memory map in Figure 10-13 (the complete map is several hundred text lines) and the run-time error message in Figure 10-14. Based on the instruction address in the error message and the starting addresses of the modules listed in the memory map, a programmer can determine that the error occurred somewhere within the function _main. But the programmer can't tell exactly which instruction caused the error or the contents of program variables at the time the error occurred.

```
TemperatureConversion

Preferred load address is 00400000
   Location              Publics by Value         Address   Lib:Object

0001:00000000    ?fahrenheit_to_celsuis@@YAMM@Z   00401000   Main.obj
0001:00000020    ?celsuis_to_fahrenheit@@YAMM@Z   00401020   Main.obj
0001:00000040    _main                            00401040   Main.obj
0001:000001b0    ??6ostream@@Z@Z                  004011b0   Main.obj
0001:000001d0    ?flush@@YAAAVostream@@AAV1@@Z    004011d0   Main.obj
0001:00009242    _ExitProcess@4                   0040a242   KERNEL32.dll
0002:000000f0    __real@8@3ffe8e38e38e38e39000    0040b0f0   Main.obj
0002:000000f8    __real@4@40048000000000000000    0040b0f8   Main.obj
0002:00000100    __real@8@40048000000000000000    0040b100   Main.obj
0002:00000108    __real@8@3fffe666666666666800    0040b108   Main.obj

entry point at           0001:00001ac2
```

TemperatureConversion.exe - Application Error

The instruction at "0x00401176" referenced memory at "0x42000000". The memory could not be "read".

Click on OK to terminate the program
Click on CANCEL to debug the program

OK   Cancel

A **symbolic debugger** is an automated tool for testing executable programs. It provides a number of features, including the ability to:

► Trace calls to specific source code statements or subroutines

► Trace changes to variable contents

► Execute source code instructions one at a time

► Detect run-time errors and report them to the programmer in terms of specific source code instructions and variables

A symbolic debugger uses the symbol table, memory map, and source code files to trace memory addresses to specific source code statements and variables. It also inserts debugging checkpoints after each source code instruction so program execution can be paused. An executable program that contains symbol table entries and debugging checkpoints sometimes is called a **debugging version**. In contrast, the

---

production version, or **distribution version,** of a program omits the symbol table and debugging checkpoints to reduce program size and increase execution speed.

Interpreted programs have an inherent advantage over compiled programs in detecting and reporting run-time errors because the contents of the symbol table and program source code always are available to the interpreter at run time. If an error occurs, it can be reported in terms of the most recent source code line translated. Memory addresses also can be converted to source code names by looking them up in the symbol table. Most interpreters directly incorporate symbolic debugging capabilities. In other words, they don't need a stand-alone symbolic debugging program.

INTEGRATED APPLICATION DEVELOPMENT TOOLS

Programming languages, program translation tools, and debugging tools address only the latter half of the SDLC implementation phase. Other tools are needed to address other implementation tasks and earlier SDLC phases. Figure 10-15 summarizes the application development tools discussed to this point, and the additional tools and the stages of the SDLC to which they apply.

Figure 10-15 ▶

Application development tools

Programmer's Workbenches

A **programmer's workbench** is an integrated set of automated support tools to speed development and testing. Programmer's workbenches generally provide the following components:

► Smart program editor
► Compiler and/or interpreter

- Link editor and a large library of classes or subroutines
- Interactive facility for prototyping and designing user interfaces
- Symbolic debugger
- Integrated window-oriented graphical user interface

The key feature of a programmer's workbench is the level of integration among tools, not the specific tools provided. A programmer's workbench leverages programmer effort and skill so programs can be written and tested in minutes, hours, or days. The workbench automates tedious and time-consuming tasks, streamlines other tasks, and maximizes reuse of existing source code. Using a programmer's workbench, a programmer can rapidly develop and test many program versions.

Editors included in the workbench assist the programmer in writing syntactically correct code by performing rudimentary syntax checking as the code is being typed and visually identifying errors. Compiler and link editor errors displayed in one window are linked directly to specific source code instructions in another window. Some compilers provide suggested corrections or context-sensitive help in addition to error messages.

Workbenches also can provide or generate program templates and skeletons that can be stored in a library or generated based on input from the programmer or from other tools. Workbenches usually contain large libraries of predefined subroutines or classes which address user interface, database manipulation, network resource access, and other common functions. Library routines can be imported directly into program source code or added via function or method calls and linked into the executable program. Libraries allow programmers to reuse source and executable code, thus speeding up application development.

Workbenches include extensive symbolic debugging support. Method calls, function calls, and changes to program variables or object data can be traced during program execution. Programs can be started and stopped at programmer specified locations so testing can be performed on specific code segments. Debugging and tracing also can be performed on machine code and operating system service calls. Testing and debugging occur in a simulated run-time environment so program crashes can't disable the workbench or the entire workstation.

Workbench hardware and operating system requirements are high. Complex tools and simulated execution environments require substantial CPU power and memory. Large libraries, complex tools, and extensive documentation require significant amounts of secondary storage. Large monitors are needed to display all of the windows used in the development process. Workbenches typically require two to three times the CPU power, memory, and disk storage of a typical business workstation.

Technology
Focus

Microsoft Developer Studio

Microsoft Developer Studio is a programmer's workbench for developing software that runs under various Microsoft Windows operating systems. It supports several programming languages including C, C++, Visual Basic, FORTRAN, and Java.

A Developer Studio workspace is a container for application components. A programmer interacts with a user-oriented view of the workspace in a window (see the upper-left subwindow in Figure 10-16). This view organizes all of the program components so the programmer can see or access them at a glance.

Figure 10-16 ▶

Microsoft
Developer Studio

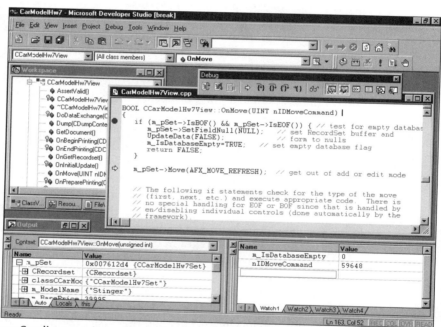

Developer Studio contains many software development tools, including compilers, link editors, program text editors, symbolic debuggers, and interactive tools for designing user interface elements such as menus, toolbars, icons, and forms. Developer Studio keeps track of which tools are used to create and modify each programmer-defined object and opens the appropriate tool automatically when the programmer selects an application component. A set of overlapping menus, toolbars, and icons provides access to various tool features and commands.

Separate tools are integrated automatically whenever possible. For example, a command to build a complete program automatically executes user interface resource compilers, program compilers for each component programming language, and the link editor, in the appropriate order. Developer Studio keeps track of file dependencies so a change to one source code file causes all dependent files to be rebuilt automatically. Developer Studio manages data movement among tools and displays progress and results within an output window (see the bottom of Figure 10-16).

Developer Studio also provides extensive symbolic debugging capabilities. A programmer can pause program execution at specified source code lines. Further execution can be directed on a line-by-line or function-by-function basis. A programmer can examine variable content while the program is paused and trace variable changes during execution.

Developer Studio provides over 1 GB of documentation that can be accessed from a hard disk, compact disc, or the Web. A sophisticated documentation browser provides extensive search and indexing capabilities. The documentation contains many cross-references and hyperlinks to glossary terms, examples, and tutorials.

CASE Tools

The term **computer assisted software engineering (CASE) tool** usually refers to a tool that supports the SDLC analysis and design phases. The term "CASE tool" is a bit of a misnomer because any tool that assists system or program development fits within the literal definition of a CASE tool. The singular form of the term also is a misnomer because CASE tools normally are highly integrated tool suites.

The key feature of a CASE tool is support for a broad range of system development activities with particular emphasis on model development. CASE tools support most or all system development activities up to, and sometimes including, program translation. CASE tools usually support a specific system development methodology such as structured system development, object-oriented development, or Information Engineering. A tool suite that primarily supports model development sometimes is called a **front-end CASE tool**. A tool suite that primarily supports program development based on specific analysis and design models sometimes is called a **back-end CASE tool**, or **code generator**.

CASE tools can perform very specialized model development and analysis tasks due to high integration levels and specialization to particular methodologies and models. CASE tools can perform consistency and other types of error checking on models. For example, a CASE tool that supports structured analysis can verify that all data flows on a data flow diagram have corresponding definitions in the data dictionary. Such a tool also can verify that the data outputs of each data flow diagram process can be produced from the data inflows based on the process logic specification.

CASE tools with extensive back-end capabilities can automate much of the software development process. For example, a CASE tool that supports object-oriented development can generate class declarations in Java or C++ source code files. Such a tool also can generate database schema descriptions and component registration entries for specific database and component management systems.

The most comprehensive CASE tools automate the entire process of generating working systems from fully specified analysis and design models. Such tools generate program and other source code from models, compile and link the programs, create databases, and create, register, and install all components. CASE tools also can provide comprehensive facilities for planning and managing development projects and for maintaining and modifying existing systems. Comprehensive CASE tools usually are expensive, require weeks or months of training, and support very specific target deployment environments.

Business
Focus
Building the Next Generation of Application Software

Southwestern Gifts, Incorporated (SGI) is a small catalog retailer of southwestern art, jewelry, and gifts. SGI distributes two catalogs per year, in May and October. Most orders are received by telephone and the rest by mail. SGI employs up to 50 people, including 15 warehouse staff and 30 order-processing clerks during the Christmas shopping season, 4 accountants and bookkeepers to handle payroll and accounts payable, 3 buyers, a catalog designer, 3 IS staff, a warehouse manager, and the owner-operator.

SGI currently has a Compaq minicomputer running the Linux operating system, a small Dell server running Windows 2000 Server, 6 microcomputers, and 12 Wyse WY series video display terminals. The Wyse terminals are connected to the minicomputer by serial cables and are used only for order entry. The Dell server provides file and printer sharing for the PCs and printer sharing for the minicomputer. The PCs are used by the accountants, warehouse manager, buyers, catalog designer, and IS staff. A 100 Mbps Ethernet LAN connects the PCs, Dell server, and Compaq minicomputer.

Application software on the personal computers includes Quicken (an accounting package), various productivity tools such as Microsoft Office, and graphical design software. SGI purchased an order entry and inventory control package eight years ago from an out-of-state vendor. The package stores data in ordinary Linux files and uses a proprietary application development and operating environment called DERQS (Data Entry, Retrieval, and Query System).

DERQS provides forms-based data entry, a forms-based data query facility, a file definition tool, and a simple report generator. Screen forms are defined using an interactive layout tool that compiles and stores screen layout and content in a screen definition file (SDF). Files also are defined using an interactive tool, and descriptions of file format and content are stored in a data description file (DDF). Data content is stored in ordinary UNIX files.

Application programs are written in a simple interpreted script language to display and manipulate screens, manipulate file content, and generate simple reports. Most of the existing application programs are DERQS scripts but some are written in C. DERQS provides a compiler library with functions that allow C programs to display DERQS screens and interact with DERQS files.

One IS staff member specializes in DERQS data entry functions and script-based applications. A second IS staff member specializes in the DERQS query language and develops and maintains C applications. A third IS staff member manages the LAN and personal computers.

SGI would like to make several improvements to their information systems, including:

▶ A Web site to provide catalog information and order entry

▶ A direct interface to financial information in DERQs data files so data can be uploaded to Quicken and Excel

▶ An automated system to support the buyers, and which interacts directly with accounts payable

▶ Modern graphical user interfaces for existing and newly developed application programs using modern thin client terminals and full-featured microcomputers

SGI recently learned that the company that developed and supported DERQS has filed for bankruptcy. No company has shown an interest in purchasing the rights to DERQS so it appears that no further upgrades or technical support will be available.

Questions

1. Should SGI develop any new software using DERQS? If not, what tools should they acquire for new system development?

2. Should SGI re-implement the functions of their existing DERQS-based applications using more up-to-date development tools?

3. What risks and hidden costs might SGI face if they acquire new tools to implement old software functions and/or develop new software?

SUMMARY

► Application systems are developed by following the steps of the systems development life cycle (SDLC). Application development follows a development methodology, develops a set of models, and uses automated tools. Most software development cycles include two sets of model: a requirements model and a design model.

► Executable software consists entirely of CPU instructions. Modern programmers write programs in higher-level programming languages that are translated into CPU instructions. Programming languages have evolved through six generations, the most recent of which is object-oriented programming. Later generations have incorporated improvements and capabilities such as instruction explosion, database access, support for graphical user interfaces, and nonprocedural programming.

► All programming language generations other than the first must be translated into CPU instructions prior to execution. Two commonly used translation approaches are compilation and interpretation. A compiler translates an entire source code file, building a symbol table and object code file as output. An interpreter translates, links, and executes source code programs iteratively, one source code instruction at a time.

► Compiled and interpreted programs must be linked to libraries of executable functions or methods. A link editor statically links external reference calls in object code to library functions and combines them into a single file containing executable code. A link editor also can dynamically link external calls by statically linking them to an operating system service function that loads and executes dynamically linked library (DLL) functions at run time. Interpreters always use dynamic linking.

► Application development is more efficiently supported by integrated suites of automated tools. A programmer's workbench integrates tools such as a program editor, screen and report designers, an intelligent compiler and link editor, a symbolic debugger, and extensive documentation. A CASE tool provides a tool suite to support earlier SDLC phases. A front-end CASE tool supports development of analysis and design models. A back-end CASE tool generates program source code from models.

In the next two chapters, we'll look at operating systems in detail. Chapter 11 describes the internal architecture of an operating system and discusses how it manages processes and memory. Chapter 12 concentrates on file and secondary storage device management.

Key Terms

applet
assembler
assembly language
back-end CASE tool
call
code
code generator
coder
compiler
compiler library
computer assisted software
    engineering (CASE) tool
control structure
data declaration
data operation
debugging version
design model
distribution version
dynamic linking
dynamically linked library
    (DLL)
early binding
executable code
external function call

fifth generation language
    (5GL)
first generation language
    (1GL)
fourth generation language
    (4GL)
front-end CASE tool
function
instruction explosion
interpretation
interpreter
Java
Java Virtual Machine (JVM)
label
late binding
link editor
link map
machine independence
machine language
memory map
message
methods
native application
nonprocedural language

object
object code
object-oriented programming
    (OOP)
procedure
production version
programmer
programmer's workbench
programming language
return
sandbox
second generation language
    (2GL)
source code
static linking
subroutine
symbol table
symbolic debugger
system requirements model
third generation language
    (3GL)
unresolved reference
variable

Vocabulary Exercises

1. A compiler allocates storage space and makes an entry in the symbol table when it encounters a(n) _____ in source code.

2. A(n) _____ is produced as output of the SDLC analysis phase.

3. A link editor searches object code for _____ .

4. A(n) _____ is produced as output of the SDLC design phase.

5. A 4GL has a greater degree of _____ than a 3GL.

6. _____ contains CPU instructions and external function calls.

7. A(n) _____ produces a(n) _____ to show the location of functions or methods in executable code.

8. A(n) _____ is a model or system component that combines a set of related data and the methods that manipulate that data.

9. The compiler adds the names of data items and program functions to the _____ as they are encountered in source code.

10. A 2GL is translated into executable code by a(n) _____.

11. A(n) _____ translates an entire program before linking and execution. A(n) _____ interleaves translation and execution.

12. A Java _____ runs within the _____ of a Web browser.

13. _____ in a source code program are translated into machine instructions to evaluate conditions and/or transfer control from one program module to another.

14. Relationships among objects that are modeled in an object-oriented system model include _____ and _____.

15. A(n) _____ uses the content of the symbol table to aid the programmer in tracing memory locations to program variables and instructions.

16. FORTRAN, COBOL, and C are examples of _____.

17. A(n) _____ tool supports system model development. A(n) _____ tool generates program source code from systems models.

18. A link editor performs _____ linking. An interpreter performs _____ linking.

19. Java programs are compiled into object code for a hypothetical hardware and system software environment called the _____.

Review Questions

1. Describe the relationships between application development methodologies, models, and tools.

2. Compare and contrast the various generations of programming languages, including object-oriented programming languages.

3. What is instruction explosion? What types of programming languages have the most instruction explosion? What types of programming languages have the least instruction explosion?

4. What are the differences between source code, object code, and executable code?

5. Compare and contrast assemblers, compilers, and interpreters.

6. What does a compiler do when it encounters data declarations in a source code file? Data (manipulation) operations? Control structures?

7. Compare and contrast the execution of compiled programs to interpreted programs in terms of CPU and memory utilization.

8. What is a link editor? What is a library? How and why are they useful in program development?

9. What types of programming language statements are likely to be translated into machine instructions by a compiler? What types are likely to be translated into library calls?

10. How does interpretation differ from compilation?

11. Compare and contrast the error detection and correction facilities of interpreters and compilers.

12. Compare and contrast static and dynamic linking.

13. With respect to the requirements of modern applications, what are the shortcomings of 3GLs?

14. What happens to application program portability when a programmer uses non-ANSI programming language extensions? Why do so many programmers use these extensions?

15. What are the primary differences between object-oriented programming languages and more traditional programming languages?

16. What components normally are part of a programmer's workbench? In what ways does a workbench improve programmer productivity?

17. What is a CASE tool? What is the relationship between a CASE tool and a system development methodology?

18. What is the difference between a front-end CASE tool and a back-end CASE tool?

Problems and Exercises

1. Develop a set of machine instructions to implement the following source code fragment:

```
a = 0;
i = 0;
while (i < 10) do
    a = a+i;
    i = i+1;
endwhile
```

2. Some RISC processors do not provide PUSH and POP instructions. The rationale is that they are complex instructions that can be built from simpler components. Write a short machine language program to implement the PUSH and POP operations using only data movement and branching instructions.

Research Problems

1. Investigate a modern CASE tool such as those offered by Sterling Software (www.sterling.com) or Oracle (www.oracle.com). On what system development methodology or methodologies is the tool based? What types of system models can be built with the tool? How is an analysis model translated into an implementation model? What programming languages, operating systems, and database management systems are supported by the back-end CASE tool? How can the tool(s) be used to support rapid prototyping?

2. Investigate a modern programmer's workbench such as Oracle JDeveloper (www.oracle.com) or Borland Delphi (www.borland.com). Are application programs interpreted, compiled, or both? What program editing tools are provided? What tools are available to support run-time debugging? What DBMSs can be accessed by application programs?

3. Investigate the latest Java-oriented microprocessors offered by Sun (www.sun.com). In what ways are they optimized to execute Java programs? What architectural features are similar to microprocessors of competitors such as Intel? Is the Java orientation of Sun microprocessors a fact or just marketing hype?

4. Investigate the graphical user interface (GUI) features of different operating systems such as Windows and MacOS, and the features of a later 3GL such as C or an object-oriented language such as SmallTalk, C++, or Java. How would you go about developing a suite of interactive test programs to ensure compliance with a programming language standard? What GUI features couldn't be tested because they're not embedded within the programming language? How would nonstandard extensions to the language affect program portability between the two operating systems? Could a language standard be constructed that would use equivalent instructions or statements to implement an interactive interface under both Windows and MacOS? Could the behavior and appearance of the GUI be exactly the same when a single program is translated and executed under both operating systems?

Chapter 11

Operating Systems

Chapter Goals

- ▶ Describe the functions and layers of an operating system

- ▶ List the resources allocated by the operating system and describe the complexities of the allocation process

- ▶ Describe how an operating system manages programs, processes, and threads

- ▶ Compare and contrast alternative CPU scheduling methods

- ▶ Explain how the operating system manages memory

- ▶ Describe signals and pipes and how they are used to coordinate threads and processes

Operating systems are critical components of all modern information systems. Users and application programs rely on operating systems to provide many services and to manage software, data, and hardware resources. As an information systems professional, you often will be asked to select, install, configure, and upgrade operating systems. Successfully performing those tasks requires a detailed understanding of how operating systems work. Figure 11-1 shows some of the topics that you will learn in this chapter.

Figure 11-1 ▶

Topics covered in this chapter

OPERATING SYSTEM OVERVIEW

The operating system is the most important system software component in an information system. It plays a dual role as a high-level manager and a tireless low-level worker. As a manager, the operating system oversees all hardware resources and allocates them to users and applications as needed. It also shifts resources as needed to meet rapidly changing demands while enforcing resource security and sharing policies. With respect to resources, the operating system is similar to "Big Brother"—an omniscient government with unquestioned authority.

At the same time, the operating system performs many low-level tasks on behalf of users and application programs such as accessing files and directories, creating and moving display windows, and accessing resources over a network. The operating system provides these services for two reasons. First, it is efficient. If many application programs perform similar tasks, it makes sense to create services accessible to all applications. That way, application programmers are free to address tasks unique to their application area. Operating system services are a form of code reuse.

The second reason operating systems provide so many low-level functions is because in doing so, the operating system maintains absolute control over hardware resources. For example, the operating system provides file access services so it can maintain complete control over secondary storage devices. With complete control, the operating system can enforce system-wide security policies, prevent programs from accidentally or maliciously destroying one another's data, and efficiently allocate disk and I/O capacity to meet the needs of all users and programs.

Operating System Management Functions

Operating system management functions (see Figure 11-2) can be divided loosely between those oriented to hardware resources and those oriented to users and their programs. Each hardware resource type (CPU, memory, secondary storage, and I/O devices) has an associated operating system management function, as do users and programs. This chapter explores CPU management, memory management, and process management.

All programs require CPU and memory to execute. Managing programs, CPUs, and memory is complex because a modern computer system can have hundreds or thousands of programs executing at the same time. The operating system tracks each program from start to finish and processes all requests for resource allocation and communication. All executing programs compete for a limited memory pool and access to a small number of CPUs. The operating system controls which programs access which resources and for how long they control those resources.

Most computer systems have gigabytes of secondary storage organized into tens of thousands or hundreds of thousands of files. Keeping track of where files are located on disk requires a complicated set of directories and allocation tables. The operating system updates directories and allocation tables as files are created, modified, and deleted. It performs all low-level file access functions on behalf of users and programs so it can maintain tight control over directories and allocation tables. The operating system also enforces access controls and provides utility functions such as backup and recovery. Secondary storage functions are described in detail in the next chapter.

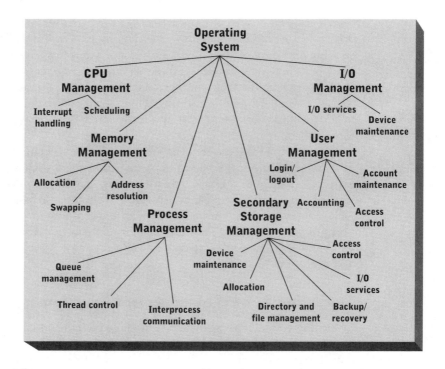

Figure 11-2 ▶

Operating system management functions

The operating system manages all I/O devices and accesses those devices on behalf of users and programs. It also manages the movement of data among I/O devices, memory buffers, and programs. As the user moves from program to program, the operating system reconnects interactive devices such as keyboards and sound cards. The operating system provides many interactive I/O services, including output to printers and display devices. By providing these services, the operating system can efficiently manage shared resources, relieve application programmers of many user interface development tasks, and provide consistent interactive user input forms across programs. The operating system creates and manages user accounts and tracks ownership of files and other resources. The operating system enforces access controls based on user identity, resource ownership, and explicit access authorizations.

Operating System Layers

Like other complex software, operating systems are organized internally into layers as shown in Figure 11-3. Using layers makes the operating system more maintainable because functions within one layer can be modified without affecting other layers. The outermost layers provide services to application programs or directly to end-users. The innermost layer encapsulates hardware resources, preventing users and application programs from directly accessing them.

Figure 11-3 ▶

Operating system layers (shaded)

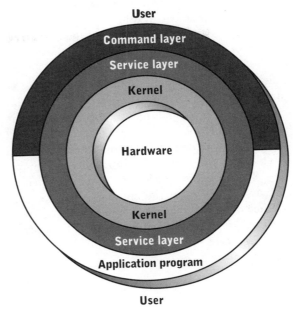

The **command layer**, sometimes called the **shell**, is the user's interface to the operating system. Through the command layer, a user or system administrator can execute programs, including operating system utility programs, and directly manage files and directories.

The command layer can be implemented with a textual or a graphical user interface. Textual interfaces accept user input from the keyboard via a command prompt. The set of commands and their syntax requirements are called a **command language**, or sometimes a **job control language** (**JCL**). Command languages tend to be difficult to learn and use. The user or programmer must know the syntax and semantics of the language in order to issue commands. Examples of command languages and command sets are MS-DOS, IBM MVS JCL, and UNIX Bourne Shell.

Modern operating systems provide a graphical user interface (GUI), which reduces or eliminates the need for users to learn a command language or JCL. With GUI-based command layers, users can execute many operating system commands by manipulating graphical images on a video display using a pointing device such as a mouse. For example, programs can be executed by double-clicking their image. The visual metaphor for files, programs, and commands is easier to learn and manipulate for most users.

The **service layer** contains a set of functions called by application programs and the command layer. A request to execute a service layer function is called a **service call**. Most operating systems provide thousands of service calls, including many to implement graphical user interfaces. Available service functions can number in the hundreds or thousands. The service layer is also an intermediary between programs and the kernel.

The **kernel** is the portion of the operating system that manages resources and directly interacts with computer hardware. Most service functions directly or indirectly access hardware resources on behalf of programs. For example, a service request to open a file is a direct request for access to secondary storage and an indirect request to allocate memory buffers or cache for file I/O. Because the kernel directly interacts with hardware, service layer functions rely on the kernel to allocate and interact with hardware resources.

The kernel includes a set of interface programs called device drivers for each hardware device in the computer system (see Figure 11-4). Examples include device drivers for keyboards, video display devices, disk drives, tape drives, and printers. Using device drivers makes the operating system modular and flexible. As hardware is added or upgraded, device drivers are added or modified to reflect the changes.

Figure 11-4 ▶

Components of the kernel

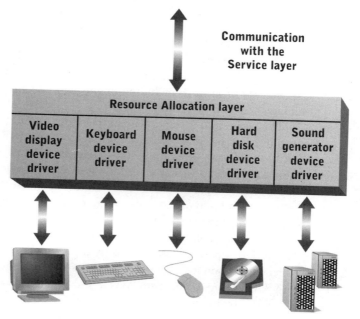

RESOURCE ALLOCATION

Early computer systems were relatively simple devices with few hardware resources. Because hardware resources were limited, complex resource allocation methods were unnecessary. An application program "took control" of the entire computer system when it executed. Because only one application program executed at a time, there was no need for operating system software to resolve competing demands for hardware resources.

Over time, computers became more powerful and their hardware resources became more numerous and diverse. Extra processing power and storage capacity made it possible for multiple users and programs to use a computer at the same time. Operating system support for multiple executing programs is called **multitasking**. Multitasking operating systems had to implement elaborate procedures to share resources among processes and users and to prevent them from interfering with one another.

Single Process Resource Allocation

Resource allocation in a single-tasking operating system involves only two executing programs—an application and the operating system. There is rarely any contention between the operating system and the application program for resources. The operating system reserves whatever resources it requires when it first boots up. All resources not allocated to the operating system are available to an application program.

In a single-tasking environment, when an application program begins executing, the operating system grants it control of all unused hardware resources. The program controls those resources until it terminates, an error occurs, or it needs service from the operating system. Errors and service calls normally are processed through interrupts (interrupts were discussed in Chapter 6). When an interrupt is detected, control of the CPU is passed back to the operating system. If an error has occurred, the operating system can attempt to correct it so the program may continue. If this is not possible, or if the operating system is not programmed to correct errors, then the application program is terminated and the operating system reclaims control of all hardware resources.

Most service calls are actually indirect requests for system resources. For example, when an application program asks the operating system to retrieve data from a file, it indirectly is asking for access to memory buffers, the system bus, the disk controller, and the disk that holds the file. An I/O service call is an implicit request for access to an I/O device and the resources needed to communicate with it. In a single-tasking operating system, no elaborate mechanisms are needed to check for competing requests because there are no other active processes.

MS-DOS is a single-tasking operating system. When MS-DOS first is loaded, it reserves enough memory for its kernel, service layer, and command layer. The command layer then awaits an explicit request by the user to execute an application program. When that request is received, the application program is loaded into memory. Control of the CPU then is turned over to the application program. The application program retains control unless an error occurs or a service request is made. Errors and service requests are signaled by an interrupt. Interrupt handlers return control of the CPU to the operating system. CPU control passes back and forth between the operating system and the application program as services are requested and provided.

Single-tasking operating systems are small and efficient because they do not need complex resource allocation procedures. MS-DOS typically requires less than 100 KB of memory and consumes relatively few CPU cycles during its execution. Most hardware resources are available to execute application programs.

Multiple Process Resource Allocation

Multitasking operating systems are the norm for general-purpose computers. Under multitasking, application and system software can be more flexible because large software programs can be built from smaller independent modules or processes. Small modules are easier to develop and update and they can be loaded, executed, suspended, and resumed as needed.

For example, a typical word processing program often needs to format output and send it to a printer. With a multitasking operating system, the word processor's printer formatting and interface functions can be a separate module. That module can be loaded and executed when printing is required, and suspended when not in use. The printing module can execute independently of other word processing modules, so the user can perform other tasks while a document is being printed.

System software also benefits from multitasking capabilities. Network interface units, for example, typically have a dedicated operating system process that detects and responds to incoming data. With multitasking, this process is loaded into memory when the operating system is booted and placed in a suspended state until data arrives. Other I/O devices such as printers and terminals also are managed by independent processes. Dividing operating system tasks among multiple independent processes makes the operating system easier to build and upgrade. It also results in more efficient use of hardware resources.

Resource Allocation Goals

A multitasking operating system manages hardware resources to achieve the following goals:

► Meet the resource needs of processes
► Prevent processes from interfering with one another
► Efficiently use hardware and other resources

Related goals include maximizing the amount of work performed per time unit, ensuring reliable program execution, protecting resource security, and minimizing the resources consumed by the operating system itself.

Resource Allocation Processes

Consider the resource allocation tasks performed by the audio visual support service for a school. The service maintains a pool of equipment, accepts equipment requests from instructors and faculty, allocates equipment to satisfy requests, and moves equipment from place to place as needed. Managing the equipment inventory and responding to user requests is a complex task that requires detailed knowledge of what equipment is available and when it is available. The service maintains detailed records of equipment capabilities, operational status, commitments, and movement schedules. The records are input data for future scheduling decisions and are modified as a result of those decisions. The service follows well-defined policies and procedures to prioritize conflicting demands for service.

An operating system's resource allocation functions are similar to the audio visual service scenario. The operating system keeps detailed records of available resources and knows which resources are used to satisfy which requests. It schedules resources based on specific allocation policies to meet present and anticipated demand, and it continually updates records to reflect resource commitment and release by processes and users.

Resource allocation goals and policies must be defined as a set of procedures or algorithms implemented in software. The records that support the procedures must be implemented as a set of data structures. The resource allocation functions of an operating system are similar to any other program—they are a set of algorithms and data structures organized to accomplish a particular purpose.

Computer scientists have long studied data structures and the most efficient ways to manipulate them. They have created a large body of knowledge that describes which algorithms and data structures are most efficient for particular types of processing tasks. This body of knowledge finds extensive application in the design and implementation of operating system resource allocation procedures.

Operating system designers take great care to implement resource allocation procedures that consume as few machine resources as possible. The reason is simple—every resource consumed by operating system resource allocation is unavailable to application programs. Think of the resource allocation function as similar to middle management functions in a business. Every business has a limited set of resources. Every resource devoted to a management function is unavailable for other business functions such as producing and selling products.

The resources consumed by the resource allocation procedures sometimes are referred to as **system overhead**. An important design goal for most operating systems is to minimize system overhead while achieving acceptable levels of throughput and reliability. Note that these goals are in direct conflict. Allocation decisions that provide higher levels of resource availability to application programs typically require more elaborate and complex allocation procedures. Ensuring higher levels of reliability also requires more extensive oversight procedures. Yet, increases in the complexity of allocation and oversight procedures consume more hardware resources when those procedures are implemented in software. Operating system designers must determine the right balance among these competing objectives.

Real and Virtual Resources

A computer system's physical devices and associated system software are called **real resources**. As allocated by the operating system, the resources that are apparent to a process or user are called **virtual resources**. Modern operating systems make the set of virtual resources available to active processes, and users appear to be equal to or greater than the available set of real resources.

Application design and programming is much simpler if the application is not concerned with resource availability. Programs can be written under the assumption that whatever resources are requested will be provided. The programmer doesn't need to implement schemes to lock and unlock resources for exclusive use. Processes are oblivious to the complex procedures used to satisfy their resource requests.

Providing virtual resources that meet or exceed real resources is accomplished by:

► Rapidly shifting resources unused by one process to other processes that need them

► Substituting one type of resource for another when possible and necessary

Although each process "thinks" it has control of all hardware resources in the computer system, it is rare for any one process to need them all at the same time. The operating system shifts resources among processes as demand rises and falls. In this way, the sum of virtual resources for all active processes can substantially exceed the real resources of the computer system.

Certain types of resources can be substituted for each other. The best example of this is storage. Recall from Chapter 5 that available storage devices of a computer system form a hierarchy with faster expensive devices such as SRAM cache at the top, and slower inexpensive devices such as magnetic tape at the bottom. Every process would "like" to have access to a large supply of the fastest storage available. Because the number of running processes is limited to the number of physical CPUs, only one or a few processes actually need fast storage at any point in time. Storage needs of inactive or suspended processes and unused storage of active processes can be shifted from expensive high-speed devices to less expensive, slower devices. The most common example of this shifting is virtual memory management, which is described in more detail later in this chapter.

PROCESS MANAGEMENT

A **process** is a unit of executing software that is managed independently by the operating system. A process can request and receive hardware resources and operating system services. Processes can be stand-alone entities or part of a group of processes that cooperate to achieve a common purpose. Processes can communicate with other processes executing on the same computer system or with processes executing on other computer systems.

Process Control Data Structures

The operating system keeps track of each process by creating and updating a data structure, called a **process control block (PCB)**, for each active process. The operating system creates a PCB when a process is created, updates the PCB as process status changes, and deletes the PCB when the process terminates. Using information stored in the PCB, the operating system can perform a number of functions, including resource allocation, secure resource access, and protecting active processes from interference by other active processes.

PCB content varies among operating systems and also can vary within a single operating system depending on whether certain functions such as resource accounting and auditing are enabled. The following data items typically are included:

► A unique process identification number
► The current state of the process; for example, executing or suspended
► Events for which the process is waiting
► Resources allocated exclusively to the process, including memory, files, and I/O devices
► Machine resources consumed; for example, CPU seconds consumed and bytes of data transferred to/from disk
► Process ownership and/or access privileges
► Scheduling priority or data from which scheduling priority can be determined

PCBs normally are organized into a larger data structure such as a linked list. The list of PCBs sometimes is called a **process queue** or **process list**. The process queue often is searched by operating system components. For example, a user control process might search the process queue for all processes owned by a certain user so they can be terminated when that user logs off. Search and update speed is an important characteristic of PCBs and the process list. Process list and PCB contents are changed frequently as processes are created and terminated and as resources are allocated and released. Using efficient data structures and processing algorithms minimizes system overhead.

Processes can create, or **spawn,** other processes and communicate with them. The original process is called the **parent process** and the newly created process is called the **child process**. Parent processes can spawn multiple child processes, and the child processes collectively are called **sibling processes**. Child processes can, in turn, spawn children of their own. A group of processes descended from a common ancestor, including the common ancestor itself, is called a **process family**.

Application programs often spawn a child process to execute general-purpose utility programs that are not part of the operating system service layer. For example, a word processor can spawn a child process to sort a list or table. The process can be another program within the same application, a utility program supplied with an application suite, or a utility program supplied with the operating system. Subdividing application programs into multiple cooperating processes can improve execution speed, especially if the computer system has multiple CPUs because multiple processes can execute simultaneously.

Even if only one CPU is available, subdividing application programs into cooperating processes increases execution speed due to more efficient resource allocation. For example, consider a word processor user who wants to edit one document while printing another. If the word processor spawns a separate printing process, then both tasks can execute concurrently. Because the two tasks use different mixes of computer resources (heavy I/O and light CPU used for the print process and the reverse for the editing process), the operating system can allocate resources efficiently so both processes move quickly toward completion.

Threads

There is a diminishing return associated with subdividing large processes into smaller ones. As the number of active processes increases, the operating system overhead required to track and manage them also increases. The process list grows and operating system processes that search and update the list become less efficient. This increases system overhead and reduces the resources available to execute application processes.

In some operating systems, processes subdivide themselves into more easily managed subunits called threads. A **thread** is a portion of a process that can be scheduled and executed independently. Process threads can execute concurrently on a single processor or simultaneously on multiple processors. Threads share all resources allocated to their parent process, including primary storage, files, and I/O devices.

The advantage of organizing an application program into a family of threads instead of multiple processes is that operating system overhead for resource allocation and process management is reduced. Because all threads of a process share storage and I/O resources, the operating system can track allocation of those resources on a per-process basis. This reduces the number of PCBs and speeds up searching and updating the process queue.

The operating system keeps track of thread-specific information in a **thread control block (TCB)**. Each PCB contains a set of pointers to its related TCBs. All active TCBs are organized into a data structure called a **run queue** or **thread list**. As with the process list, the run queue is implemented with efficient data structures that can be searched or updated quickly.

A process or program that divides itself into multiple threads is said to be **multithreaded**. An operating system that supports threads automatically creates a thread for each process, but subdividing processes into multiple threads is not automatic. The programmer or compiler specifically must identify code segments that can execute independently and create a thread for each segment. Operating systems that support multithreaded processes include Windows 2000 and later versions of UNIX.

CPU ALLOCATION

Threads[1] progress toward completion only when they have CPU cycles with which to execute their instructions. A multitasking operating system can execute dozens, hundreds, or thousands of threads within the same time frame. Most computer systems have only one or two CPUs so threads must share CPUs.

The operating system makes rapid decisions about which threads receive CPU control and for how long that control is retained. Typically, a thread controls the CPU for no more than a few milliseconds before it relinquishes control and the operating system gives another thread a turn. Figure 11-5 shows three threads sharing a single CPU using small time slices. This method of CPU sharing is called **concurrent execution** or **interleaved execution**.

---

[1] We will use the term thread throughout this chapter, though all references to threads also apply to processes in operating systems that don't support threads.

Figure 11-5 ▶

Concurrent
(interleaved)
thread execution
on a single CPU

	Time slice 1	Time slice 2	Time slice 3	Time slice 4	Time slice 5	Time slice 6	Time slice 7
Thread 1	Running	Idle	Idle	Idle	Idle	Running	Idle
Thread 2	Idle	Running	Idle	Running	Idle	Idle	Idle
Thread 3	Idle	Idle	Running	Idle	Running	Idle	Running

Thread States

An active thread can be in only one of the following states:

▶ Ready
▶ Running
▶ Blocked

Figure 11-6 shows the three thread states and the events that move a thread from one state to another. Threads in the **ready state** are waiting for access to a CPU. There can be many threads in the ready state at any point in time. When a CPU becomes available, the operating system chooses a ready thread to execute on that CPU.

Figure 11-6 ▶

Thread
movement
among states

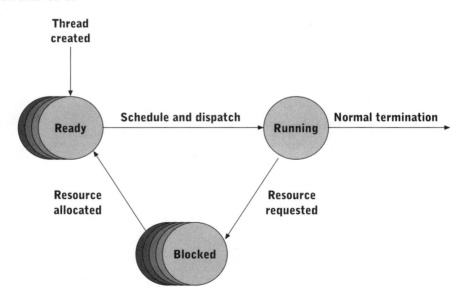

The act of giving control of a CPU to a ready thread is called **dispatching**. The operating system dispatches a thread by loading the instruction pointer with an instruction address that belongs to the dispatched thread. During the next fetch and execute cycle, the thread takes control of the CPU and its directly related resources, such as general-purpose registers.

Once dispatched, a thread has entered the **running state** and retains control of the CPU until one of the following events:

▶ The thread or its parent process terminates (halts) normally

▶ An interrupt occurs

When a process or thread terminates its own execution, control returns to the operating system. The exact way in which the operating system regains CPU control varies across operating systems and CPUs. The most common method is for the process or thread to execute an exit service call. The exit call triggers a software interrupt.

Interrupts can occur for a variety of reasons, including:

▶ Executing another service call such as a file I/O request

▶ A hardware generated interrupt indicating an error such as overflow or a critical condition such as a power failure alarm from an uninterruptible power supply

▶ An interrupt generated by a peripheral device such as a network interface unit when a packet arrives

Recall from Chapter 6 that when any interrupt is received, the CPU automatically suspends the currently executing thread, pushes current register values onto the stack, and transfers control to the operating system master interrupt handler. The suspended thread remains on the stack until interrupt processing is completed. During this period of time the thread is in a **blocked state**. Once the interrupt has been processed, the operating system can leave the suspended thread in the blocked state, move it to the ready state, or return it to the running state.

Interrupt Processing

A blocked thread is waiting for an event to occur, such as allocation of a requested resource or correction of an error condition. Error conditions might or might not be correctable. If they are not correctable, such as overflow or memory protection fault, the thread is halted. If the error can be corrected, the thread remains in the blocked state until the error condition is resolved.

For example, if a thread attempts to access a file on a removable disk and the disk is not in the drive, the disk drive controller will generate an interrupt to signal the error. The CPU will automatically suspend the thread and pass CPU control to an operating system interrupt handler. In an operating system such as

Windows, the interrupt handler displays an error message asking that a disk be inserted. If the user complies, the error condition is resolved and the thread leaves the blocked state. If the user does not comply, the error is not corrected and the requesting thread is terminated involuntarily.

Most service calls directly or indirectly require resource allocation. For example, a service call that requests additional memory for a thread is a direct request for resource allocation. A request to open a file is an indirect request for memory buffers to hold file data as it is transferred to or from secondary storage. If the request is for read/write access, it also indirectly requests exclusive access to the file. A subsequent request to read data from the file is an indirect request for access to the secondary storage device, its controller, and the I/O channel(s) that connect the secondary storage device to its memory buffers.

Some service calls, such as requesting additional memory, usually can be performed immediately. But many service calls, such as requesting input from a file or device, require a period of time to complete. In that case, the thread remains in the blocked state until the request is satisfied.

For example, the following event sequence is typical when a process or thread reads from a file:

1. The thread requests file input by issuing an appropriate software interrupt.

2. The thread automatically is pushed onto the stack and the appropriate interrupt handler is called.

3. The interrupt handler executes. It checks to see if the requested data is already in a memory buffer. If it is, the interrupt handler transfers the data from the memory buffer to the data area of the requesting process. The interrupt handler exits, and the operating system places the suspended thread in the ready or running state.

4. If the requested data isn't already in a memory buffer, the interrupt handler generates an appropriate read command and sends it to the secondary storage controller. The interrupt handler then exits without waiting for the data to be returned. The operating system places the suspended thread in a blocked state pending arrival of the requested data.

5. When the data is read from disk, the secondary storage controller sends an interrupt to the CPU. The interrupt handler transfers the data from the secondary storage controller to a memory buffer over the system bus and then copies it to the data area of the requesting process. The interrupt handler exits, and the operating system places the blocked thread in the ready or running state.

Scheduling

The decision-making process used by the operating system to determine which ready thread moves to the running state is called **scheduling**. The portion of the operating system that makes scheduling decisions is called the **scheduler**. Operating systems vary widely in their scheduling methods though the following methods are typical:

► Preemptive scheduling
► Priority-based scheduling
► Real-time scheduling

Preemptive Scheduling In **preemptive scheduling**, a thread can be removed involuntarily from the running state. A running process or thread controls the CPU by controlling the content of the instruction pointer. CPU control is lost whenever an interrupt is received and the CPU pushes the current thread onto the stack and transfers control to the operating system. The portion of the operating system that receives control is called the **supervisor**.

The supervisor performs two important functions:

► Calls the appropriate interrupt handler
► Transfers control to the scheduler

The supervisor uses the value in the interrupt register as an index into the interrupt table. It extracts the corresponding address, pushes itself on the stack, and transfers control to the interrupt handler by placing its address in the instruction pointer. When the interrupt handler finishes, it passes control back to the supervisor by popping the stack. The supervisor then passes control to the scheduler by executing an unconditional branch instruction.

The scheduler performs four tasks:

► Updates the status of any process or thread affected by the last interrupt
► Decides which thread to dispatch to the CPU
► Updates thread control information and the stack to reflect the scheduling decision
► Dispatches the selected thread

Processing an interrupt usually changes the state of at least one thread. If the interrupt resulted from a request for a resource that couldn't be provided immediately, the thread currently on top of the stack must be moved from the running state to the blocked state. If processing the interrupt cleared an error condition or supplied a resource for which some thread was waiting, the thread must be moved from the blocked state to the ready state. The scheduler makes updates to the appropriate TCBs to reflect these state changes. Figure 11-7 illustrates the processing sequence for both a service request and a delayed satisfaction of that request. The processing steps on the left occur after Thread 1 makes an I/O service call. The processing steps on the right occur after the I/O device completes the I/O operation.

Figure 11-7 ►

Interrupt
processing

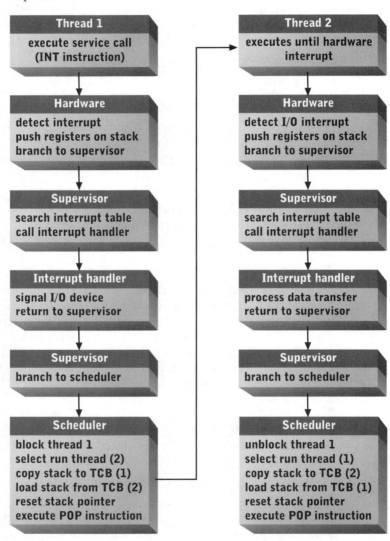

Next, the scheduler examines the run queue to determine which threads are in the ready state. It chooses a ready thread based on a predefined scheduling algorithm. Scheduling algorithms are discussed in detail in the next section.

Once the scheduler decides which thread will run next, it updates the thread's TCB and manipulates the stack contents, if necessary. If the suspended thread is to become the running thread, then the scheduler pops the stack and the thread resumes execution. If another thread is selected as the running thread, then the scheduler must save the entire stack contents to the appropriate TCB because elements below the topmost element represent suspended subroutines or functions of the interrupted thread.

For example, assume that a thread called main begins execution, calls one subroutine (A), that subroutine calls another subroutine (B), and an interrupt is received before any of the subroutines completes execution. After the interrupt is received and control is passed to the supervisor, there are three threads on the stack (see Figure 11-8). The stack contents are main, which was pushed when subroutine A was called; A, which was pushed when subroutine B was called; and B, which was pushed when the interrupt was detected. The entire stack must be saved to the thread's TCB so it can be restored when the thread is dispatched at some future point in time.

Figure 11-8 ▶

Subroutine calls (left) and associated stack changes (right)

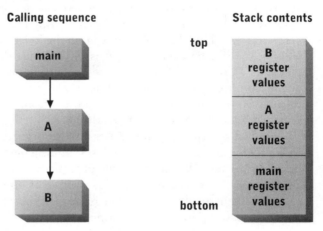

Calling sequence

main

A

B

Stack contents

top

B register values

A register values

main register values

bottom

After the suspended thread's stack is saved to its TCB, the operating system copies the saved stack of the next running thread from its TCB to the stack. The scheduler then transfers control to the thread by issuing a POP instruction.

Timer Interrupts Because interrupt arrival is unpredictable, the time that a thread remains in the ready state also is unpredictable. Theoretically, it is possible for a thread to control the CPU indefinitely if it makes no service calls, generates no error conditions, and receives no external interrupts. Under these conditions a thread stuck in an infinite loop would execute forever.

Most CPUs periodically generate an interrupt to give the scheduler an opportunity to suspend the currently executing thread. A **timer interrupt** is generated at regular intervals of between several dozen and several hundred CPU cycles. As with other interrupts, the currently executing process is pushed onto the stack and control is transferred to the supervisor. Because a timer interrupt isn't a "real" interrupt, there is no interrupt handler to call so the supervisor passes control to the scheduler.

Timer interrupts are an important CPU hardware feature for multitasking operating systems. They guarantee that no thread can hold the CPU for long periods because the scheduler has an opportunity to place another thread in the running state. This ensures that all threads make progress toward completion at regular intervals.

Priority-Based Scheduling Methods **Priority-based scheduling** methods can determine which ready thread should be dispatched to the CPU, including:

► First come first served
► Explicit priority
► Shortest time remaining

Under **first come first served (FCFS)** scheduling, the scheduler always dispatches the ready thread that has been waiting the longest. The simplest way to determine which thread has been waiting the longest is to order the run queue by waiting time. Each time a thread is moved into the ready state the scheduler places its TCB at the back of the run queue. The scheduler then searches the run queue from front to back until it finds a ready thread and that thread is the next to run. Implementing the run queue in this fashion is relatively simple so FCFS scheduling has minimal system overhead.

Explicit priority scheduling uses a set of priority levels and assigns a level to each process or thread. A priority level can be assigned to a newly created thread based on the default priority of the process owner, a priority stated explicitly in an operating system command, or a number of other methods. Many operating systems automatically assign high priority levels to their own internal threads under the assumption that operating system threads must be completed quickly to maintain total system throughput.

The scheduling method can use priority levels in two ways:

► Always dispatch the highest priority ready thread
► Assign larger time slices to high priority threads

The first strategy ensures that high priority processes always will be dispatched before lower priority processes. However, scheduling decisions based solely on initial priority level can result in extremely long or infinite idle time for long-running threads with low priority. Such threads always are moved to the back of the run queue and are unable to obtain sufficient CPU time to complete. To prevent this, many operating systems automatically increase the priority level of "old" threads. Priority-based scheduling using variable-length time slices also addresses this problem though that method is seldom implemented.

Shortest time remaining scheduling chooses the next process to be dispatched based on the expected amount of CPU time needed to complete the process. This can be implemented directly by tracking time to completion for each thread and ordering the run queue with shortest time remaining threads first. Shortest time remaining scheduling can be accomplished indirectly by increasing the explicit priority of a thread as it nears completion.

In either case, the scheduler must know how much CPU time is required for each thread to execute and how much time already has been used. This implies that the required CPU time was provided to the scheduler when the thread was created and that this information was stored in the TCB. It also implies that the amount of CPU time used by a thread is stored in its TCB and is updated each time the thread leaves the ready state. Creating, updating, and using time to completion information increases system overhead, so shortest time remaining scheduling is rarely implemented.

Real-Time Scheduling Real-time scheduling guarantees a minimum amount of CPU time to a thread if the thread makes an explicit request when it is created. Real-time scheduling is used when a thread must have enough resources to complete its function within a specified time. Real-time scheduling often is used in transaction processing, data acquisition, and automated process control. For example, in a transaction processing application, an automated bank teller machine typically will expect a response to a query of a customer's balance within a specified amount of time.

Data acquisition and process control applications use data that arrives at a constant rate from one or more hardware devices. Data acquisition programs, such as those used to record radio astronomy observations, copy incoming data to a storage device such as tape or disk. Data analysis is performed later by a separate process. Process control programs actively process the data as it arrives. For example, a process control program in a chemical manufacturing plant processes data from sensors within pipelines and reaction vessels.

Data inputs flow from data collection devices or process sensors into memory buffers. A data acquisition or process control program must extract data from the buffers for analysis quickly enough to keep the buffers from overflowing. The program must be allocated sufficient resources to ensure that it can process or store incoming data at least as quickly as it arrives.

When a thread with a real-time scheduling requirement is created, it informs the operating system of the maximum CPU time needed to complete one thread cycle and the frequency of the cycles. A **thread cycle** can execute instructions to process a single transaction, retrieve and store data from an I/O device, or retrieve and analyze one set of process variables. For example, data acquisition equipment might send data every second and the thread might need a maximum of 150 milliseconds to complete processing the data.

The thread signals the operating system at the start of each thread cycle and the scheduler starts a timer. The scheduler tracks thread progress by updating CPU time used in the TCB each time it enters and leaves the ready state. The scheduler checks the timer, CPU time used, and maximum thread cycle time each time it makes a scheduling decision to ensure that the thread completes its cycle before the timer expires. The scheduler can manage multiple real-time threads though real-time scheduling overhead can quickly consume available CPU resources.

Technology
Focus
Windows 2000 Scheduling

The Windows 2000 operating system supports a number of advanced features including multitasking, multithreaded processes, and preemptive priority-based scheduling and dispatching. Scheduling, dispatching, and interrupt handling are performed by an operating system component called the microkernel, which acts as both supervisor and scheduler.

Each process managed by the microkernel has a base priority level that is an integer with a value from 1 to 31. Higher values represent higher scheduling priority. Priority levels are grouped into four categories called base priority classes. The classes are Idle (priority levels 1-6), Normal (priority levels 6-10), High (priority levels 11-15), and Real-time (priority levels 16-31). A process is assigned a base priority class when it first is created.

Most user processes start in the Normal base priority class. The microkernel raises the base priority class to High when a process is moved from the background to the foreground; in other words, when the process is receiving input from the keyboard and mouse. It is important to note that the Real-time base priority class does not meet the definition of real-time scheduling given earlier in the chapter. It simply describes the highest group of priority levels.

Threads have base priority levels that normally vary within the range of priority levels for their parent processes' base priority class. Thread priority classes are defined as Idle, Lowest, Below Normal, Normal, Above Normal, Highest, or Time Critical. The precise meaning of these classes within process base priority classes is summarized in Table 11-1.

The current raw priority level of a thread is called its dynamic priority. The initial value of a thread's dynamic priority level is based on the base priority class of the thread and parent process. The microkernel can alter dynamic priority to improve system performance and to reflect high or low demand for system resources by a thread. Dynamic priority levels can be adjusted over a wide range of possible values.

Table 11-1 ▶

Thread priority
classes and levels

Base priority class	Thread priority class	Priority level
Real-time	Time Critical	31
	Highest	26
	Above Normal	25
	Normal	24
	Below Normal	23
	Lowest	22
	Idle	16
High	Time Critical	15
	Highest	15
	Above Normal	14
	Normal	13
	Below Normal	12
	Lowest	11
	Idle	1
Normal	Time Critical	15
	Highest	10
	Above Normal	9
	Normal	8
	Below Normal	7
	Lowest	6
	Idle	1
Idle	Time Critical	15
	Highest	6
	Above Normal	5
	Normal	4
	Below Normal	3
	Lowest	2
	Idle	1

Figure 11-9 shows a Windows 2000 utility called Process Viewer. Process Viewer shows information about active processes, including the process's base priority level and the priority class and dynamic priority of its threads. This example shows information for a process named *explorer* and its five threads. The base priority class of *explorer* is Normal. The initial base priority class for the highlighted thread (0) is Above Normal. The initial dynamic priority for this thread was 9. Note that the current value shown at the bottom of the display is 14. Some time after the thread was created the microkernel adjusted its dynamic priority upward.

Figure 11-9 ►

Windows 2000
Process Viewer

The microkernel implements a strict priority-based scheduling algorithm. A higher priority thread in the ready state always is executed before a lower priority thread. When a new thread is created with a higher priority level than the currently executing thread, the microkernel moves the currently executing thread to the ready state where it remains until all higher priority threads terminate or become blocked.

Business
Focus
Which Windows for the Business Desktop?

Microsoft currently sells two competing versions of Windows for desktop computers—Windows Millenium Edition (Me) and Windows 2000 Professional. Windows Me, marketed to home users, is the latest product in a line that began with Windows 95. Windows Me includes many features useful to home users, including support for digital cameras and home networks. Windows 2000 Professional, marketed to business users, is the latest product in a line that began with Windows NT 3.5. Windows 2000 Professional includes many "industrial strength" features, including better reliability, portability, and security. Although the user interface (command layer) of Windows Me and Windows 2000 Professional is nearly identical, there are many differences "under the hood."

One of the most significant differences is compatibility with older programs designed for DOS and Windows 3.1. Windows Me was designed for maximum compatibility with older programs. Many older programs execute directly under Windows Me. Those that don't execute directly can execute in a special DOS compatibility mode, and Windows Me automates the process of installing and executing such programs. Many older programs will not execute under Windows 2000 Professional because it routes all hardware device accesses through its own device drivers and prohibits programs from bypassing them. Because many older programs directly access hardware such as video controllers, they are incompatible with Windows 2000 Professional.

Both Windows products implement multitasking but do so using different techniques. Applications designed specifically for Windows Me use preemptive multitasking, but older programs use a method called cooperative multitasking. Under cooperative multitasking, a program must relinquish CPU control voluntarily. Programs that fail to do so can crash the entire system. Under Windows 2000 Professional, all programs use a highly reliable method of preemptive multitasking that prevents one program from crashing others or the entire system.

Memory protection in Windows Me is not completely reliable because significant portions of the operating system memory space can be accessed by application programs. Applications designed for Windows Me are protected from one another, but older applications share memory space and can interfere with one another. Program lockups and system crashes are more common with Windows Me. Windows 2000 Professional programs each run in their own protected memory space and all operating system code is protected from applications. The result is a highly reliable execution environment in which crashes are rare.

Both products incorporate extensive support for networking, sufficient for the majority of typical networking needs. The primary difference is in the methods used to share resources among computers. Resource sharing among Windows Me computers is less secure than resource sharing among Windows 2000 Professional computers. Network security also is easier to manage centrally with Windows 2000 Professional than with Windows Me.

Hardware requirements are higher for Windows 2000 Professional than for Windows Me. A minimal usable configuration for Windows Me is a Pentium 150 MHz processor, 32 MB of memory, and several hundred megabytes of disk space. A minimal usable configuration for Windows 2000 Professional is a Pentium 200 MHz processor, 64 MB memory, and at least 1 GB of disk space.

Questions

1. Which operating system would you recommend for a home user?
2. Which operating system would you recommend for a small business? Assume that the business has 10 PCs connected to a file and print server.
3. Which desktop operating system would you recommend for a local office of a Fortune 500 company? Assume the office has 125 PCs, 5 network servers, and dedicated high-speed network connections to a nationwide corporate network.

MEMORY ALLOCATION

Processes need memory to hold instructions and data during execution. The operating system allocates memory to processes when they are created and responds to requests for additional memory during a thread's lifetime. The operating system also allocates memory to itself and for other needs such as buffers and caches. The complexity of the memory allocation process depends on the number of supported processes and the type of memory addressing and allocation.

Note that most of what is said about memory allocation in this section also applies to managing secondary storage devices. Keep this in mind as you read this section because many of the terms and concepts will be used again in the next chapter.

Single-Tasking Memory Allocation

Memory allocation in a single-tasking operating system is fairly simple. A memory map for a single-tasking operating system is shown in Figure 11-10. The bulk of the operating system normally occupies lower memory addresses and the application program is loaded immediately above it. For the application program in Figure 11-10, memory allocation is **contiguous**. That is, all portions of the program and operating system are loaded into sequential locations within memory.

Figure 11-10►

Contiguous memory allocation in a single-tasking operating system

To conserve physical memory, some operating system components are loaded into memory only when needed. When loaded, they usually are placed in upper memory, as shown in Figure 11-11. The memory allocation for the operating system shown in Figure 11-11 is **noncontiguous**, or **fragmented**. Although most of the operating system is stored contiguously in low memory, a portion is stored in high memory, with intervening space that is not allocated to the operating system.

Figure 11-11 ▶

Noncontiguous
memory
allocation

OS extensions	16 MB
	15 MB
	14 MB
Free space	13 MB
	12 MB
	11 MB
	10 MB
	9 MB
	8 MB
Application process	7 MB
	6 MB
	5 MB
	4 MB
Operating system	3 MB
	2 MB
	1 MB
	0

As discussed in Chapter 5, most CPUs use some form of offset addressing. With offset addressing, programs can be located in physical memory locations other than those assumed at the time of compilation and linking. Normally, programs are compiled and linked assuming that the first instruction will be located at the first address (zero) in physical memory.

When a CPU instruction uses a memory address as an operand, the operand is called a memory reference. Examples of CPU instructions that contain memory references include load, store, and branch. The process of determining the physical memory address that corresponds to a memory reference is called **address mapping**, or **address resolution**. Address resolution is a simple process when memory is allocated contiguously. Each memory reference is mapped to its equivalent physical memory address by adding the offset value. For the application program shown in Figure 11-11, all memory references are mapped to physical memory addresses by adding the value 4,194,305 (4 MB + 1) to the reference.

Programs can grow or shrink during execution by calling operating system service functions to allocate or deallocate memory. To service such requests, the operating system must keep track of allocated and free memory. Each time memory is allocated to a program or the operating system, the allocation must be recorded so the same memory region is never allocated more than once.

Multitasking Memory Allocation

Memory allocation is much more complex when the operating system supports multitasking. The operating system finds free memory regions in which to load new processes and threads, and reclaims memory when processes or threads terminate. The goals of multitasking memory allocation are:

► Allow as many active processes as possible

► Respond quickly to changing memory demands of processes

► Prevent unauthorized changes to a process's memory region(s)

► Implement memory allocation and addressing as efficiently as possible

Multitasking operating systems use partitioned memory, which divides memory into equally sized regions, each of which can hold all or part of a process or thread. Figure 11-12 shows 16 MB of physical memory divided into 1 MB partitions, each of which can hold an operating system component, a process, or nothing at all (free space).

Figure 11-12 ►

Partitioning main memory into fixed-size regions

Region	
Region 15	16 MB
Region 14	15 MB
Region 13	14 MB
Region 12	13 MB
Region 11	12 MB
Region 10	11 MB
Region 9	10 MB
Region 8	9 MB
Region 7	8 MB
Region 6	7 MB
Region 5	6 MB
Region 4	5 MB
Region 3	4 MB
Region 2	3 MB
Region 1	2 MB
Region 0	1 MB
	0

Figure 11-13 shows three processes and the operating system loaded into fixed-size memory partitions. Process 1 occupies three memory regions and Process 2 occupies two memory regions. Each process uses all memory within its allocated regions. Process 3 occupies four complete regions and part of a fifth. Partitions 14 and 15 are free space.

The operating system maintains a memory partition table and updates it each time a partition is allocated or freed (see Table 11-2). When a process is ready to be loaded, the operating system searches the table to find a sufficient number of contiguous free partitions. If they are found, the table is updated and the process is loaded into the free partition(s). When a process terminates, the table is updated to free the memory partition(s).

Figure 11-13 ▶

Several processes and the operating system loaded into 1MB fixed-size memory partitions

Free space {	15	16 MB
	14	15 MB
	13	14 MB
	12	13 MB
Process 3 {	11	12 MB
	10	11 MB
	9	10 MB
Process 2 {	8	9 MB
	7	8 MB
	6	7 MB
Process 1 {	5	6 MB
	4	5 MB
	3	4 MB
Operating system {	2	3 MB
	1	2 MB
	0	1 MB
		0

Table 11-2 ▶

Fixed-size memory partitions maintained by the operating system

Partition	Starting address	Status	Allocated to
0	0 MB	Allocated	OS
1	1 MB	Allocated	OS
2	2 MB	Allocated	OS
3	3 MB	Allocated	OS
4	4 MB	Allocated	Process 1
5	5 MB	Allocated	Process 1
6	6 MB	Allocated	Process 1
7	7 MB	Allocated	Process 2
8	8 MB	Allocated	Process 2
9	9 MB	Allocated	Process 3
10	10 MB	Allocated	Process 3
11	11 MB	Allocated	Process 3
12	12 MB	Allocated	Process 3
13	13 MB	Allocated	Process 3
14	14 MB	Free	
15	15 MB	Free	

Contiguous program loading, coupled with fixed-size memory partitions, usually results in wasted memory space. For example, Process 3 uses only part of its last partition. The remainder is wasted because the operating system can't allocate less than a full partition. Wasted space can be reduced by reducing partition size. In general, the smaller the size of the partitions, the less wasted space. However, smaller partitions require a larger partition table, which increases operating system memory requirements and the time needed to search and update the table.

Memory Fragmentation

As processes are created, executed, and terminated, memory allocation changes accordingly. Over time, memory partition allocation and deallocation leads to an increasing number of small free partitions separated by allocated partitions. Figure 11-14 shows changes in memory allocation over time as processes are created and terminated. Figure 11-14(a) shows a starting point with four processes in memory. Figure 11-14(b) shows memory allocation after Process 3 has terminated and Process 5 has been loaded into memory. A new free space area has been created because Process 5 is smaller than Process 3. In Figure 11-14(c), Process 2 has terminated and its former partition is now free.

Figure 11-14 ▶

Changes over time in the allocation of memory partitions as processes execute and terminate

(a)   (b)   (c)

There is now 4 MB of free memory but only two contiguous free partitions. Although there might be other processes waiting to execute that require 4 MB or less, only those that require 2 MB or less can be loaded due to the fragmentation of available free space.

Over time, memory will become more and more fragmented as shown in Figure 11-15(a), and larger processes will have increasing difficulty finding enough contiguous partitions. One way to address the problem of fragmented free space is to periodically relocate all programs in memory in a process called **compaction**. After compaction, all free partitions form a contiguous block in upper memory, as shown in Figure 11-15(b). Compaction is a time-consuming process because entire programs are moved within memory and many partition table entries are updated. The overhead required to perform compaction generally is larger than the overhead required to implement a more common strategy—noncontiguous memory allocation.

Figure 11-15 ▶

Multiple fragments of free space (a) are collected into a single contiguous fragment (b) by compaction

(a)   (b)

Noncontiguous Memory Allocation

In an operating system that supports noncontiguous memory allocation, portions of a process can be allocated to free partitions anywhere in memory. Noncontiguous memory allocation uses small fixed-size partitions, usually no larger than 64 KB, though we will continue to assume 1 MB partitions to simplify the current discussion and figures.

Examine Figure 11-15(a) and assume that a new process (Process 6) awaits loading and needs 4 MB of memory. Under contiguous memory allocation Process 6 can't be loaded unless memory first is compacted. Under noncontiguous memory allocation, Process 6 can be loaded into the four free partitions, as shown in Figure 11-16. The process is divided into four 1 MB partitions that are allocated to available free memory partitions.

Figure 11-16 ▶

Process 6 is
allocated to
noncontiguous
memory
partitions

Partition 15	Process 6d	16 MB
Partition 14		15 MB
Partition 13	Process 4	14 MB
Partition 12		13 MB
Partition 11	Process 6c	12 MB
Partition 10		11 MB
Partition 9	Process 5	10 MB
Partition 8		9 MB
Partition 7	Process 6b	8 MB
Partition 6	Process 6a	7 MB
Partition 5		6 MB
Partition 4	Process 1	5 MB
Partition 3		4 MB
Partition 2		3 MB
Partition 1	Operating system	2 MB
Partition 0		1 MB
		0

Noncontiguous memory allocation is much more flexible than contiguous memory allocation, but flexibility comes at a substantial price. Process 6 was compiled and linked under the assumption that it would occupy contiguous memory locations starting at address 0. Memory references within Process 6 now must cross noncontiguous partitions, for example a branch instruction in partition 7 to an instruction in partition 15. What offset value should be used to resolve the memory reference to a physical address?

Under noncontiguous memory allocation, each process partition has its own offset value. To resolve addresses correctly, the operating system must keep track of each program partition and its offset and make appropriate adjustments to memory references that cross partition boundaries. The partition tables and address calculations that support noncontiguous memory allocation are much more complex than for contiguous memory allocation. Table content and address resolution are described in the next section.

Virtual Memory Management

The only portions of a process that must be in memory at any given time during execution are the next instruction to be fetched and any operands stored in memory. Only a few bytes of any process must reside in memory at any one time. Most operating systems minimize the amount of process code and data stored in memory at any one time. The benefit of this is freeing large quantities of memory for use by other processes and greatly increasing the number of processes that can execute concurrently.

Virtual memory management divides a program into partitions called **pages**. Each page is a small fixed-size portion of a program, normally between 1 and 4 KB. Memory is also divided into pages of the same size. Each memory page is called a **page frame**. During program execution, one or more pages are allocated to page frames and the rest are held in secondary storage. As pages in secondary storage are needed for current processing, they are copied into page frames by the operating system. If necessary, pages currently in memory are written to secondary storage to make room for pages being loaded.

Each memory reference is checked to see if the page it refers to is currently in memory. A reference to a page held in memory is called a **page hit** and a reference to a page held in secondary storage is called a **page fault**. Tables, called **page tables**, store information about page locations and allocated page frames and secondary storage space. Each active process has a page table or portion of a page table dedicated to it. Table contents include page numbers, a status field indicating whether the page currently is held in memory, and the page frame number in main memory or on secondary storage. A sample page table is shown in Table 11-3.

Table 11-3 ►

Portion of a process's page table

Page number	Memory status	Frame number	Modifications status
1	In memory	214	no
2	On disk	101	n/a
3	On disk	44	n/a
4	In memory	110	yes
5	On disk	252	n/a

Because page size is fixed, memory references can easily be converted to the corresponding page number and offset within the page. The page number can be determined by dividing the memory address by the page size. The whole portion of the result is the page number and the remainder is the offset into that page. For example, if page size is 1 KB, a reference to address 1500 is equivalent to an offset of 476 (1500 − 1024) into page number 2 (1500/1024 + 1). If the table is stored sequentially, the corresponding entry in the process's page table can be computed quickly as an offset into the table.

If the reference is to an address in a page held in memory, the corresponding memory address is an offset into the memory page indicated in the table. Using Table 11-3 as an example, a reference to address 700 (offset of 700 into page 1) would resolve to an offset of 700 into memory frame 214. Once again, fixed page size is used to calculate the corresponding memory address as 219836 (214 * 1024 + 700). Note that these calculations are similar to those used for addressing array contents, as described in Chapter 3.

A secondary storage region, called the **swap space,** is reserved exclusively for the task of holding pages not held in memory. The swap space is divided into page frames in the same manner as memory. A memory reference to a page held in the swap space results in that page being loaded into a page frame in memory. As with address resolution, page location within the swap space can be computed quickly by multiplying the page number by the page size.

If all page frames currently are allocated, a page currently in memory, known as the **victim,** must be written to the swap space before the reference page is loaded into a page frame. A page fault in the currently executing process implies not only a page swapped into memory but also an existing page swapped out of memory. Some commonly used methods for selecting the victim are:

► Least recently used

► Least frequently used

Both methods require the operating system to maintain information about accesses to pages in memory. Searching and updating this information is a part of the system overhead associated with virtual memory management.

When a victim has been selected, it might or might not be copied back to the swap space. The sample data in Table 11-3 shows an entry that indicates whether a page has been modified since it was swapped into memory. If a page has not been modified, the copy held in the swap space is identical to the copy in memory, which means that the page does not have to be copied to the swap space if the page has been selected as the victim.

Memory Protection

Memory protection refers to the protection of memory allocated to one program from unauthorized access by another program. This can apply to interference between programs in a multitasking environment or interference between a program and the operating system in either a single or multitasking environment. If memory is not protected, errors in one program can generate errors in another. If the program being interfered with is the operating system, this might result in a system crash.

In the simplest form of memory protection, the operating system checks each write to a memory location to ensure that the address being written is allocated to the program performing the write operation. Complicating factors include various forms of indirect addressing, virtual memory management, and cooperating processes, which are two or more processes that "want" to share a memory region. Memory protection adds overhead to each write operation.

Memory Management Hardware

Early types of multitasking, protected memory access, and virtual memory management were implemented exclusively by the operating system, which imposes severe performance penalties. Consider, for example, the overhead required to map program memory references using virtual memory management. Each reference requires the operating system to search one or more tables to locate the appropriate page and to determine the corresponding memory or disk location of that page. A memory reference that should consume only one or a few CPU cycles consumes many additional cycles for paging, swapping, and address mapping functions.

The benefits of advanced memory addressing and allocation schemes are offset quickly by reduced performance when they are implemented in software. Because of this, modern CPUs and computer systems incorporate advanced memory allocation and address resolution functions in hardware. For example, all Intel microprocessors since the 80386 have included hardware support for virtual and protected memory management.

Technology
Focus
Intel Pentium Memory Management

Intel Pentium family CPUs dedicate six registers to hold data structures called segment descriptors. The registers are named CS, DS, ES, FS, GS, and SS. Segment descriptors contain various information about a segment, including:

► The physical memory address of the first byte of the segment, that is, its base address

► The size of the segment

► The segment type

► Access restrictions

► Privilege level

A memory address used in an operand consists of a reference to one of the segment registers and an offset value. For example, the operand DS:018C refers to byte 018C (hexadecimal) within the segment pointed to by the DS segment descriptor. The CPU converts this to a physical memory address by adding the segment base address and the offset.

Segments range in size from 1 byte to 4 GB. Segment types include data segments and code segments (executable instructions). The data type is further subdivided into "true" data and stack data. The CS register always contains a segment descriptor for a code segment. The DS, ES, FS, and GS registers always contain segment selectors for a data segment, or a null descriptor. The SS registers always contain the segment descriptor of a stack.

Pentium CPUs protect process memory by checking instruction operands that reference segment descriptors. The simplest check is a limit check on the offset value that prevents processes from reading or writing data outside their own segments. For each memory reference, the CPU compares the offset value to the size parameter stored in the segment descriptor. If the offset is larger than the segment size, an error interrupt is generated.

Access types must be specified for each segment. Data segments can be marked as read-only (RO) or read-write (RW). Code segments can be marked as execute-only (EO) or execute-read (ER). A segment's access type restricts the types of instructions that can execute using memory references within the segment. The CPU generates an error interrupt if an instruction attempts to write to a segment marked RO or read from a segment marked EO. Write operations to code segments always generate an error interrupt. The CPU also checks segment types when loading segment descriptors into a register. Segment descriptors for data segments cannot be loaded into the CS register. A segment descriptor for a code segment cannot be loaded into the DS, ES, FS, GS, or SS registers.

Pentium CPUs use two different methods to prevent access to code segments not allocated to a process. The first method specifies a privilege level for each process and segment. Privilege levels are numbered 0 through 3, with 0 being the most privileged. Every process is assigned a privilege level and it cannot access a segment with a privilege level numbered lower than its own. This method protects operating system components from interference by application programs. Segments allocated to the kernel typically have a privilege level of 0. Segments allocated to other operating system components typically have a privilege level of 1 or 2. Application processes typically have a privilege level of 3.

Memory segments can be completely hidden from a process. The processor maintains a global descriptor table (GDT) containing descriptors for all segments. If an application process uses the GDT, then it can "see" all segments. The operating system can execute an instruction that allocates a local descriptor table (LDT) to a process. The operating system defines the content of the LDT prior to issuing the instruction. Typically, the operating system populates the LDT only with descriptors allocated to the process. It is impossible for a process using an LDT to know of any segments other than its own.

Pentium processors implement virtual memory management tables in hardware. Page size normally is 4 KB. The CPU uses two types of tables—page directories and page tables. A page directory is a table of pointers to page tables. By convention, the entries in the page directory correspond directly to the segment's descriptors contained in a process's LDT (see Figure 11-17). Each entry in a page directory points to one entry in a page table. Each entry in a page table contains descriptive information about one 4 KB page.

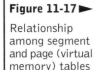

Figure 11-17 ►

Relationship
among segment
and page (virtual
memory) tables

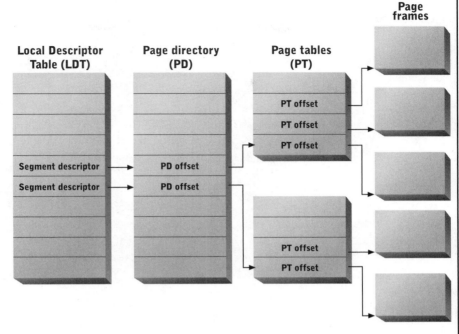

A page table entry includes the physical memory address of the page if it is loaded into memory. The entry also contains a number of bit field flags including a flag that indicates whether the page is in memory, a flag that indicates whether or not the page has been accessed since it was loaded, and a flag that indicates whether the content of a page in memory has been written.

Virtual memory management responsibility is split between the CPU and the operating system. When virtual memory management is enabled, the CPU converts addresses containing a segment descriptor and offset into addresses containing a page directory offset, a page table offset, and an offset into the 4 KB page. The CPU automatically performs a lookup in the page directory and page table based on these offsets to find the base physical address of the page (see Figure 11-18). It then adds the page offset to generate a physical address. If the page table entry indicates that the page is not in memory then the CPU generates an interrupt. The CPU also sets the access and write flags in the page table as page contents are read or written.

The operating system maintains its own table that maps pages to page frames in the swap space, and clears the access and write flags when a page is loaded into memory. The operating system sets and clears page table "in memory" flags each time pages are loaded or swapped. It also provides an interrupt handler to implement page swaps in response to page fault interrupts.

Figure 11-18▶

Address resolution with virtual memory management

The performance improvements gained by implementing memory protection and virtual memory management within the CPU are substantial. A few versions of UNIX were developed to run on the Intel 80286 CPU, which had no hardware support for advanced memory management. The result was execution so slow that it was unusable.

The 80386 CPU was introduced in 1987 and provided memory management functions similar to those of Pentium family CPUs. When 80286 UNIX versions were rewritten for the 80386 system, throughput was improved by a factor of 10 to 20 times. The 80386 word size was double that of the 80286 and clock rates were marginally faster, which normally would have improved software performance by only 2 to 3 times. The extra performance gain resulted from transferring many advanced memory management functions from software to hardware.

INTERPROCESS COMMUNICATION

Cooperating processes and threads might need to synchronize their activities or pass data among themselves. Synchronization often is needed with operating and database management systems and occasionally is needed with application software. For example, a distributed database management system might need to synchronize updates to redundant data stores, or an application thread that prints a document might signal another thread when it is finished.

Interprocess Signals

A **signal** is a message, usually an integer value, used to synchronize the actions of two processes or threads. Signals can be sent between threads of the same process, between independent processes, or between the operating system and a process or thread. Processes and threads use an operating system service call to create or wait for signals. The operating system movement keeps track of which processes or threads are waiting for which signals.

Consider the three threads shown in Figure 11-19. Threads A and B execute independently, but thread C cannot execute until both A and B have terminated.

Figure 11-19 ►

Process containing three threads

Process (word processor)

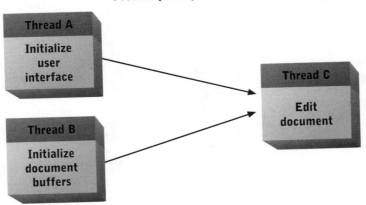

When Thread C creates Threads A and B it executes operating system service calls such as:

```
create_thread("Init_UI",thread_ID_A)
create_thread("InitDocBuffers",thread_ID_B)
```

The operating system uses the first parameter to determine what executable file or function to load and execute. When each thread is created, the operating system assigns it a unique integer number and returns it to Thread C through the

---

second parameter. Thread C uses the thread numbers to tell the operating system that it wants to wait for the threads to complete by executing service calls such as:

```
wait(thread_ID_A,exit_code_A)

wait(thread_ID_B,exit_code_B)
```

When the operating system processes the calls, it suspends Thread C pending completion of Threads A and B. When each thread finishes executing, the operating system passes a signal containing its exit code back to Thread C through the second parameter of the wait service call. Thread C can test either or both exit codes to determine whether the other threads executed successfully.

More complex signals are required when processes coordinate execution of individual process cycles. Consider, for example, the two cooperating processes shown in Figure 11-20. Process 1 retrieves a record from an I/O or secondary storage device and stores it in a shared memory region. Process 2 formats the record and sends it to an output device. Each process has an interlocking repetitive cycle. Process 2 cannot format the record until Process 1 has read and stored it. Similarly, Process 1 cannot store the next record until Process 2 has retrieved the previous record.

Figure 11-20 ▶

Two cooperating processes using interprocess signals

```
Process 1                          Process 2

read(record)
while not end-of-file              signal(process 1)
    wait(process 2)                wait(process 1)
    store(record)                  retrieve(record)
    signal(process 2)              while (record = 'eof')
    read(record)                       signal(process 1)
end while                              transform(record)
store('eof')                           write(record)
                                       wait(process 1)
                                       retrieve(record)
                                   end while
```

Both processes can coordinate their actions with a pair of signals. Process 1 sends a signal to Process 2 after each record is stored in shared memory. Process 2 sends a signal to Process 1 each time a record is retrieved from shared memory, which prevents Process 1 from overwriting shared memory with a new record before Process 2 has retrieved the old record. The operating system suspends each process pending the arrival of a signal from the other process.

The processes also need to coordinate their activities if Process 1 reads an end of file. Another signal could be defined to handle this condition or a single multivalued signal can be used for both purposes. Figure 11-21 illustrates the latter alternative.

Figure 11-21▶

Two cooperating
processes using
a multivalued
signal

```
Process 1                          Process 2

read(record)                       signal(process 1, 'ready')
while not end-of-file              wait(process 1, condition)
     wait(process 2)               while (condition = 'ready')
     store(record)                      retrieve(record)
     signal(process 2, 'ready')         signal(process 1, 'ready')
     read(record)                       transform(record)
end while                               write(record)
signal(process 2, 'eof')                write(process 1, condition)
                                   end while
```

Managing signals is an important operating system function because signals are used by operating system components to control access to shared resources. Note that in the previous two examples, signals not only synchronize the processes, but also coordinate access to their shared memory region. The operating system implements similar access controls for many shared resources such as the CPU, memory, secondary storage, and I/O devices.

Interprocess Data Communication

It is unusual for interprocess data communication to be implemented entirely through signals and process-controlled shared memory. In general, implementing shared memory, as shown in Figures 11-20 and 11-21, requires both processes to be combined during link editing. But processes that are truly independent must each occupy a separate region of memory.

Operating systems provide a mechanism called a **pipe** so there can be communication via shared memory between independent processes. A conceptual representation of a pipe is shown in Figure 11-22.

Figure 11-22▶

Pipe between
two processes
with one-way
data flow

Pipes are manipulated by operating system service calls such as:

▶ open_pipe(process_id,direction,pipe_id)

▶ close_pipe(pipe_id)

▶ read_pipe(pipe_id,data)

▶ write_pipe(pipe_id,data)

The open_pipe call asks the operating system to allocate a shared memory region for communication between the calling process and the process identified by process_id. The operating system returns pipe_id as an identifier for use in subsequent pipe operations. The argument "direction" in the open_pipe call tells the operating system whether data will flow to or from the calling process. It also can be possible to use the pipe for bidirectional data communication, in which case possible values of direction are "to", "from", or "both". The close_pipe call tells the operating system that the shared storage is no longer needed.

The write_pipe call directs the operating system to move the information stored in the data structure "data" to the storage region allocated to the pipe. The read_pipe call directs the operating system to move data from the storage region to the data structure referenced in the "data" argument. Note that the location of the storage region, the movement of data into and out of that region, and the data structure(s) used to implement the pipe are entirely under operating system control.

In some operating systems such as UNIX, input and output to and from pipes is implemented in a manner similar to input and output to and from files. Once created, a process interacts with a pipe in essentially the same manner as it interacts with a file (in other words, it uses file I/O service calls). Normal functions such as reading, writing, buffering, and end of file conditions are the same for pipes as for files. The advantage to such an approach is that it can be used both for data communication and for synchronization.

Figure 11-23 shows the example from Figure 11-21, modified to use pipes for both data communication and process synchronization. Note that most of the explicit signal and wait commands have been eliminated. The signal to coordinate reading and writing to and from shared memory is replaced by I/O service calls to read or write data to or from the pipe. If Process 1 tries to write to a full pipe, the operating system suspends it until Process 2 has consumed some or all of the data in the pipe. Similarly, if Process 2 tries to read from an empty pipe, the operating system suspends it until data is written into the pipe by Process 1. An end of file condition is handled by explicitly read and writing the end of file marker to and from the pipe.

Figure 11-23 ▶

Two cooperating processes using a pipe for synchronization and data communication

```
Process 1

open_pipe(process 2,"to",pipe_id)
signal(process 2,pipe_id)
read(record)
while not end-of-file
    write_pipe(pipe_id,record)
    read(record)
end while
close(pipe_id)
```

```
Process 2

wait(process 1,pipe_id)
read_pipe(pipe_id,record)
while (not eof(pipe_id))
    transform(record)
    write(record)
    read_pipe(pipe_id,record)
end while
close(pipe_id)
```

SUMMARY

► An operating system is the most complex component of system software. Its primary purpose is to manage hardware resources and to provide support services to users and application programs. In functional terms, the operating system manages the CPU, memory, processes, secondary storage (files), I/O devices, and users. In architectural terms, the operating system consists of the kernel, service layer, and command layer.

► The operating system allocates hardware resources to individual user processes on demand. In multitasking environments, the operating system coordinates access to shared resources by many processes while ensuring that each process receives the resources it needs. The resource allocation function requires extensive recordkeeping and complex procedures to make allocation decisions. An important design goal for any operating system is to minimize system overhead while ensuring that resource allocation goals are met.

► Application software is simpler to develop if programs are unaware of resource allocation functions. Operating systems make a large set of virtual resources appear to be available to each application program. The sum of these apparent virtual resources generally exceeds the real resources that exist within the computer system. The operating system implements virtual resources by rapidly reallocating real resources among programs and by substituting one resource for another where appropriate and necessary.

► The operating system stores information about each process in a process control block (PCB). PCB contents are updated continually as resources are allocated to processes. In some operating systems, processes can create executable subunits called threads, which can be independently scheduled. Threads share all resources with their parent process. Execution speed is increased when multiple threads execute concurrently or simultaneously.

► An active thread is always in one of three states—ready, running, or blocked. In a single CPU machine, only one thread may be in the running state at any one time. Ready threads are waiting only for access to the CPU. Blocked threads are waiting for some event to occur, such as completion of a service request or correction of an error condition. Processes can be scheduled by many methods, including first come first served, shortest time remaining, and explicit prioritization.

► Memory is divided into fixed-size partitions and processes are allocated one or more memory partitions to store instructions and data. The operating system maintains tables to track partition allocations and free space. Memory references are mapped to physical addresses through table lookups and address calculations.

▶ Modern operating systems implement a form of memory allocation and management called virtual memory management. Portions of processes, called pages, are allocated to small memory partitions called page frames. Pages are swapped between memory and secondary storage as needed. Complex memory management procedures incur substantial overhead. To reduce operating system overhead, modern CPUs implement many of these functions in hardware.

▶ Processes and threads often need to synchronize their actions or transfer data among themselves using signals and pipes. Signals are flags, or single-valued data items, passed among processes. Pipes are shared memory regions written by one process or thread and read by another. The operating system manages signals and pipes and blocks and unblocks processes and threads as they create, use, and wait for signals or pipe I/O.

This chapter presented an overview of internal operating system architecture and described CPU and memory allocation. In Chapter 12, we'll examine secondary storage allocation and file management. You will see that many of the issues addressed in memory allocation must also be addressed in secondary storage allocation. In Chapter 13, we'll turn our attention to external resources, and explore how an operating system cooperates with other operating systems to provide local users access to distant resources.

Key Terms

address mapping
address resolution
blocked state
child process
command language
command layer
compaction
concurrent execution
contiguous
dispatching
explicit priority
first come first served (FCFS)
fragmented
interleaved execution
job control language (JCL)
kernel
multitasking
multithreaded
noncontiguous
page

page fault
page frame
page hit
page table
parent process
pipe
preemptive scheduling
priority-based scheduling
process
process control block (PCB)
process family
process list
process queue
ready state
real resources
real-time scheduling
run queue
running state
scheduler

scheduling
service call
service layer
shell
shortest time remaining
sibling process
signal
spawn
supervisor
swap space
system overhead
thread
thread control block (TCB)
thread cycle
thread list
timer interrupt
victim
virtual resource
virtual memory management

Vocabulary Exercises

1. A(n) _____ operating system supports multiple active processes or users.

2. Under virtual memory management, the location of a memory page is determined by searching a(n) _____.

3. A(n) _____ occurs when a process or thread references a memory page not currently held in memory.

4. A(n) _____ is a shared memory region, managed by the operating system, that is used to send data between independent processes.

5. Dispatching a process moves it from the _____ to the _____.

6. The CPU periodically generates a(n) _____ to provide the scheduler an opportunity to allocate the CPU to another ready process.

7. A process in the _____ requires only access to the CPU to continue execution.

8. In the _____ scheduling method, processes are dispatched in order of their arrival.

9. A(n) _____ process contains subunits that can be executed concurrently or simultaneously.

10. _____ scheduling guarantees that a process will receive sufficient resources to complete one _____ within a maximum time interval.

11. Hardware resources consumed by an operating system's resource allocation functions are called _____.

12. _____ scheduling refers to any type of scheduling in which a running process can lose control of the CPU to another process.

13. The act of selecting a running process and loading its register contents is called _____ and is performed by the _____.

14. To achieve efficient use of memory and a large number of concurrently executing processes, most operating systems use _____ memory management.

15. When a process makes an I/O service request, it is placed in the _____ until processing of the request is completed.

16. Memory pages not held in primary storage are held in the _____ of a secondary storage device.

17. On a computer with a single CPU, multitasking is achieved _____ execution of multiple processes.

18. Under _____, all portions of a process must be loaded into sequential physical memory locations.

19. The _____, _____, and _____ are the primary layers of an operating system.

20. A(n) _____ operating system can support multiple active processes at any one time.

21. A(n) _____ is the unit of memory read or written to the swap space.

22. A(n) _____ resource is apparent to a process or user, although it might not physically exist.

23. Under a(n) _____ memory allocation scheme, portions of a single process might be located physically in scattered segments of main memory.

24. Under virtual memory management, memory references by a process must be converted to an offset within a(n) _____.

25. Information about a process's execution state such as register values and process status are stored in a(n) _____.

26. Under the _____ scheduling method, processes requiring the least CPU time are dispatched first.

27. A(n) _____ causes the currently executing process to be _____ and control passed to the _____.

28. The process of converting an address operand into a physical address within a memory partition or page frame is called _____.

29. A(n) _____ can be used for interprocess _____ when no data needs to be passed.

30. A(n) _____ is an executable subunit of a process that is scheduled independently but shares memory and I/O resources.

Review Questions

1. Describe the functions of the kernel, service, and command layers of the operating system.

2. What is the difference between a real resource and a virtual resource?

3. What hardware resources are allocated to processes by the operating system?

4. What are the goals of an operating system resource allocation function? Describe the conflict among them.

5. How and why does a process move from the ready state to the running state? How and why does a process move from the running state to the blocked state? How and why does a process move from the blocked state to the ready state?

6. What is a process control block and for what is it used?

7. What methods does an operating system provide for interprocess synchronization and communication? Under what conditions would each be most useful?

8. What is a thread? What resources does it share with other threads within the same process? What resources doesn't it share?

9. Briefly describe the most common methods for making scheduling decisions.

10. What scheduling complexities are introduced by real-time processing requirements?

11. Describe the operation of virtual memory management.

12. What is memory protection and why is it needed? Where is it most efficiently implemented or enforced?

Research Problems

1. The RISC approach to processor design described in Chapter 4 avoids internal CPU complexity to gain raw processor speed. Strict adherence to this design philosophy might be construed to eliminate CPU support for memory protection and virtual memory management. Investigate the memory management techniques of a RISC CPU such as the Sun UltraSPARC (www.sun.com) or IBM PowerPC (www.chips.ibm.com). What support is provided for memory protection and virtual memory management?

2. Investigate the memory protection, multitasking, and interprocess communication capabilities of Sun Microsystems' Solaris operating system (see the textbook Web site for a set of Web resources). How are application programs isolated from the operating system? How are application programs isolated from each other? Is it possible for one application to cause another application or the operating system to crash? How are CPU cycles allocated to application programs? Can a thread or process lock up the system by never releasing the CPU? How are pipes and signals implemented?

File Management Systems

Chapter Goals

- ► Describe the components and functions of a file management system

- ► Compare the logical and physical organization of files and directories

- ► Explain how secondary storage locations are allocated to files and describe the data structures used to record those allocations

- ► Describe file manipulation operations, including open, close, read, delete, and undelete operations

- ► List access controls that can be applied to files and directories

- ► Describe security, backup, recovery, and fault tolerance methods and procedures

File management is an important system software function because stored programs and data are important individual and organizational resources. The entire collection of system software that performs file management and access functions is known as a **file management system (FMS)**. The FMS usually is part of the operating system though it sometimes is supplemented by additional software such as database management systems. Figure 12-1 shows the FMS functions described in this chapter.

Figure 12-1 ►

Topics covered in this chapter

FILE MANAGEMENT SYSTEMS

An FMS is implemented in layers similar to those of the operating system as a whole (see Figure 12-2):

► Application program or command layer
► File control
► Storage I/O control
► Secondary storage devices

Figure 12-2 ▶

Operating system
layers compared
to file
management
system layers

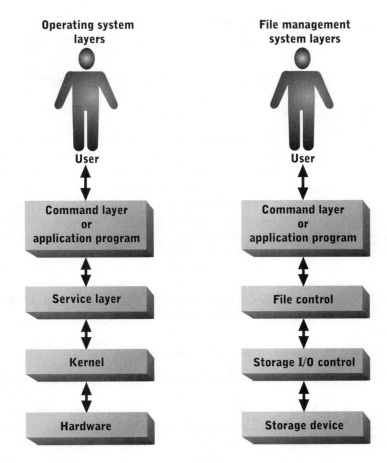

The secondary storage device level that corresponds to the operating system hardware level includes both storage devices and storage device controllers. Storage devices physically store bits, bytes, and blocks within a storage medium, and storage device controllers interact with the bus and with operating system device drivers to transfer data between storage devices and memory. As discussed in Chapter 6, the device controller presents a logical view of the storage device or media to the device driver. The logical view is a linear sequence of storage locations or linear address space.

The **storage I/O control layer** accesses storage locations and manages data movement between storage devices and memory. The storage I/O control layer is part of the operating system kernel. Software modules within this layer include:

► Device drivers for each storage device or device controller
► Interrupt handlers
► Buffers and cache managers

The **file control layer** provides a set of service functions for manipulating files and directories. It processes service calls from the application or command layer and issues commands to the storage I/O control layer to interact with hardware. The file control layer maintains the directory and storage allocation data structures used to locate files and their associated physical storage locations.

An FMS provides command layer functions and utility programs that users and system administrators use to manage files, directories, and secondary storage devices. With the command layer, users perform common file management functions such as copying, moving, and renaming. Utility programs address more complex functions such as creating text files, formatting storage devices, and creating backup copies of files, directories, and storage volumes on storage media such as magnetic tape or compact disc.

Logical and Physical Storage Views

The file control layer is the bridge between logical and physical views of secondary storage. Users and applications logically view secondary storage as a collection of files organized within directories and storage volumes. The physical view of secondary storage is a collection of physical storage locations organized as a linear address space. A typical computer system has up to a few dozen storage volumes, hundreds to thousands of directories, tens of thousands to millions of files, and millions to billions of physical secondary storage locations (see Figure 12-3).

Figure 12-4 shows the logical structure of a typical data file. The file is subdivided into multiple **records** and each record is composed of multiple **fields**. A record usually contains information about a single person such as a customer or employee, a thing such as a product held in inventory, or an event such as a transaction. A field contains a single data item that describes the record subject.

Figure 12-3 ►

Logical and physical secondary storage for a typical PC

Volumes (4)

Directories (1,000)

Files (200,000)

Disk blocks (80,000,000)

Logical view

Physical view

Figure 12-4 ►

Logical structure of a data file

Fields

Records

99635	Smith	George	R	1414 Oak Road	Los Angeles	...
04667	Jones	Alice	L	12 Main Street	Buffalo	...
52046	Torres	Maria	H	9823 7th Avenue	Chicago	...
34421	Miller	Fritz	S	451 Quincy Road	Atlanta	...
63752	Chang	Wendy	B	310 G Street	Fort Worth	...

The logical file structure is independent of the physical device on which it is stored. A number of physical structure characteristics are simplified or ignored in the corresponding logical file structure, including:

► Physical storage allocation
► Data access method(s)
► Data encoding method(s)

Physical storage allocation considerations include the placement of fields and records within a file, and the distribution of the file across storage locations, media, or devices. Physical data access considerations include whether file content is accessed sequentially, using an index, or by some other method. Data encoding issues include the data structures and coding methods used to represent individual fields. Related issues include data encryption and data compression.

File Content and Type

A file can store many different data types including text, numbers, complex data structures, and executable instructions. It is difficult to design an FMS that accounts for all possible variations in file content and organization. Such an FMS would be very complex and have a very large number of file-oriented commands and service routines. To avoid these problems, most FMSs directly support only a limited number of file types, including:

► Executable programs
► Operating system commands
► Textual or unformatted binary data

Among other things, file type determines:

► Physical organization of data items and data structures within secondary storage
► Operations that may or may not be performed upon the file
► Filename restrictions

For example, the Windows FMS supports executable code stored within executable (EXE) files and dynamically linked library (DLL) files. EXE files are stored in a format that simplifies loading them into memory for execution. DLL files contain subroutines that can be called from executable files and an index that allows the operating system to locate subroutines quickly by name. Operating system batch command (BAT) files are stored as ordinary text files and the text commands are passed automatically to a command interpreter when the user double-clicks a BAT file icon. If the user double-clicks a text (TXT) file icon, the operating system automatically calls the Notepad text editing utility, which opens the file for editing. The relationship between file types and the programs or operating system utilities that manipulate them is called **file association**.

Modern FMSs provide a framework to support additional file types. In this framework, users and application programs can register new file types and install programs to perform common file manipulation operations such as printing, editing, and error checking. Figure 12-5 shows a partial listing of application file types registered on a computer running Windows 2000. Figure 12-6 shows registration details for Microsoft Word document files including the associated programs and commands used to print a document. Once a file type is registered, the operating system loads and executes the registered program each time the user prints, edits, or otherwise manipulates the file.

File type normally is declared when a file is created. In some FMSs, such as UNIX, file type is stored within the directory. In other FMSs, such as Windows, the file type is declared through a filename restriction or convention. For example, in Windows, executable filenames must end in .EXE, dynamically linked library filenames must end in .DLL, and the default text filenames end in .TXT. Windows registered file types also use naming conventions such as .DOC for Microsoft Word document files and .CDR for CorelDraw graphics files (see the Extensions column in Figure 12-5).

Figure 12-5 ▶

Registered
Windows 2000
file types

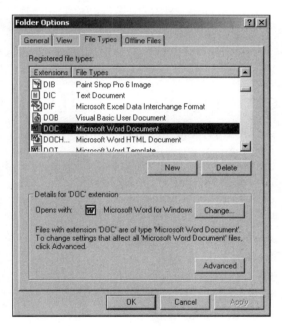

Figure 12-6 ►

Sample
association
details of a
Windows 2000
file type

DIRECTORY CONTENT AND STRUCTURE

A **directory** contains information about files and other directories. Directory information normally is stored in a table, though more complex data structures are sometimes used. Users and system administrators can view most directory information using command layer utilities. Some directory information, such as disk location, is hidden from users because it is only used by the FMS. Typical directory contents include:

► Name
► File type
► Location
► Size
► Ownership
► Access controls
► Time stamp(s)

All files and directories must have a unique name within their own directory but names can be reused in other directories. Operating systems vary in requirements for valid names. Many older operating systems restrict both the length of a name and the characters used. For example, MS-DOS uses a two-part name with a maximum of eight characters in the first part, three characters in the second part, and a mandatory dot (.) symbol separating the parts, for example SAMPLE.DOC (this format is sometimes called 8.3). Embedded space characters are not allowed nor are most nonalphabetic and nonnumeric characters. Modern operating systems are far less restrictive in their requirements for filenames.

File type can be stored implicitly through a file naming convention, such as .EXE for an executable file, or directly in a coded field within the directory. The type field can be displayed in several formats including icons in a graphical display or special characters appended to the filename in a text-oriented display. Most FMSs store information about embedded directories in the same way as information about files. An FMS uses a special directory code in the file type field to distinguish directories from files.

The location field content varies considerably across FMSs. In simpler FMSs, the location field usually contains the disk address of the file's first disk block. In more complex FMSs, it contains or points to a data structure such as an index that contains the addresses of all disk blocks.

The size field can contain the number of bytes allocated to the file, the actual number of bytes stored in the file, or both. The number of allocated bytes usually is larger than the actual number of bytes because a portion of the last disk block typically is not used. For example, the directory listing of a file with 513 actual bytes might list the file size as 1024 bytes if the allocation unit size is 512 bytes, because only one byte of the last disk block actually is used.

The file owner field contains the account name or identifying number of the file creator or the last account to take ownership. Most complex FMSs grant special permissions to the file owner. Access controls list the accounts that have been granted or denied rights to the file. Rights vary among operating systems, but the following are typical:

- List
- Read
- Modify
- Change

With the List right, an account or group can view a file in a particular directory listing or list the contents of a directory. With the Read right, an account or group can view the content of a file which also implies the right to copy it. With the Modify right, an account or group can add new content, alter or delete existing content, rename the file, move the file to a new directory, or delete the file.

With the Change right, which usually is granted to the owner by default, an account or group can alter permissions for other accounts or groups.

Most FMSs store one or more time stamps in the directory for each file. Time stamps can include:

► When the file was created
► When the file most recently was read
► When the file most recently was written
► When the file last was backed up

Hierarchical Directory Structure

In a **hierarchical directory structure**, directories can contain other directories but a directory cannot be contained within more than one parent. A hierarchical directory structure sometimes is called a **tree directory structure** because directory diagrams resemble upside-down trees. The left pane of Figure 12-7 shows a portion of a hierarchical directory structure as displayed by Windows Explorer. There is a root directory for each secondary storage device (drives A:, C:, F:, and G:, as shown in Figure 12-7). Each root directory can contain other directories, files, or both, as can directories below the root directory. The number of recursively descending directory levels theoretically is unlimited.

The right pane of Figure 12-7 shows the content of a single directory named First. For each process or user, the operating system maintains a pointer to the directory that currently is being accessed. That directory is called the **current directory** or **working directory**. First is the current directory in Figure 12-7.

Figure 12-7 ►

A hierarchical directory structure

In a multiuser operating system, each user normally has a default working directory, or **home directory**. When a user logs on interactively or runs a batch process, his or her home directory is the default current directory. The user or process subsequently can change the current directory by issuing a command through the command layer or by making the appropriate service call.

Within the hierarchy of directories, names of access paths can be specified in two ways. A **complete path**, also called a **fully qualified reference**, begins at the root directory and proceeds through all directories along a path to the desired file. Directory names are separated by a special character such as "\" in Windows or "/" in UNIX. A fully qualified reference to the current directory is shown in the Address text box near the top of Figure 12-7:

F:\Textbooks\Systems Architecture\Third Edition\Drafts\First

The fully qualified reference to the first file in that directory is:

F:\Textbooks\Systems Architecture\Third Edition\Drafts\First\Chapter01.pdf

A **relative path** begins at the level of the current directory. For example, the name:

Chapter01.pdf

is a relative path to the file Chapter01.pdf in the current directory displayed in Figure 12-7.

Graph Directory Structure

A **graph directory structure** (see Figure 12-8) is more flexible than a hierarchical directory structure. A graph directory relaxes two of the restrictions enforced within a hierarchical directory structure:

▶ Files and subdirectories can be contained within multiple directories

▶ Directory links can form a cycle

Figure 12-8 ▶

Graph directory structure (links shown as dashed lines)

In Figure 12-8, subdirectories contained within multiple directories are shaded. The directories Images and Documents (on the left) are each contained in two directories, and WWW Pages is contained in three directories. The dashed lines in Figure 12-8 are called **links** in UNIX and **shortcuts** in Windows.

Links create the possibility of cycles or loops in the directory structure. There are two cycles in Figure 12-8—one from WWW Pages to MGT 331 to Documents and then back to WWW Pages, and the other from WWW Pages to MGT 337 to Documents and then back to WWW Pages. Cycles create special problems for programs that recursively search directories because the search program can become stuck in an endless loop as it traverses the cycle over and over again. Special processing procedures are required to avoid this situation. Programs that recursively access directory contents can avoid infinite loops by not traversing links or by keeping a list of directories already "visited" and never rereading a directory already on the list.

STORAGE ALLOCATION

The storage I/O control layer allocates secondary storage locations to individual files and directories. Because there are so many directories, files, devices, and secondary storage locations, storage allocation data structures and the functions that maintain their content are complex. The data structures and procedures used to manage secondary storage allocation to files are similar to those used to allocate memory to processes, as described in the previous chapter. The most significant differences are the number of storage locations, which are much larger for secondary storage, and the frequency of change in allocation, which is much lower for secondary storage.

Allocation Units

An **allocation unit** is the smallest number of secondary storage bytes that can be allocated to a file. Allocation units cannot be smaller than the unit of data transfer between the storage device and controller, which normally is called a **block**. Typical secondary storage device block size ranges from 512 bytes to 4 KB in multiples of 512 bytes.

Allocation unit size can be a multiple of block size. For example, a disk might use a 512 byte block size but the system administrator might decide to use 16 KB allocation units, each containing 32 blocks. Allocation unit size usually is set when an operating system or storage device is installed. Some operating systems such as DOS and early Windows versions[1] set unit size automatically based

---

[1] All MS-DOS versions and many Windows versions use similar storage allocation and file management methods. For the remainder of this chapter, all references to Windows Me also apply to Windows 98, Windows 95, Windows 3.1, and all MS-DOS versions unless otherwise stated.

on storage device or media capacity. In other operating systems, such as UNIX and Windows 2000, the installer selects the unit size. Once allocation unit size is set it is difficult to change.

Allocation unit size is a tradeoff among:

▶ Efficient use of secondary storage space for files
▶ Size of storage allocation data structures
▶ Efficiency of storage allocation procedures

Smaller allocation unit size results in more efficient use of storage space. For example, a file containing a single byte must be stored in a full allocation unit. If allocation unit size is 512 bytes, 511 bytes are empty and wasted. If the allocation unit size is 16 KB and only one byte is stored, 16,383 bytes are empty and wasted. If a storage device holds many small files, a large allocation unit size can waste a great deal of storage capacity.

The advantage of larger allocation unit sizes is that storage allocation data structures can be smaller. As allocation unit size decreases, the number of allocation units increases. For example, consider a 100 GB disk. If unit size is set to 512 bytes, there are 209,715,200 (100 x 1024^3 ÷ 512) allocation units in the device. If unit size is set to 16 KB, there are 6,553,600 (100 x 1024^3 ÷ 16,386) allocation units in the device.

Storage Allocation Tables

A **storage allocation table** is a data structure that records which allocation units are free and which belong to files. A storage allocation table contains one entry for each allocation unit. A smaller allocation unit size increases the number of entries in the storage allocation table. As the storage allocation table grows larger, the time required to search and update the table increases, which slows down any processing function that creates, deletes, or changes the size of a file.

Storage allocation table format and content vary across FMSs. Operating systems such as Windows Me use relatively simple data structures such as tables and linked lists. More complex operating systems such as UNIX and Windows 2000 use more complex data structures such as bit maps and B+ trees. We'll stick to simpler data structure examples in this chapter.

Figure 12-9 shows a hypothetical storage device with 36 allocation units. Allocation units shaded the same color all belong to the same file or to free space. Table 12-1 shows directory entries for the files stored in Figure 12-9. Free allocation units are assigned to a hidden system file called SysFree. A file's allocation units can be stored in any order on the device. The linear address of each file's first allocation unit is stored in the directory. The addresses of other file allocation units are stored in the storage allocation table.

Figure 12-9 ▶

Storage blocks
allocated to
three files

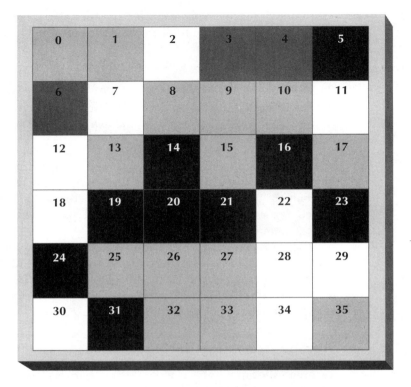

Table 12-1 ▶

Directory
content for
the files shown
in Figure 12-9

Filename	Owner name	First allocation unit	Length (in allocation units)
File1	Smith	0	14
File2	Jones	3	3
File3	Smith	5	9
SysFree	System	2	10

Table 12-2 shows a storage allocation table for the files in Figure 12-9. The format and content of this table are nearly identical to the **file allocation table (FAT)** used in Windows Me. The table contains an entry for each allocation unit. All of a file's allocation units are "chained" together in sequential order by a series of pointers. The table entry for each allocation unit contains a pointer to the next allocated unit in the file. For example, the entry for unit 0 (the first unit allocated to File1) contains a pointer to the table entry for unit 1 (the second unit allocated to File1). The pointers form a linked list that ties together the entries of all allocation units assigned to a specific file. The entry for a file's last allocation unit contains a special code to indicate that it is the last unit. All unallocated units are linked into a single chain, which simplifies finding storage units to allocate to new or expanded files.

Table 12-2 ▶

A storage
allocation table
matching the
storage alloca-
tions shown in
Figure 12-9

Unit	Pointer	Unit	Pointer	Unit	Pointer	Unit	Pointer
0	01	9	10	18	22	27	32
1	08	10	13	19	20	28	29
2	7	11	12	20	21	29	30
3	04	12	18	21	23	30	34
4	06	13	15	22	28	31	End
5	14	14	16	23	24	32	33
6	End	15	17	24	31	33	35
7	11	16	19	25	26	34	End
8	09	17	25	26	27	35	End

Sequential access to a file's allocation units is efficient when the storage allocation table uses linked lists. However, random access is much less efficient, particularly if the file is large. Some FMSs define a separate file type for random access files and store an index or similar data structure in the file's first allocation unit. The contents of this index are redundant with the content of the storage allocation table but the index allows specific units of a file to be located and accessed efficiently.

Blocking and Buffering

Some application programs access files by logical records. A **logical record** is a collection of data items, or fields, that is accessed by an application program as a single unit. A **physical record** is the unit of storage transferred between the device controller and memory in a single operation. For disk and other devices that use fixed size data transfer units, a physical record is equivalent to a block. For storage devices with variable size data transfer units, such as tape drives, block size might differ from physical record size or might be undefined.

If logical record size is less than physical record size, a single physical record might contain multiple logical records, as shown in Figure 12-10(a). If logical record size is larger than physical record size, multiple physical records are required to hold a single logical record, as shown in Figure 12-10(b). Logical record grouping within physical records is called **blocking**. Blocking is described by a numeric ratio of logical records to physical records called the **blocking factor**. The blocking factor in Figure 12-10(a) is 4:3 and the blocking factor shown in Figure 12-10(b) is 2:3. If a physical record contains just one logical record, then the file is said to be **unblocked**.

Figure 12-10 ▶

Blocking of
logical records
into physical
records

Read and write operations to or from an unblocked file can be implemented with simple and efficient algorithms because of the one-to-one correspondence between logical and physical records. File I/O is much more complex when logical and physical records have different sizes because the FMS must coordinate physical record I/O and extract logical records on behalf of the requesting program.

The FMS uses buffers in primary storage to store data temporarily as it moves between programs and secondary storage devices. Buffers are allocated automatically when the file is first accessed and are managed by the operating system on behalf of application programs. Each buffer is the size of one allocation unit. As physical records are read from secondary storage, they are stored in the buffer(s). The FMS extracts logical records from the buffers and copies them to the data area of the application program (see Figure 12-11). The procedure is reversed for write operations.

Figure 12-11 ▶

Input from
secondary
storage to an
application
program using a
buffer

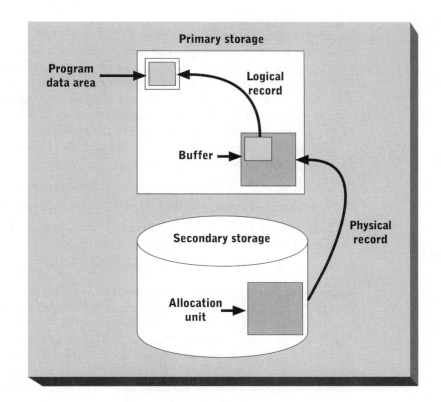

A buffer is a scratchpad for extracting logical records from physical records. As discussed in Chapter 6, buffering also improves I/O performance if a sufficient number of buffers are used. For high blocking factors, a small number of buffers can improve performance dramatically. For example, if each physical record contains 10 logical records, reading a single physical record provides sufficient data for 10 logical read operations. Reading the first logical record results in copying 10 logical records, which is one physical record, into the buffer. The next nine sequential read operations can be satisfied from the buffer without further input from secondary storage.

Low blocking factors, with larger logical records, require more buffers to achieve significant performance improvements. The FMS usually allocates enough buffers to hold at least one physical record. If enough buffers can't be allocated to hold an entire physical record, then a logical record must be moved to the program data area in a series of physical read and buffer copying operations.

An Example

The following example is typical of storage allocation and file I/O procedures in a simple FMS. Assume that the directory entries in Table 12-1 represent actual files stored on a disk drive and that the storage allocation unit size used by the disk drive and FMS is 512 bytes. Allocation units are assigned to files as shown in Figure 12-9, and the file allocation table is shown in Table 12-2. Assume that 55 byte logical records are stored sequentially in File3.

In response to any read operation performed by the application program, the FMS performs the following tasks:

1. Determines which allocation unit contains the requested record.

2. Loads that allocation unit into the buffer if it is not already there.

3. Copies the portion of the allocation unit that contains the desired data to the application program's data area in memory.

4. Increments a pointer to the current position in the file.

Sequential Access The first allocation unit contains the first logical record or the first part of the logical record if it is larger than an allocation unit. When the first logical record is read, the FMS looks up the File3's first allocation unit in the directory (5), issues a read request to the disk controller, and then loads the physical record into the buffer. Once loaded into the buffer, the first logical record's first byte is stored in byte 0. The FMS copies the first 55 bytes from the buffer to the application program. At the conclusion of the operation, the FMS sets the file pointer to 55 to point to the start of the second logical record.

During subsequent read operations, the FMS calculates the allocation unit and offset for each logical record. For the second record, the calculation is as follows:

$$\frac{(\text{Pointer} - 1) \times \text{Record.Size}}{\text{Block.Size}} = \frac{(2-1) \times 55}{512} = 0, \text{remainder } 55$$

The second record begins in allocation unit 0 at byte 55. Because that block is already in the buffer, the FMS copies 55 bytes starting at offset 55 and adds 55 to the file pointer.

Direct Access

Assume that the first read operation requests the 37th logical record. Using the same formula as before, the calculation is:

$$\frac{(\text{Pointer} - 1) \times \text{Record.Size}}{\text{Block.Size}} = \frac{(37-1) \times 55}{512} = 3, \text{remainder } 499$$

The 37th record begins in the fourth allocation unit allocated to File3 at byte 499. Allocation unit 3 is the fourth allocation because allocation unit 0 is the first allocation unit.

To find File3's third allocation unit, the FMS follows the chain of entries in the file allocation table. The first allocated unit (5) is recorded in the directory. File3's second allocation unit is the pointer field of entry 5 in the file allocation table, which is allocation unit 14. File3's third allocation unit is the pointer field of entry 14 in the file allocation table, which is allocation unit 16. File3's fourth allocation unit is the pointer field of entry 16 in the file allocation table, which is allocation unit 19. Therefore, the 37th record starts in allocation unit 19 at byte 499.

The FMS loads allocation unit 19 into the buffer and begins transferring 55 bytes to the application program. While copying, the FMS reaches the end of the buffer before it has finished copying 55 bytes. The FMS recognizes this condition, looks in the file allocation table to determine File3's next allocation unit (20), issues a read request for allocation unit 20, loads it into the buffer, and copies the first 42 bytes to the application program.

FILE MANIPULATION

All FMSs provide service layer functions so application programs can create, copy, move, delete, read, and write files. The exact set of service layer functions varies widely among FMSs. Application programs interact directly with the FMS through the service layer and users interact with the FMS indirectly through the command layer.

File Open and Close Operations

The FMS must perform several tasks, collectively called a **file open operation**, before an application program can read or write a file's contents. The application program executes a file open service call to inform the FMS that intends to read or write a file. In response to the service call, the FMS performs the following steps:

1. Locates the file within the directory structure and reads its directory entry.
2. Searches an internal table of open files to see if the file already is open.
3. Ensures that process has sufficient privileges to access the file.
4. Allocates one or more buffers.
5. Updates an internal table of open files.

The FMS maintains a table of open files to prevent application programs from interfering with each other's file I/O activities. If a file already is open for read-only access and another program tries to open the file, the FMS normally will open the file but allocate a separate set of buffers to each program. If another program attempts to open the file for writing, the request normally will be denied because buffer content is difficult to manage if one program can change content that another program already has read. In some complex FMSs, multiple programs can read or write a file by employing complex schemes for locking physical records.

When an application program finishes reading or writing a file it executes a file close service call. The FMS completes the **file close operation** by performing the following steps:

1. Flushing the program's file I/O buffers to secondary storage.
2. Deallocating buffer memory.
3. Updating the file's directory entry time stamps.
4. Updating the open file table.

If the program that issued the request is the only program accessing the file, the FMS deletes the file's entry in the open files table. If other programs still are accessing the file, the FMS deletes the program from the list of programs actively using the file.

Delete and Undelete Operations

In most FMSs, files are not removed immediately from secondary storage when they are deleted. Instead, the file's storage allocation units are marked as free and its directory entry is marked as unused. As new files are created or existing files are expanded, the deleted file's allocation units are reassigned to other files and overwritten with new content. Some portion of the deleted file's content remains on secondary storage until all of its allocation units have been reassigned and overwritten. When a new file is created in the deleted file's former directory, the deleted file's directory entry is overwritten.

Implementing file deletion as just described is efficient but it has two important consequences:

▶ Files can be undeleted by reconstructing appropriate storage allocation table contents

▶ File content can be visible to intruders who can bypass the storage allocation table and read allocation units directly

A user might be able to recover a deleted file by performing an **undelete operation**. For example, if File2 in Figure 12-9 and Table 12-1 is deleted, the second row of the directory table is marked as deleted and allocation units 3, 4, and 6 are added to the chain of allocation units for SysFree. If the user executes a file recovery utility before performing any other file operations, File2 can be recovered based on the information still in the directory entry and the chained contents of the storage allocation table in Table 12-2.

Giving a user the ability to undelete a file usually is considered a good thing, but an intruder who breaks system security can perform exactly the same operation. In addition, an intruder also might be able to access deleted file contents using low-level CPU instructions to read disk sectors directly. In some environments such as law enforcement and defense research, users need to know that a deleted file can never be recovered. In some FMSs, users or system administrators can configure the FMS so directory entries and allocation units immediately are overwritten with blanks. This provides additional security but slows the deletion process, particularly for large files. Removable magnetic media might need to be destroyed because it is possible to recover old data after a single overwriting.

ACCESS CONTROLS

Because data is an important and valuable organizational resource, an FMS helps prevent loss, corruption, and unauthorized access to files. An FMS relies on the operating system to identify and authenticate users and their processes. In operating systems that enforce access controls, each user has a unique account name or identification number and users must authenticate their identity through passwords or other means. Once user identity is authenticated, the user's name or ID is passed to the FMS with every service request.

By default, individual users are the owners of files they create and can grant or deny file access privileges to other users or groups of users. Different FMSs provide different sets of access controls. For example, UNIX defines three access control types:

► Read
► Write
► Execute

With read access, a user or process can view the contents of a file. With write access, a user or process can alter the contents of a file or delete it altogether. With execute access, a user or process can execute a file, assuming that the file contains an executable program or set of operating system commands. A file owner can reserve these access privileges to himself or herself, thereby denying those access privileges to all other users except the system administrator. The file owner can grant any access privilege to other members of a user group or to all users. For example, a file owner might grant read and write access to one workgroup but only read access to other users. A file owner can revoke his or her own access privileges. For example, a file owner might deny write access to prevent accidental deletion of an important file.

Most FMSs use similar access controls for files and directories. Users are the owners of their home directories and any directories that they create below the home directory in a hierarchical directory structure. Access controls for reading (listing directory contents) and writing (altering directory contents) are defined.

Access controls are enforced automatically in FMS service routines that access and manipulate files and directories. Although access controls are a necessary part of file manipulation, in most FMSs they impose additional processing overhead. In some FMSs, the system administrator might choose among several different levels of enforcement for file access controls to balance overall FMS performance with the relative need for file security.

An FMS places access restrictions on secondary storage devices, storage allocation tables, and root directories to prevent users and processes from bypassing security controls built into FMS service routines. File and directory accesses also can be logged for later review by the system administrator. As with other security measures, logging imposes a performance penalty on all file accesses.

Technology
Focus

Windows NTFS

Early Microsoft operating systems including MS-DOS and Windows versions through Windows Me use a file system called the file allocation table (FAT) file system. When Microsoft developed Windows NT, which evolved into Windows 2000, it decided to develop an entirely new file system, the NT File System (NTFS). Windows NT was targeted to high-performance and "mission critical" application software that required several file system features, including:

► High-speed directory and file operations

► Ability to handle large disks, files, and directories

► Secure file and disk content

► Reliability and fault tolerance

Windows NTFS organizes secondary storage as a set of volumes. A volume normally corresponds to one partition of a disk drive but volumes can span multiple partitions and drives. Volumes contain a collection of storage allocation units called clusters. Cluster size can be 512, 1024, 2048, or 4096 bytes. Each cluster is identified by a 64-bit logical cluster number (LCN) within a linear address space. A volume can be as large as 4096×2^{64} bytes.

A volume's master directory is stored in a data structure called the master file table (MFT). The MFT contains a sequential set of file records, one for each file on the volume. All volume contents are stored as files, including user files, the MFT itself, and other volume management files such as the root directory, storage allocation

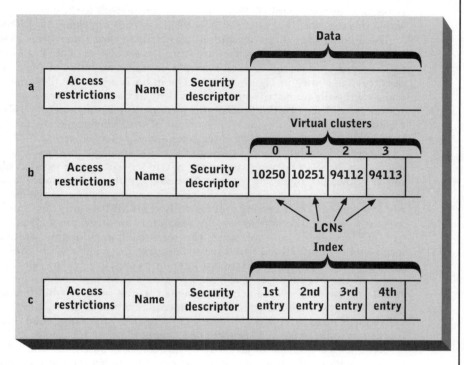

Figure 12-12 ▶

NTFS MFT records for a small file (a), a large file (b), and a small directory (c)

table, bootstrap program, and bad cluster table. The first 16 MFT entries, numbered 0 through 15, are reserved for the MFT and volume management files. All subsequent MFT entries, numbered 16 and higher, store records about user files.

Conceptually, a file is an object with a collection of attributes including name, global access restrictions such as read only, and security descriptor such as owner and owner-defined access controls). A file's data content is just another attribute though it is usually much larger than the other attributes. Each attribute type is assigned a numeric code and file attributes are stored in ascending code order in the file's MFT record, as shown in Figure 12-12(a). MFT record size is 1, 2, or 4 KB and is determined by the operating system when a volume is formatted.

Each file attribute contains a header and a data value. The header contains the attribute name, a resident flag, the length of the header, and the length of the attribute value. Attributes can be either resident or nonresident. A resident attribute is a short value stored within an MFT record immediately following its header such as access restrictions and filename.

A nonresident attribute is stored in clusters elsewhere in the storage device and the cluster addresses are stored in the file's MFT record. Data attributes are the most common nonresident attributes though normally resident attributes such as the security descriptor can grow too large to fit within an MFT record. The clusters assigned to a nonresident attribute are called virtual cluster numbers (VCNs). The area immediately following the nonresident attribute's header stores a sequence of

LCNs, as shown in Figure 12-12(b). The VCN corresponds to the position of an LCN within this sequence. For example, LCN 94112 is VCN 2 in Figure 12-12(b).

The volume root directory and all user-defined directories are stored in the same manner as files. That is, they are stored as sequences of attributes within an MFT record. A directory's data attribute contains an index of files within the directory, as shown in Figure 12-12c. The index is sorted by filename and also contains the file number (which is equivalent to the MFT record number), time stamp(s), and size. By duplicating this information in the directory index, directories can be listed more quickly. The index of small directories is stored sequentially in the MFT record. Larger directories are stored as B+ trees.

File security is implemented through the object manager facilities of the operating system. Files, I/O devices, and many system services are managed as objects by the Windows NT operating system. Accessing an object automatically invokes a security subsystem that compares the object's security descriptor against the security descriptor of the accessing process or user. An MFT security descriptor has the same structure and content as security descriptors for other object types.

NTFS has several fault tolerance features, including redundant storage of critical volume information, bad cluster mapping, logging of disk changes, and optional RAID. (RAID is discussed later in this chapter.) The MFT always is stored at the beginning of a volume but a partial second copy is stored in the middle of the disk in case a block assigned to the primary MFT becomes corrupted or unreadable. The FMS tracks unreadable blocks during formatting and subsequent read and write operations. Clusters containing bad blocks are marked as unreadable in a separate bad cluster file.

NTFS uses a delayed, or lazy, write protocol. Disk blocks are cached in memory and write operations are confirmed immediately to the requesting process. Cache flushing is performed by a background process. Write operations that affect volume structure, such as file creation, file deletion, and directory modification, are written to the cache and also written immediately to a log file stored on disk. In the event of a system crash, the contents of the log file always are current and can be used to restore the volume structure to a consistent state.

FILE MIGRATION, BACKUP, AND RECOVERY

Most file management systems provide utilities and embedded features to protect files against damage or loss. Specific functions supported by most large scale FMSs include:

► File migration (version control)
► Automatic and manual file backup
► File recovery

Smaller scale FMSs such as those within LAN and personal computer operating systems typically do not support file migration but do support backup and recovery.

File Migration

When a user alters a file, the original file version usually is overwritten by the new version. However, there are advantages to maintaining the original file version, including allowing "undo" operations and having the original available as a backup. Many commonly used application programs such as word processors and text editors automatically save original file copies. For example, Microsoft Word can be configured to create a backup of the original version automatically each time it saves a document file. The most recently saved file is stored with a DOC filename extension and the original backup file is stored with a WBK filename extension.

Many transaction processing programs also preserve original versions of input files. For example, the original version of a bank's master account file usually is copied prior to processing daily batch transactions such as checks, interest, and monthly fees. The transaction processing program then reads the master and transaction files and updates the master with the transactions as shown in Figure 12-13. The original account master commonly is called the **father** and the copy that has been updated to reflect new transactions is called the **son**. After another set of transactions is processed, the father becomes the **grandfather**, the son becomes the father, and the new copy of the account master file becomes the new son.

In mainframe file management systems such as IBM's OS/390, the process of naming and storing original versions of altered files can be automated. That is, the original version of a file can be archived automatically by the file management system whenever the file is modified. In such systems, it is common to attach a version number to each filename. When a file is created, it is assigned a version number of 1. The first time it is altered, the altered copy is assigned a version number of 2, and so forth. The file management system automatically generates and stores copies, as well as tracks version numbers. A user or application program can access an older version of a file by explicitly referring to its version number.

Figure 12-13 ►

Batch account
transaction
update

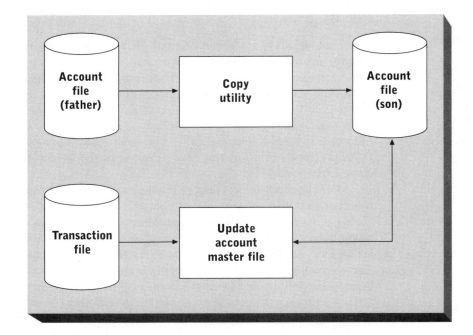

As files are modified, older versions accumulate on secondary storage, which can rapidly consume secondary storage capacity, especially when files are large or frequently altered. To compensate for this, the file management system automatically performs file migration. **File migration** is a file management technique that balances the storage cost of each file version with anticipated user demand for that version. As a file version becomes outdated, the probability of an access decreases. The file management system migrates older file versions from local disks to remote disks and eventually to offline backup storage such as tape or CD-RW.

A simpler file migration method is based on file age rather than multiple file versions. The FMS periodically checks the last access time stamp of each file and migrates older files. The file still appears in directory listings and is reloaded automatically from distant or backup storage if accessed. This file migration method frequently is used in file management systems that manage files for many users with infrequent accesses. It often is coupled with robotic devices to access offline tape or removable optical disc volumes.

File Backup

Most FMSs provide general-purpose utility programs to create backup copies of files and directories on removable or remote storage media. Backup utilities can be executed manually or automatically at set periods. Backups protect against data loss due to many causes including storage device or media failure, accidental erasure, or malicious attack. Backup utilities can protect file content, directory content, and storage allocation tables.

Backup copies should be stored on a different storage device or volume than the original. For example, it makes little sense to back up files on one partition of a hard disk to another partition of the same disk drive. Both the original and the backup would be lost in the event of total drive failure. For backups created on removable media, the media should be stored in a different physical location. Large computer centers typically store backup copies in a separate building or site to minimize the probability of a disaster destroying both copies, for example, a fire that destroys an entire building. Backup copies can be transmitted directly to remote storage facilities via a high-speed network, eliminating the cost, delay, and risk inherent in physically transporting storage media.

Types of periodic backups include:

► Full backup
► Incremental backup
► Differential backup

When a **full backup** operation is performed, the FMS copies all files and directories for an entire storage volume. A full backup operation also can include storage allocation tables, partition tables, and other important disk management data structures. A full backup is time consuming due to the large number of files copied and the slow write speeds of most backup storage devices. Full backups usually are performed at long time intervals, such as weekly, and during off-peak hours.

An **incremental backup** operation archives only files that have been modified since the previous incremental or full backup. To make incremental backup operations possible, the FMS must keep track of when backups are performed and when files are modified. The backup utility compares the most recent update and backup time for each file and directory. Only files and directories modified since their last backup operation are copied to backup storage. Incremental backups usually are much faster than full backups because many files are not modified frequently.

Most large-scale FMSs use both full and incremental backups. For example, incremental backups might be created at the end of each business day and full backups created each weekend. Mixing backup types reduces the hardware resources consumed by backup operations but it complicates the recovery process. If files must be recovered, the full backup first must be restored, then the incremental backups created since the last full backup must be recovered in the order they were created.

A **differential backup** is similar to an incremental backup in that only files changed since the most recent backup are archived. It differs from an incremental backup because only portions of the files that have been changed since the last backup are archived. Differential backup utilities are complex because they require the FMS to log changes to files at the logical or physical record level. The differential backup utility uses this log to determine which records to write to the archive. Recovery operations also are more complex. Changed records stored in a differential backup must be merged with the contents of a complete file backup; that is, from an earlier incremental or full backup operation. This process is far more complex than restoring a complete file from an incremental or full backup.

Transaction Logging

Transaction logging, also called **journaling**, is a form of automated file backup. The term transaction in this context should not be confused with the more generic meaning of the term, for example, a business transaction such as a customer purchase. To an FMS, a **transaction** is any single change to file contents or attributes such as a newly added record, modified field, or changed access controls. In an FMS that supports transaction logging, all changes to file content and attributes are recorded automatically in a separate storage area in addition to being written to the file's I/O buffer. Log entries are immediately or frequently written to a physical storage device.

Transaction logging provides a high degree of protection against data loss due to program or hardware failure. When an entire computer system fails, the contents of file I/O buffers are lost. If these buffers were not written to physical storage prior to the failure, file content becomes corrupted and content changes are lost. With transaction logging, the FMS can recover most or all of the lost changes and repair corrupted files. When the system is restarted, the contents of the transaction log are reviewed and compared to the file content on disk. Lost updates are identified and written to the files.

Transaction logging imposes a performance penalty because every file change requires two write operations—one to the file and another to the transaction log. Transaction logging commonly is used only where the costs of data loss are high, such as for large scale e-commerce sites.

File Recovery

Backup procedures and utility programs must be supplemented by a reliable set of recovery procedures to form a complete file protection mechanism. Typically, recovery procedures have both automated and manual components. For example, transaction log replay and subsequent file repair usually is fully automated. Recovery procedures based on full or incremental backups stored on removable media usually rely at least to some degree on manual procedures.

The FMS maintains backup logs to aid in locating backup copies of lost or damaged files. Recovery programs can search these logs for particular files or groups of files. Backup logs record the storage device or medium on which the backup copies are located. At the time of the backup, the backup utility writes an identification number or code to the backup storage medium and sometimes to a label that is manually applied to the medium. With the external label, the system administrator can locate the medium and mount it in the appropriate device. The recovery utility reads the embedded identification number or code to verify that the correct medium has been mounted before beginning recovery operations.

Recovery procedures for a crashed system or physically damaged storage device usually are more sophisticated and highly automated. Damage might have occurred to files, directories, storage allocation tables, and other important disk management data structures. The recovery utility reconstructs as much of the directory and storage allocation data structures as possible and makes a consistency check to ensure that:

► All storage locations appear within the storage allocation table and other data structures

► All files have correct directory entries

► All storage locations of a file can be accessed through the storage allocation table

► All storage locations can be read and/or written

Consistency checking and repair procedures consume a great deal of time— anywhere from a few minutes to several hours. But they mitigate the need to do large amounts of data recovery from backup copies and minimize the amount of current data that is lost. They also can eliminate the need to reinstall system and application software.

Fault Tolerance

As applied to FMSs, **fault tolerance** describes methods of securing file content against hardware failure. Magnetic and optical drives are complex devices with many mechanical parts. To improve performance, manufacturers employ high spin rates and small distances between read/write heads and recording media. The result is devices that provide high performance at the cost of occasional catastrophic failure.

Common causes of disk failure include head crashes, which are contact between a read/write head and a spinning platter, and burned out motors and bearings. Repairing a failed disk drive is prohibitively expensive due to the nature of modern manufacturing methods. The mean time between failures (MTBF) of a modern magnetic disk drive is approximately 10 years of continuous use. But the large number of disks in use guarantees that some failures will occur before 10 years.

File backup, recovery, and transaction logging are forms of protection against disk failure but they all require time to implement and the data is unavailable to users during recovery operations. In many processing environments, an occasional down time period for file recovery is acceptable. In other processing environments such as banking, retail sales, e-commerce, and production monitoring and control, down time is unacceptable or very expensive. In general, any business or organization that performs continuous updates and queries against files and databases is a candidate for advanced methods of fault tolerance such as mirroring or RAID.

Mirroring

Disk mirroring is a fault tolerance technique in which all disk write operations are made simultaneously or concurrently to two different storage devices. In some cases, the two devices might be located physically in different cabinets, rooms, or buildings. If one device fails, the other device contains a duplicate of all data. Data is available continuously because either device can respond to a read request.

Disk mirroring can be implemented through the FMS by configuring it to perform duplicate writes to duplicate storage devices, but that can reduce system performance substantially. Software-based mirroring is required if duplicate disks are not attached to the same disk controller.

When duplicate disks are located within the same cabinet, mirroring usually is implemented within hardware by the device controller, which reduces CPU and system bus overhead. Multiple disk drives are attached to the controller and write operations are duplicated automatically by the controller to each drive. Read operations are split between the drives to improve performance. Special utility programs are executed to configure the disk controller for mirroring and to initialize a new duplicate drive if a failure occurs.

Disk mirroring provides a high degree of protection against data loss with no performance penalty if implemented in hardware. The primary disadvantages of mirroring are the cost of redundant disk drives and the higher cost of disk controllers that implement mirroring. Mirroring at least doubles the cost of data storage.

Technology
Focus

RAID

Redundant Array of Inexpensive Disks (RAID) is a disk storage technique used to improve performance or fault tolerance. The original RAID version, now known as RAID 0, was developed at the University of California, Berkeley in the late 1980s. RAID has evolved considerably since then, and a large number of products now are available commercially. A flurry of RAID development in the early 1990s resulted in many incompatible approaches and products. The RAID Advisory Board (RAB), http://www.raid-advisory.com, was formed in 1992 to define standard methods of implementing RAID levels through 5 (see Table 12-3). RAID 1 is disk mirroring, as described in the previous section.

Level	Description
0	Data striping without redundancy
1	Mirroring
2	Data bit striping with multiple error check sums
3	Data byte striping with parity check data stored on a separate disk
4	Data block striping with parity check data stored on a separate disk
5	Data block striping with parity check data stored on multiple disks

All RAID levels except RAID 1 use some form of data striping. **Data striping** breaks a unit of data into smaller segments and stores those segments on multiple disks. For example, a 16 KB block of data can be divided into four 4 KB segments and each segment written in parallel to a separate disk as shown in Figure 12-14. A subsequent read of the original 16 KB block would access all four disks in parallel.

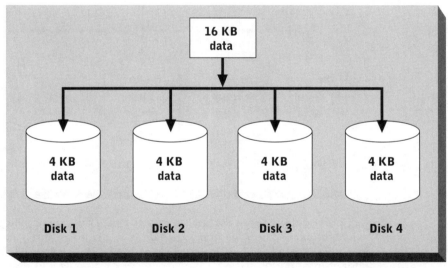

Data striping improves read performance because a large read operation to one disk is broken into multiple smaller parallel read operations. The elapsed time to perform the entire read operation is reduced because disks can perform a small read operation faster than a large read operation. Write performance also improves because smaller write operations occur more quickly. For both read and write operations, however, overhead is incurred to disassemble and reassemble larger data segments and to issue read or write requests to multiple disks.

RAID levels 1 through 5 achieve fault tolerance by generating and storing redundant data during each write operation. RAID levels 3 through 5 generate parity bits for data bytes or blocks and store those parity bits on one of the disks. If a single drive fails, no data is lost because missing bits can be reconstructed from the data and parity bits on the remaining drives.

For example, assume that even parity is used and that four one-valued bits are stored on four different disks. Also assume that a fifth disk holds parity data as shown in Figure 12-15. The parity bit for four one-valued data bits also is a one-valued bit because four is an even number of one-valued bits. If the parity disk fails, then the other disks still retain the original data bits. If any of the data disks fails there is a single missing bit that must have held either a one or zero value. The missing bit value is determined by comparing the parity bit to the remaining bit values. Because there are three one-valued bits and the parity bit is one-valued, the missing bit must be one-valued.

Figure 12-15 ▶

A RAID write operation to multiple disks

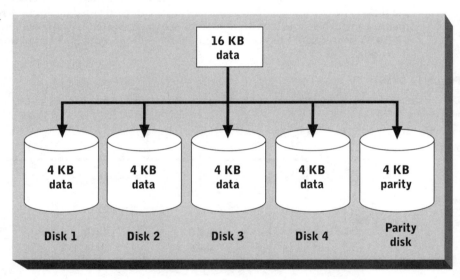

Storing parity data reduces the usable capacity of a disk array. The proportional reduction depends on the number of disks in the array. In Figure 12-15, five disks are used resulting in 20 percent of the available disk space being used for redundant information. Using a larger number of disks would decrease the portion of space used for parity bits at the expense of a slight increase in the probability of data loss due to multiple drive failures.

RAID can be implemented in software or hardware. Hardware implementations are common for a number of reasons, including:

▶ The ability to configure all RAID components in a single cabinet

▶ Extended hardware fault tolerance through redundant power supplies and disk controllers

> ▶ Reduced load on the host CPU
>
> ▶ Reduced system software complexity
>
> A RAID storage device appears as a single large disk drive to an operating system. A dedicated controller performs all RAID-related processing including segmenting read and write operations, generating parity data, reconstructing missing data if a drive fails, and repopulating data to a replacement drive. Hardware-based RAID systems for LAN and small WAN servers typically are based on the SCSI bus. RAID systems for larger computer systems usually use other high-capacity communication channels such as Fibre Channel.

SUMMARY

▶ The file management system (FMS), usually a part of the operating system, manages all aspects of user and program access to secondary storage. The FMS presents a logical view of stored data and programs to users as a set of files organized into directories and storage volumes. As users and programs manipulate the logical view, the FMS translates those operations into appropriate commands to physical storage devices. FMSs directly support only a limited number of file types, including executable programs, executable operating system command files, and user data files. In modern FMSs, users and programs can define new file types and install utility programs to manipulate them.

▶ With directories, users can organize the thousands of files stored in a typical computer system. Each storage device or volume has a root directory. In hierarchical directory structures, directories can be stored in other directories, creating a tree structure in which each file belongs to only one directory. In graph directory structures, files can belong to more than one directory and there is the possibility of loops or cycles in the directory structure. Directories normally are stored as tables and contain descriptive information about files and directories, such as name, owner, file type, access controls, and time stamps.

▶ Secondary storage units are divided into allocation units, which are typically a few kilobytes in size. The FMS assigns allocation units to files and directories as they are created or expanded and reclaims allocation units as files and directories shrink or are deleted. The FMS uses complex data structures, called storage allocation tables, to track the assignment of allocation units to files and directories. Entries within the table can be linked lists in simpler FMSs or indices or other complex data structures in more complex FMSs.

▶ The FMS allocates buffers to support program file I/O. A program executes a file open service call before reading or writing a file, which causes the FMS to find the file, verify access privileges, allocate buffers, and update an internal table of open files. Multiple programs can open the same file for reading but in most FMSs, only one program can open a file for writing. When a program is finished with a file it executes a file close service call which causes the FMS to flush buffer content to the storage device, deallocate the buffers, update the file time stamps, and update the table of open files.

► The FMS enforces access controls when accessing files on behalf of a user or program. File owners and system administrators can grant or deny access privileges for reading, writing, and executing files. When a user or user program attempts to access a file, the FMS checks the user identification against the access controls stored in the file's directory entry to determine whether access is permitted. Enforcing access controls provides security at the expense of additional FMS overhead.

► FMSs provide utilities to make backup copies of files and directories and to recover them if needed. Backups can be performed manually or may be automated fully. Different types of backups include full, incremental, and differential. Some FMSs support automatic storage and backup of old file versions in a process called file migration. As file contents are changed, the original contents are archived in another file. Older file versions are migrated automatically to slower or offline storage as they become further out of date.

In this chapter and the previous chapter, we covered allocation and management of the CPU, primary storage, and secondary storage. In both chapters, it is assumed that the resources being managed were part of the local computer system. In the next chapter, we turn our attention toward resources outside the local computer system. We'll examine how users and applications interact with external resources including files, I/O devices, and programs.

Key Terms

---

allocation unit
block
blocking
blocking factor
complete path
current directory
data striping
differential backup
directory
disk mirroring
father
fault tolerance
fields
file allocation table (FAT)
file association

file close operation
file control layer
file management system (FMS)
file open operation
file migration
full backup
fully qualified reference
grandfather
graph directory structure
hierarchical directory structure
home directory
incremental backup
journaling
link
logical record

physical record
record
Redundant Array of Inexpensive Disks (RAID)
relative path
shortcut
son
storage allocation table
storage I/O control layer
transaction
transaction logging
tree directory structure
unblocked
undelete operation
working directory

Vocabulary Exercises

1. A(n) _____ is the unit of file I/O to and from an application program. A(n) _____ is the unit of file I/O to and from a secondary storage device.

2. The term blocking factor describes the number of _____ contained within a single _____

3. A file _____ releases allocated buffers and flushes their content to secondary storage.

4. A(n) _____ operation allocates buffers for file I/O and updates a table of files in use.

5. The content of a logically, but not physically, deleted file may be recovered in a(n) _____ operation.

6. _____ describes the tracking of old file versions and their movement to offline and archival storage devices.

7. Under _____ , changes to files are written to a log file as they are made.

8. The _____ layer presents a service layer interface to application programs and the command layer. The _____ layer manages the movement of data between secondary and primary storage.

9. A(n) _____ specifies a storage device or volume and all directories leading to a specific file. A(n) _____ specifies file location with respect to the current or working directory.

10. In a(n) _____ directory structure, a file can be located within no more than one directory. This restriction does not apply in a(n) _____ directory structure.

11. A(n) _____ consists of a master (root directory), one or more subdirectories, and a filename.

12. A(n) _____ only copies portions of the files that have been changed since the last backup.

13. MS-DOS and some Windows versions record storage allocation information in a(n) _____ .

14. An FMS can implement _____ through disk mirroring or _____ .

15. A(n) _____ records the allocation of storage locations to specific files.

16. When an old version of a master file is saved, the current version can be called the _____ , the previous version the _____ , and the version before that the _____ .

Review Questions

1. List the layers of a file management system (FMS) and describe their functions.

2. What is the difference between the logical and physical structure of a file? What advantages are realized by *not* having an application program interact directly with physical file structure?

3. What file types usually are supported directly by a file management system?

4. What is an allocation unit? What are the advantages of using small allocation units? What are the disadvantages?

5. Describe the use of buffers in file I/O operations. When are buffers allocated? When are they released?

6. Describe the structure of a hierarchical directory. What are its advantages and disadvantages as compared to graph structure directories?

7. How is file deletion normally accomplished? What security problems might arise from this method?

8. What levels of access privilege can exist for a file?

9. What is transaction logging, or journaling? Describe the performance penalty that it imposes on file update operations.

10. Describe various methods of implementing hardware fault tolerance. What are their comparative advantages and disadvantages?

Problems and Exercises

1. Modify the directory in Table 12-1 and the storage allocation table in Table 12-2 to store a new file containing seven allocation units.

2. Assume that the first character of the filename in a deleted file's directory entry is overwritten with an ASCII 0 to mark the file as deleted. Write a step-by-step procedure for undeleting a deleted file, assuming that no file operations have been performed since the deletion.

Research Problems

1. To increase efficiency of application program execution, some FMSs use a large number of file organization and access methods, each optimized to a very specific type of processing. The MVS (IBM mainframe) operating system contains such an

FMS. Investigate this FMS to determine the various methods of file organization and access that are supported. For what type(s) of application program file manipulation is each method intended?

2. The UNIX operating system, and its embedded file management system, originally were not designed to support large scale transaction processing systems. For this and other reasons, UNIX lacks many aspects of file management, such as transaction logging, indexed files, and file migration, that would support such an environment. Identify the specific deficiencies of the UNIX operating system with respect to large scale online transaction processing. How can application developers overcome these deficiencies?

Chapter

13

Internet and Distributed Application Services

Chapter Goals

- Describe client-server and multi-tier application architecture and discuss their advantages compared to centralized applications

- Explain how operating systems and network protocol stacks cooperate so users and programs can access remote resources

- Describe low-level protocols for interprocess communication across networks, including sockets, named pipes, RPC, and DCE

- List and describe standard Internet protocols used to access distributed resources

- Discuss component-based application development and describe the protocols and standards that support component-based applications

- Explain the role and function of directory services and the LDAP standard

Users of modern computers and information systems interact with a variety of resources located on computer systems all over the world. In this chapter, we'll examine the complex set of network protocols, infrastructure, and services that make it possible for users to interact with geographically dispersed resources as if they were all located on their own computer. Figure 13-1 shows the topics covered in this chapter.

Figure 13-1 ▶

Topics covered in this chapter

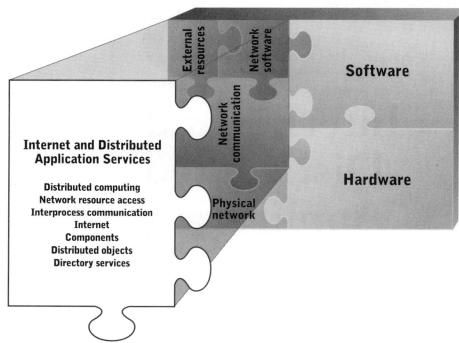

DISTRIBUTED COMPUTING

Modern information systems often are distributed across many computer systems and geographic locations. For example, an organization's corporate financial data might be stored on a mainframe computer in its central office. Minicomputers in regional offices might periodically generate accounting and other reports based on data stored on the mainframe, while microcomputers in branch offices might access and view periodic reports as well as query and update the central database. Similar tasks might be performed with laptop computers or personal digital assistants (PDAs) using wireless networks. Distributing parts of an information system across many computer systems and locations is called **distributed computing** or **distributed processing**.

Client-Server Architecture

Client-server architecture is currently the dominant architectural model for distributing information system resources. **Client-server architecture** divides software into two classes—client and server. A **server** manages one or more system resources and provides access to those resources through a well-defined communication interface. A **client** uses the communication interface to request resources and the server responds to those requests.

The client-server architectural model can be applied in many different ways. The architecture by which workstations access a shared printer on a LAN, as shown in Figure 13-2, is one approach to client-server architecture. An application program on a workstation sends a document to a server computer that dispatches it to a management process for the specified printer. The server acknowledges the client request and notifies the client when the document is sent to the printer.

Figure 13-2 ▶

Network printing services implemented with client-server architecture

Service request →

← Response

Document →

N-Layer Client-Server Architecture

Modern variants of client-server architecture divide application software into the following set of client and server processes called layers or tiers:

▶ The **data layer** manages stored data, usually in one or more databases
▶ The **business logic layer** implements the rules and procedures of business processing
▶ The **view layer** accepts user input and formats and displays processing results

This approach to client-server architecture sometimes is called **three-layer architecture** or **three-tier architecture**. Figure 13-3 illustrates how the three layers interact. The view layer acts as a client of the business logic layer, which in turn acts as a client of the data layer.

Figure 13-3 ▶

Three-layer architecture

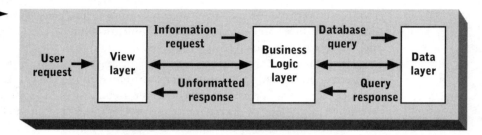

Three-layer architecture simplifies distributing or replicating application software across a network. Interactions between the layers are service requests and responses. Layers can be placed in different processes on the same computer system or on different computer systems. Layers that overload a single computer's capacity can be replicated on multiple machines.

Additional layers can be used when processing requirements or data resources are complex. Architectures that employ more than three layers are called **n-layer architectures** or **n-tiered architectures**. For example, corporate databases often are stored using multiple database management systems. An additional data layer is interposed between the business logic layer and the database management systems to present a unified view of data resources from all corporate databases.

Layers also can be replicated in different forms. A single business logic layer might interact with multiple view layers from different applications. For example, business logic for validating inventory items might be used by an inventory ordering application and by a customer merchandise return application. A single application might use multiple view layers that interact with a single business logic layer. For example, order entry business logic might interact with a view layer used by telephone sales representatives that runs on simple character-based terminals. The same business logic might interact with a Web-based view layer used directly by customers.

The connections between layers of client-server and multi-tier application software can be very complex. Every layer can use a unique combination of programming language, operating system, and computer hardware. Well-defined interfaces and communication protocols are needed so the layers function as an integrated whole.

The term **middleware** describes software that "glues" together multiple components in a client-server or multi-tier application. Middleware is a wide-ranging system software category because the connection and communication requirements of applications vary. For example, a simple client-server application might need only to transmit messages between a single client and server in predetermined locations. A more complex multi-tier application might need additional middleware, such as Web servers that support embedded client-side programs, database servers that support store procedures, and network operating systems that share files among clients on multiple computers.

A continuing trend in operating system development is to embed additional middleware functions within newer versions. For example, software to implement network protocols such as TCP/IP was not embedded in microcomputer operating systems until the mid-1990s. Web server software, once an optional component of most server operating systems, now is standard, and other server operating system functions such as e-mail and document distribution often rely on it.

Other middleware can be obtained as optional software for a network operating system or as completely separate packages. The Microsoft BackOffice suite is optional middleware that extends the capabilities of Microsoft Windows 2000 Server. Novell Groupware and BEA Systems Tuxedo are examples of middleware packages that work with many operating systems. International Data Corporation has estimated middleware sales for 2000 at $3 billion, with a projected rise to $9 billion by 2004.

The remaining sections of this chapter explore various forms of middleware and network-related services.

NETWORK RESOURCE ACCESS

The primary role of an operating system is to manage hardware, software, and data resources. As part of this role, the operating system accepts and processes resource access requests from users and applications via the service layer. Modern operating systems enable users to interact with resources of the local computer system and of remote computer systems. To provide distributed access, the operating system must be able to distinguish between local and remote resources and interact with distant operating systems. This section covers the operating system components that implement those functions.

Protocol Stacks

An operating system implements network I/O and services as a complex set of software layers. The Open Systems Integration (OSI) network model, covered in Chapter 9, was an early attempt to standardize the number and functions of those software layers. Software that implements the lowest five levels of the OSI model commonly is called a **protocol stack**.

Figure 13-4 shows a typical protocol stack for a workstation. Stacks 1 and 2 have different network, transport, and session layer stacks but share a single physical device driver, Ethernet connection, and physical network connection. The client can access Novell NetWare LAN servers through the Transport Layer Interface (TLI), Sequenced Packet Exchange (SPX), and Internet Packet Exchange (IPX) protocols in Stack 1, and Internet services through the sockets and TCP/IP protocols in Stack 2. With the ODI layer, which is not part of the OSI model, both upper-level protocol stacks share a common physical network interface.

Figure 13-4 ►

Two upper-level stacks sharing a single NIU and driver

Layer Description	Stack 1	Stack 2
OSI Session layer	Transport Layer Interface	Sockets
OSI Transport layer	Sequenced Packet Exchange	Transmission Control Protocol
OSI Network layer	Internet Packet Exchange	Internet Protocol
Logical network interface driver	Open Data Interface	
OSI Data Link layer (Physical Network Interface Driver)	10-Mbps Ethernet	
OSI Physical layer	Twisted-pair copper wire	

Protocol stacks provide several advantages for implementing network I/O and services:

► They divide the task of network interaction into several well-defined pieces that can be separately implemented and installed

► They provide the flexibility needed to keep up with rapid protocol standard evolution

► They insulate application programs and many portions of the operating system from details of low-level network communication protocols and physical network implementation, which ensures software portability across a wide range of network protocols and transmission media

Accessing Remote Resources

Connections to remote resources can be either static or dynamic. A **static connection** is initialized by the user or system administrator prior to accessing a remote resource. The remote resource is given a local object, resource, or service name. The window labeled My Computer in Figure 13-5 shows static resource connections to three server directories (drive letters: K:, L:, and M:) under Windows 2000. For example, the directory SharedFiles on server Averia is connected statically to the local resource name M:. The figure also shows a dialog box labeled Map Network Drive to create new static connections. Similar static connections can be created for printers and other shared resources.

Figure 13-5 ▶

Displaying and creating static directory connections in Windows 2000

Static connections are inherently difficult to initialize and maintain. The user or system administrator must configure the operating system to establish a connection each time it boots up. If the location or name of a remote resource changes, the configuration of all computers with static connections to that resource also must be changed.

Remote resource access in modern operating systems is based on the following premises:

- Operating systems, application programs, and user interfaces are simpler if there is no distinction between local and remote resource access
- All resources potentially are shared across a network
- Any computer system potentially is both a client and a server
- Resources can be moved among computer systems

Software and user interfaces can be simplified by providing a common method of accessing both local and remote resources. For example, a word processor executing on a workstation should use the same service call and parameters to access document files stored on local disks and document files stored on server disks. A Web browser should access resources on remote machines in the same manner it accesses resources on the local machine. This characteristic of software and user interfaces is called **location transparency** or **network transparency**.

The second and third premises go hand in hand. If every local resource might be needed by remote users or processes, every computer system is potentially both a client and a server. To provide remote access, all operating systems need to incorporate server-like functions. Operating systems that implement this design feature are said to implement **service-oriented resource access**.

Figure 13-6 illustrates the arrangement of software components that support service-oriented resource access. Two layers, the service provider and resource locator, are interposed between the service layer and device drivers. The service provider is a server interface to a specific resource such as a printer or file system. Service requests from local users or programs are passed down through the operating system service layer. Service requests from remote users or programs are passed through the low-level network protocol stack and directed to the service provider by the resource locator. The service request format is the same regardless of the request's origin.

Figure 13-6 ►

Software layers
to access local
and remote
resources

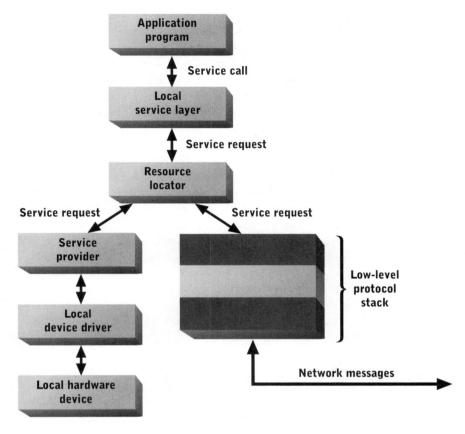

The resource locator has two important tasks:

► Locate resources referred to in service requests from local or remote users
and programs
► Route service requests to the appropriate service provider

The resource locator maintains a **resource registry** containing the names and
locations of known resources and services. The format of the names is the same
whether the resource or service is local or remote. When local service provider
processes are started, they register themselves with the resource locator, which
updates the resource registry accordingly. Resource requests from the local service layer are checked against the registry. If the requested resource or service
name is not found, the resource locator searches the network for the resource.
The exact nature of this search depends on the high-level network service protocols in use.

Most network service protocols use a distributed directory of services and resources. As services and resources are registered with a resource locator, broadcast messages are sent over the network to inform other resource locators of their existence. Some machines on the network can be designated as primary repositories of this information. When a resource locator receives a request for a resource not listed in its registry, it queries the closest primary repository to find the service. That repository can query other repositories until the requested resource is located.

The interaction between a resource locator and a primary resource registration repository is inherently dynamic. Connections established through those interactions are **dynamic connections**. A local resource locator updates its registry based on query responses from remote registries, which improves the efficiency of future accesses. Resource registry contents must change as local and remote server processes are started, stopped, moved, or reconfigured. The resource locator provides a dynamic mechanism for reconfiguring connections to services and resources.

The resource locator also acts as a router for resource access requests arriving from remote machines. The local physical network layer receives remote service requests and passes them up through the protocol stack to the resource locator. The resource locator then passes the request to the appropriate local service provider process and passes the response back to the requestor via the protocol stack (see Figure 13-6).

It should be clear by now that the traditional distinctions between client and server operating systems are blurry. All modern operating systems are organized internally as a collection of server processes that can respond to requests from any location. There still are some distinctions between client and server operating systems such as scalability and security. For example, Windows 2000 Server supports up to 32 CPUs and a sophisticated directory-based security system. Windows 2000 Professional supports one or two CPUs and uses less sophisticated security mechanisms unless it is part of a network managed by Windows 2000 Server. But differences between client and server operating systems are primarily a matter of different configuration, not fundamentally different architecture.

INTERPROCESS COMMUNICATION

When an application is split into multiple processes, those processes must communicate with one another to share data and coordinate their activities. In Chapter 11, we briefly looked at pipes and signals, which allow independent processes executing on the same computer to coordinate their activities and share data. But how do processes executing on different computers communicate and coordinate their activities?

A variety of protocols and standards have been developed over the last 20 years to address the problem of process coordination across networks. In this section, we'll concentrate on lower-level **peer-to-peer communication protocols** and standards, which allow processes to use communication synchronously across a network (see Figure 13-7). These protocols often are used by system software such as operating systems and database management systems to exchange data and coordinate activities. Distributed applications usually use higher-level protocols, which are described later in the chapter. However, those higher-level protocols often are layered above the peer-to-peer protocols.

Figure 13-7 ►

Peer-to-peer protocols layered over TCP/IP

OSI Application layer	DCE
	RPC
	Named Pipes
OSI Session layer	Sockets
OSI Transport layer	Transmission Control Protocol
OSI Network layer	Internet Protocol

Sockets

As first described in Chapter 9, a **socket** is a unique combination of an IP number and a port number. For example, the socket 129.24.8.1:53 is the network listening address for the primary DNS name server at the University of New Mexico. A port number is an unsigned 16-bit integer, so there are 65,536 possible port numbers. Some port numbers are permanently assigned for standard Internet or vendor-specific services, but many port numbers are available for other uses including client-server or peer-to-peer communication among application programs.

All modern operating systems support sockets and provide systems service calls so programs can initialize sockets, receive messages sent to a socket, and send messages to sockets anywhere on the Internet. Figure 13-8 shows a communication scenario between client and server processes on two computer systems. Client processes on Computer A are attached to sockets 129.24.8.212:2 and 129.24.8.212:6 and communicating with two server processes on Computer B attached to sockets 207.46.230.219:1 and 207.46.230.219:6. Each socket uniquely identifies a process on the Internet.

Figure 13-8 ▶

Multiple
processes
communicating
through sockets

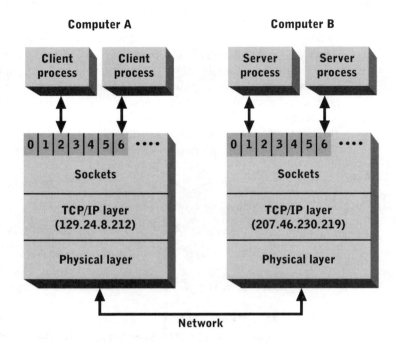

Named Pipes

In Chapter 11, a pipe was defined as a shared memory region, managed by the operating system and used to transfer data among cooperating processes. A **named pipe** is a pipe with two additional features:

▶ A name that is permanently placed within a file system directory

▶ The ability to communicate among processes on different computers

When a named pipe is created, a directory entry also is created in the local file system. Programs can read or write the named pipe as they would read or write an ordinary file. Typically, the server side of a client-server application creates the named pipe. In a peer-to-peer application, either side of the application can create the named pipe. A client or peer opens a named pipe as a network resource and reads or writes it as if it were a file on a shared directory. A server or peer on the machine where the named pipe was created reads and writes the pipe as if it were a local file.

The operating systems at both ends of the pipe manage communications to and from the pipe. Named pipes actually are a high-level interface to sockets, so the operating system assigns a free socket to the named pipe when it is created. The operating system also allocates I/O buffers and routes data flowing in and out of the pipe through the low-level network protocol stack, as shown in Figure 13-9. The operating system on the client side also allocates a socket each time a remote named pipe is opened.

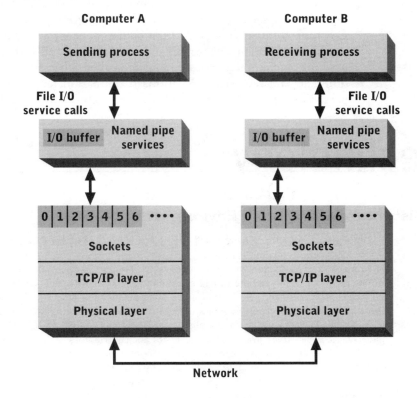

Figure 13-9 ►

Two processes communicating through a named pipe

Data can flow in either direction through a named pipe though bidirectional data flow usually is performed with two one-way pipes. In client-server applications, multiple clients can send messages to a single server by writing to the same named pipe. The server can tell which client sent which messages or data because each client has a unique socket number.

Remote Procedure Calls

With the **remote procedure call (RPC)** protocol, a process on one machine can call a process on another machine. As with function or procedure calls within a single program, the calling process:

1. Passes parameters to the called process
2. Waits for the called process to complete its task

---

3. Accepts parameters back from the called process

4. Resumes execution with the instruction following the call

Parameter passing among machines potentially is problematic because data representation varies across CPUs and sometimes across operating systems. Common differences include little endian versus big endian storage, character coding (ASCII, Unicode, or EBCDIC), and which IEEE floating point format is used for real numbers. If the calling and called process execute on machines with different data representation formats, the parameters must be converted when passed to the called process and when passed back to the calling process.

Technology
Focus

Distributed Computing Environment

Distributed Computing Environment (DCE) is a standard for distributed operating system services defined by the Open Group, formerly known as the Open Software Foundation. DCE is a wide-ranging standard covering network directory services, file sharing services, remote procedure calls, remote thread execution, system security, and distributed resource management. Its primary goal is to promote interoperability of distributed software across operating systems and middleware products. Many operating systems partially comply with DCE, including Windows 2000, many versions of UNIX, and OS/390 (MVS). IBM, Compaq, and Hewlett-Packard are principal supporters of and contributors to the standard.

DCE defines a subset of operating system services and a standard interface to those services. DCE functions are incorporated directly into an operating system or supplied as an optional component (see Figure 13-10). In theory, DCE services can replace their counterparts in an existing operating system. In practice, DCE-compliant services usually translate DCE service calls into native operating system service calls.

Figure 13-10 ▶

DCE software layers

Security is an integral part of every DCE service. DCE security is based on the **Kerberos** security model, which defines interactions between clients, services, and a trusted security service. Clients and servers authenticate one another by asking the security server to authenticate the other party. The security server provides security "tickets" to each party that they exchange to verify their identity.

The security database also maintains an access control list for each service and resource. Once a client has been authenticated, its identity is checked against the access control list for the resource it is attempting to access. If the client is part of the access control list, it is issued a ticket that it presents to the server to gain access. Tickets, passwords, and other security mechanisms are all encrypted during network transmission. Tickets also are time stamped and expire in minutes or hours.

DCE file services are based on a fully distributed file and directory structure. Every file within a DCE file system has a unique name consisting of its host name, a file system resource name within the host, and its path name within the file system resource. File systems can be distributed or replicated across multiple servers. DCE file services implement transaction logging for all updates to enable rapid recovery in the event of a crash.

Clients, servers, and all DCE services interact through remote procedure calls. RPC messages are formatted according to an Interface Definition Language (IDL). DCE-compliant client and server programs are compiled and linked with a set of DCE library routines that implement the IDL. These routines implement the low-level aspects of message passing, parameter exchange, data format conversions, and interaction with DCE security services.

The DCE standard is implemented widely in UNIX operating system variants but it also has penetrated deeply into other operating systems. The standard is used widely, though competing standards such as CORBA, which is discussed later in the chapter, are starting to supplant it. But DCE probably will be around for many years to come given its widespread operating support and large installed application base.

THE INTERNET

The Internet and World Wide Web are popular frameworks for implementing and delivering information system applications. However, there is a general lack of agreement regarding definitions of the Internet, Web, and related terms. We'll use the following definitions in this chapter:

► The **Internet** is a global collection of networks that are interconnected using TCP/IP.

► The **World Wide Web** (**WWW**), also called the **Web**, is a collection of resources (programs, files, and services) that can be accessed over the Internet by standard protocols such as the File Transfer Protocol (FTP) and Hypertext Transfer Protocol (HTTP).

► An **intranet** is a private network that uses Internet protocols but is accessible only by a limited set of internal users (usually members of the same organization or workgroup). It also describes a set of privately accessible resources that are organized and delivered via one or more Web protocols over a TCP/IP network.

The Internet is the infrastructure upon which the Web is based. In other words, Web resources are delivered to users over the Internet. An intranet uses the same protocols as the Internet and Web but restricts access to a limited set of users. Access can be restricted in many ways, including using privately registered resource names, firewalls, and user/group account names and passwords.

The Web is organized using client-server architecture. Web resources are managed by server processes that can execute on dedicated server computers or on multipurpose computer systems. Clients are programs that send requests using one or more standard Web resource request protocols. Web protocols define valid resource formats and a standard means of requesting resources. Any program, not just a Web browser, can use Web protocols to access Web resources.

Standard Web Protocols and Services

All Web resources are identified by a unique **Uniform Resource Locator (URL)** such as http://averia.unm.edu/default.htm. A URL has four components, as shown in Figure 13-11:

- ► *Protocol*—an optional header specifying the resource access protocol (http:// is the default value)
- ► *Host*—the IP number or registered name of an Internet host computer or device
- ► *Port*—an optional port number that, together with the IP address, specifies a socket as described in Chapter 9 (if omitted, a standard port number for the protocol is assumed)
- ► *Resource*—the complete path name of a resource on the host (if omitted, the host can be configured to supply a default value)

Figure 13-11 ►

URL components

Web standards define a number of protocols for resource format, content, transfer, and manipulation. The number of standard protocols is growing and their content evolves rapidly. Table 13-1 summarizes major Web protocol categories.

Table 13-1 ▶

Web protocols

Category	Sample Protocols
Formatted and linked documents	Hypertext Markup Language (HTML) and Extensible Markup Language (XML)
File and document transfer	File Transfer Protocol (FTP) and Hypertext Transfer Protocol (HTTP)
Remote login and process execution	Telnet, tn3270, and Remote Procedure Call (RPC)
Mail and messaging	Simple Mail Transfer Protocol (SMTP), Post Office Protocol (POP), and Internet Message Access Protocol (IMAP)
Executable programs	Java, JavaScript, and Visual Basic Script (VBScript)

The Web as we know it today began with the development of **Hypertext Markup Language (HTML)**. The original version of HTML defined a device-independent document-formatting language in which links to other documents could be embedded. The first Web browser, called Mosaic, displayed HTML documents on any computer and output device to which it was ported. Mosaic spawned the current generation of Web browsers such as Netscape Navigator and Microsoft Internet Explorer. Both HTML and browser software have evolved through several generations to include capabilities such as forms, style sheets, data transfer from client to server, and embedded scripts and programs. HTML will eventually be replaced by **Extensible Markup Language (XML)**, which extends HTML to describe the structure, format, and content of documents.

Hypertext Transport Protocol (HTTP) is a companion protocol to HTML and XML that specifies the language by which clients request documents and how servers respond to those requests. HTTP is an extension of an older Web protocol, called **File Transfer Protocol (FTP)**, that specifies a client-server request and response language for copying files from one Internet host to another. Because HTTP is an extension of FTP, servers that respond to HTTP requests also can respond to FTP requests.

With the **Telnet** protocol, users on one Internet host can interact with the operating system command layer of another host, as shown in Figure 13-12. Telnet emulates a character-based VDT and is limited to interacting with command-line-oriented command layers such as Windows COMMAND.COM and the UNIX Bourne shell. Tn3270 is a variant of Telnet that emulates an IBM 3270 VDT to interact with older IBM mainframe operating systems and application programs.

Figure 13-12 ►

Telnet
connection

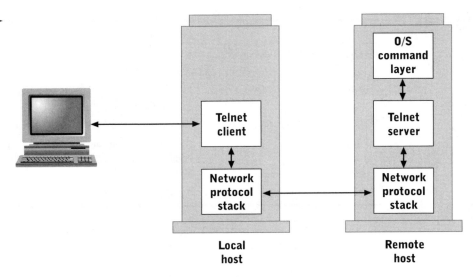

E-mail standards have been part of the Internet since the 1970s. The earliest e-mail protocol is the **Simple Mail Transport Protocol (SMTP)**, which defines how text messages are forwarded and routed among Internet hosts. E-mail client programs on each host interact with a server process to access forwarded messages and to send messages to users on other hosts. SMTP was later extended by the **Multipurpose Internet Mail Extensions (MIME)** protocol to allow non-text files to be included in mail messages. MIME also is used by HTTP.

Modern e-mail server software is based on the **Post Office Protocol version 3 (POP3), Internet Message Access Protocol version 4 (IMAP4)**, or both. POP3 standardizes the interaction between e-mail clients and servers so the client and server can execute on different Internet hosts. Under POP3, mail messages temporarily are held on the server, downloaded to the client when a connection is established, and deleted from the server as soon as the download is completed. IMAP4 extends POP3 to permanently store and manage e-mail messages on the server, which simplifies e-mail clients and gives users easy access to stored e-mail from any Internet host.

In later HTML versions, program code or scripts can be embedded within HTML documents. Java applets, first described in Chapter 11, can be called from an HTML document with parameters passed in either direction. Figure 13-13 shows sample HTML code to call a Java applet named VocabMan.class. The applet is downloaded to a Web browser along with the surrounding HTML document. Figure 13-14 shows the Web page with the embedded Java applets displayed as boxes containing asterisks. The Web browser executes the applet whenever the cursor is placed over a box and the applet displays answer text when the mouse button is pressed.

Figure 13-13 ▶

A Java applet
call embedded
within HTML

```
<li>A compiler allocates storage space and makes an entry in the symbol
    table when a(n)
    <applet width="175"
            height="25"
            code="VocabMan.class"
            codebase="http://asm.unm.edu/sa2e_student/Java/">
            <param name="answer"
                   value="data declaration">
              You must have Java enabled for this feature
    </applet>
    is encountered in source code.<br><br>
</li>
```

Figure 13-14 ▶

Vocabulary
exercise answers
displayed or
hidden by a Java
applet

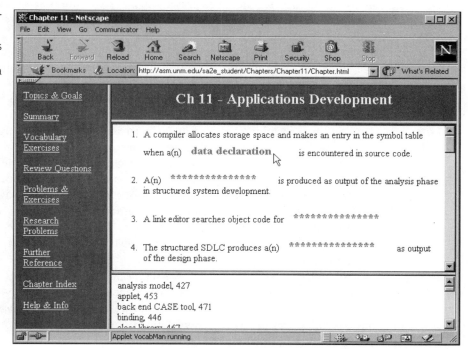

Internet **scripting languages** are slimmed-down programming languages. JavaScript and Visual Basic Script (VBScript) are two widely used scripting languages. Scripts embedded in HTML pages can perform many functions that full-fledged programs can, without the need for compilation or link editing. Typical functions performed by embedded scripts include data validation and customizing page layout or content based on Web browser configuration.

Web servers can execute programs and scripts on the server in response to requests sent by clients as a URL. Internet search engines are a common example of this approach. The Web browser downloads an HTML form in which the user

---

Chapter 13   **507** enters search criteria. When the user clicks a Send or Go button, the search criteria are transmitted back to the Web server encoded in a URL such as:

http://lycospro.lycos.com/srchpro/?query=vacation+caribbean&t=all&x=36&y=6

The URL actually is an encoded function call to a search engine program. The text after the question mark contains input data for the program. The program encodes processing results in HTML and the Web server sends them back to the Web browser for display.

The Internet as an Application Platform

Internet and Web technologies present an attractive alternative for implementing distributed applications. For example, consider a payroll system for a geographically dispersed computer consulting firm. Such an organization might have dozens or hundreds of offices and many employees traveling or on extended assignment at client locations. Employees need to update withholding, insurance, and other payroll-related information and enter time-related payroll information. How can all of the organization's employees interact with its payroll system quickly and easily?

One way to address the problem is to build client-server application software that uses a private network to connect remote clients with servers at the organization's administrative offices. The client portion of the application is installed on employees' laptop computers and employees connect to payroll servers using a modem and private telephone number. The disadvantages of this approach include the cost to build and maintain a private network, the cost to develop full-fledged self-contained client-side software, and the difficulties inherent in installing, configuring, and updating client software on many different computer systems.

An alternative is to build a client-server application that uses a Web browser interface. The application program executes on a Web server that can be accessed from any computer with an Internet connection. Employees can access the payroll server with their own laptop computer or a computer in a client office or hotel business suite.

Figure 13-15 shows the architecture of a typical Web-based three-layer payroll system. The client interacts with the Web server to download HTML pages that can include embedded scripts or Java applets. The bulk of the application code resides on the server, and program functions are called by the Web server in response to client requests encoded within URLs. The server side application code interacts with a back-end database server using lower-level Internet standards such as sockets, named pipes, or RPC. All parts of the application communicate using standard Internet and Web protocols.

Figure 13-15 ▶

A distributed Web-based application

Implementing the application via the Internet expands its accessibility and eliminates the need to install custom client software on employees' laptop computers. The application can be updated by updating software only on the Web server. The application is cheaper to develop and deploy because it is built around existing Web standards and relies on Web browser software already installed on clients.

The primary disadvantages of implementing applications via the Internet are security, performance, and reliability. If an employee can access the system via the Web, then others also might be able to gain access. System access can be restricted by various means including user accounts and passwords, but the risk of a security breach always will be present. Performance and reliability are limited by the employee's Internet connection point and the available Internet capacity between that connection point and the application server. Unreliable or overloaded local Internet connections can render the application unusable.

Each resource that participates in the payroll update application also can participate in other applications. For example, the database server hardware and software can host marketing and production databases. The Web server can store product manuals and host an order entry application for customers. The

Web browser on an employee's microcomputer can access other Web applications or information resources anywhere on the Internet.

Although Internet standards such as sockets, named pipes, HTTP, and HTML are adequate to implement distributed applications, they're not optimal for the following reasons:

► With lower-level protocols, server addresses are stored in client configuration files or source code. If server resources are moved, the clients must be reconfigured or recompiled.

► Breaking up server-side processes into small manageable pieces is difficult. Each new distributed piece requires a new set of hard-coded connections.

► Developers usually create large complex server processes to avoid the complexity of large numbers of connections between many smaller server processes. But doing so reduces the chances that server processes can be incorporated into multiple distributed applications.

Newer techniques and standards for deploying and supporting distributed applications address these problems in several ways. They make it easier to break up large server processes into small reusable pieces by providing standards for communication among server processes. They improve flexibility by providing directory services so processes can locate one another using location-independent names. They make it easier to build systems from small reusable processes by providing the infrastructure to manage large numbers of interprocess connections.

COMPONENTS AND DISTRIBUTED OBJECTS

A **component** is a standardized and interchangeable software module that:

► Is executable
► Has a unique identifier
► Has a well-known interface

Components are ready-to-use software. They are compiled and linked for the operating system and CPU on which they'll execute, or they're Java programs that can execute in an installed Java Virtual Machine (JVM).

Every component has a unique identifier (ID), which is a number or symbolic name. Like Internet names and sockets, component names must be unique. Unique IDs enable one component to find and request services from another component. Component IDs must be registered with a directory service that can be queried by other components.

A component's interface is the set of services it provides or tasks it performs. Each service or task is similar to a function or subroutine within a program—it has a name and parameters. When one component asks another to perform a task, it sends it a message containing the task or service name and the required input parameters. The component that receives the message performs the task and, if necessary, passes back results through parameters.

Component-Based Software

Components are important in modern software development because complex programs and applications can be constructed from small, previously developed parts. Other complex products have been built from components for decades. For example, a modern automobile contains tens or hundreds of thousands of parts, relatively few of which are manufactured by the companies that assemble and sell cars. Automobile manufacturers specify standards for components such as engines, bearings, and tires that often can be used in several models. Other companies manufacture parts based on their specifications. Without component-based construction, modern automobiles would be much more expensive, much less reliable, and much more difficult to repair and modify.

Component-based design and construction provides similar benefits to complex software products. For example, consider the grammar-checking function in most word processing programs. Grammar checking can be implemented as a function or subroutine that is called by other parts of the word processing program. The grammar-checking function source code is integrated with the rest of the word processor source code during program compilation and linking. The executable program then is delivered to users.

Now consider two possible changes to the original grammar-checking function:

▶ The developers of another word processing program want to incorporate the existing grammar-checking function into their product

▶ The developers of the grammar-checking function modify it to improve speed and accuracy

To integrate the existing function into a new word processor, the developers integrate the existing source code of the grammar-checking function into the new word processor program. They add appropriate grammar-checking function calls to their word processor source code and then compile, link, and distribute the program to users.

When the developers of the grammar checker revise their source code to implement the faster and more accurate function, they deliver the source code to the developers of both word processors. Both development teams integrate the new grammar-checking source code into their word processors, recompile and relink the programs, and deliver a revised word processor to their users.

What's wrong with the above scenario? Nothing in theory, but a great deal in practice. The grammar checker developers can provide their function to other developers only as source code, which creates potential problems concerning intellectual property rights and software piracy. Also, integrating the grammar-checking function is difficult or impossible if the word processor is written in a different programming language than the grammar checker.

When the grammar-checking function is updated, the developers of both word processing programs must recompile and relink their entire product to update the embedded grammar checker. The new executable program then must be delivered to users and installed on their computers—an expensive and time-consuming process.

A component-based approach to software design and construction solves all of these problems. Component developers can deliver their product as a ready-to-use executable function. Developers of the word processing programs simply plug in the component. Updating a single component doesn't require recompiling, relinking, and redistributing the entire application. Applications already installed on user machines can be updated by installing only the new component. This is exactly the mechanism used by many companies to update installed software, for example, Microsoft Windows Update and Symantec Live Update for Norton Utilities.

Components and Objects

It is difficult to talk about components without lapsing into object-oriented (OO) terminology. Components are similar to objects within an OO program because they send and respond to messages, encapsulate internal data, and interact with other components through a well-defined interface. OO program design concepts also can be applied to component-based applications. In essence, component-based design and development scales OO programming concepts up to the level of application programs and entire information systems.

Components usually are developed with OO software development tools and programming languages, but that isn't required. A component can be implemented with any programming language as long it interacts with other components via messages and a well-defined public interface. Nonetheless, the vast majority of components are implemented with OO tools because they naturally are suited to component development. A component behaves as a distributed object regardless of its internal implementation.

Connection Standards and Infrastructure

Interoperability among hardware or software components requires well-defined and widely adopted standards. For example, consider telephone connections and the services provided by telecommunication companies (see Figure 13-16). In the U.S., all telephones are connected to the public telephone grid by a four-conductor wire with an RJ-11 connector. Each wire in an RJ-11 connector carries a specific electrical signal with known voltage and other characteristics. This standard ensures that any telephone can be connected easily to any telephone interface.

Figure 13-16 ▶

Standard connectors and infrastructure enable communication between telephones

Telephones rely on basic services supplied by the public telephone infrastructure. A unique telephone number identifies every telephone or set of telephones. The infrastructure provides connection services among telephones, as well as directory services to match numbers to people, organizations, and physical locations. Switching and routing are handled by the infrastructure with no specific instructions from sending and receiving telephones. Additional services such as forwarding and voice mail also can be provided. The standards that enable any telephone to access any other telephone or service have been developed over many years through industry cooperation and occasional governmental intervention.

Software components require similar standards for connections and services. They require the software equivalent of an RJ-11 plug and an infrastructure that provides directory, routing, and forwarding services. Within a single computer, operating system interprocess communication services can connect components and configuration files can supply directory information.

Connecting components located on different machines running different operating systems requires a standard network protocol. A network protocol is desirable even when components all execute on the same machine because it provides the flexibility needed to distribute components in the future. Standard

Internet services and protocols such as TCP/IP and sockets provide part of a component connection solution. But such protocols don't address two important issues:

► Format and content of valid messages and responses
► Means of uniquely identifying each component on the Internet and routing messages to and from that component

Addressing those issues requires additional standards, protocols, and services. Many organizations have participated in developing such standards over the last two decades. Currently, only two standard families are well-developed and widely implemented: CORBA and COM+.

CORBA

In the 1980s, many computer hardware and software organizations joined forces to create an industry-wide component interoperability standard that is known today as the **Common Object Request Broker Architecture (CORBA)**. CORBA specifies the middleware used by objects to interact across networks. The two key components of CORBA are:

► **Object Request Broker (ORB)**, a service that maintains a component directory and routes messages among components
► **Internet Inter-ORB Protocol (IIOP)**, a component message-passing protocol

The ORB is a server process that can reside anywhere on a network or can be distributed across many network nodes. The ORB acts as a component registry, a message router, and a translator. Components must register themselves with the ORB before other components can connect to them. The ORB assigns a unique identifier to each registered component, so that component's methods can be invoked by other components anywhere on the Internet.

A component that wants to invoke a method within another component sends an IIOP formatted request to the nearest ORB. ORBs cooperate with one another to locate the desired component and establish a connection. Once a connection has been established, parameters can be passed in both directions. If necessary, an ORB can translate parameters from one format to another. This enables otherwise incompatible objects, such as those developed with different languages or executing on different CPUs, to exchange data.

The CORBA standard is robust, inherently scalable, and independent of programming language, operating system, and CPU architecture. These features are the result of deliberate design, a long development history, and participation by a large number of computing vendors and organizations. The standard is implemented widely, though most of those implementations are relatively new.

COM+

The **Component Object Model Plus (COM+)** is a Microsoft specification for component interoperability. COM+ has its roots in older Microsoft specifications including object link embedding (OLE) and the Component Object Model (COM). In the early and mid 1990s, Microsoft incorporated DCE services into COM and called the resulting specification the Distributed Component Object Model (DCOM). COM+ is the most recent upgrade of that standard.

Like CORBA, COM+ defines component registration, message routing services, and a component communication protocol. The COM+ services and protocol are similar to CORBA with the following key differences:

▶ Components are not assigned a permanent identifier and their internal states cannot be stored permanently. COM+ components can't remember information from one invocation to the next. COM+ components are similar to functions or subroutines. CORBA components are objects.

▶ COM+ components are registered in the Windows Registry of the client machine on which they're installed. The Windows Registry stores information other than component registrations, including hardware configuration, software configuration, and user profile information. A CORBA ORB is dedicated to component services.

COM+ is implemented widely. Most developers of general-purpose application software such as word processing and spreadsheet programs used COM+'s predecessor standards and are using or already have migrated to COM+. COM+ also is the cornerstone of Microsoft's approach to building and deploying multi-tier distributed applications. Brisk sales of Microsoft run-time tools that support this environment, for example, Internet Information Server and Microsoft Transaction Server, clearly indicate that COM+ is being used to develop and deploy significant numbers of business applications.

DIRECTORY SERVICES

When resources are distributed across network nodes, resource users and resource providers must have some way of finding one another. **Directory services** describe middleware that:

▶ Stores the name and network address of distributed resources
▶ Responds to directory queries
▶ Accepts directory updates
▶ Synchronizes replicated or distributed directory copies

Directory services are integral components of all network operating systems. Typically, network operating system directories store information about:

► Registered users and their permissions to access directory objects

► Shared hardware resources such as printers

► Shared files, databases, and programs

► Computer systems and specialized hardware devices such as network storage appliances

Modern directory services are distributed in a manner similar to Internet name services. Directories are organized hierarchically to create a single namespace for all network resources and objects. In large networks, responsibility for maintaining directory content and answering queries is distributed throughout the network. Directory content can be replicated in multiple servers in different parts of the network to reduce response time and improve fault tolerance.

Lightweight Directory Access Protocol (LDAP)

Historically, most directory services have been tied to a specific operating system such as NDS for Novell NetWare, or to a specific application type such as file sharing or e-mail distribution. Application and operating specificity is a problem in networks with multiple operating systems and many types of distributed resources and objects. Today, such networks are the norm rather than the exception, so directory services must bridge operating systems and applications.

In the 1980s, the International Telecommunications Union (ITU) developed the X.500 standard, which defines nonproprietary directory services for e-mail and network addresses. The standard was never implemented widely, though partial implementations have appeared in some commercial products such as Novell NetWare and in higher-level standards such as DCE. Most vendors considered X.500 to be cumbersome and too closely tied to the OSI network model. The X.500 implementations that have been deployed can't interoperate so the primary goal of the X.500 standard wasn't achieved.

The X.500 standard became the basis for another standard called the **Lightweight Directory Access Protocol (LDAP)**, which has been adopted by the Internet Engineering Task Force (IETF) as a formal Internet standard. The standard was updated regularly in the late 1990s and those efforts still are in progress. LDAP is implemented widely, though not all products provide all the features of the most recent standards.

An LDAP directory stores information about LDAP objects. Each object is an instance of an **objectclass**, which defines the attributes common to all member objects. For example, the objectclass Shared_Printer might define attributes named Building, Room, Manufacturer, Model, Print_Engine_Type, Color, Duplex, and Pages_Per_Minute. Each directory entry for a shared printer would be an object of type Shared_Printer and contain values for some or all of the defined attributes.

LDAP objects are organized into a hierarchical directory structure. Objects can be grouped into container objects that can be grouped into other container objects (see Figure 13-17). LDAP defines several standard container object types including Country (C), Organization (O), and Organizational Unit (OU). All objects in an LDAP schema have an attribute called distinguished name (DN), which uniquely identifies the object within an objectclass. A fully qualified distinguished name such as:

DN=Stephen Burd,O=Faculty,O=School of Management,OU=University of New Mexico,C=USA

specifies a complete path from a directory root node through one or more container objects to a specific object.

Figure 13-17 ▶

An LDAP hierarchy of objects and container objects

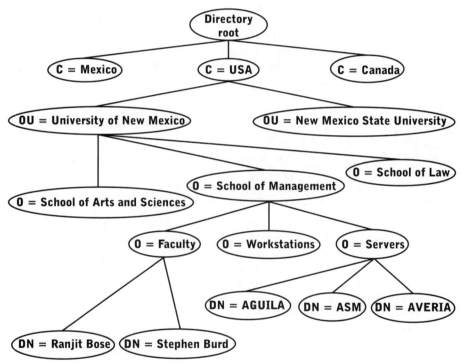

LDAP defines a standard protocol for updating and querying LDAP directories. LDAP queries contain plain text characters, though the syntax is complex and awkward for people. Many client programs such as e-mail clients and Web browsers provide a more user-friendly interface to LDAP services. To see an example, try accessing the URL ldap://ldap.whowhere.com with Microsoft Internet Explorer version 5 or higher.

Strict adherence to the LDAP standard guarantees interoperability among all LDAP clients and servers. However, there is an important area of interoperability that has yet to be addressed: standard schema names and structure. There currently are no standard objectclass or attribute names for entities and resources common to most directories such as people, shared printers, and components. As a result, there is no standard way for a directory from one vendor or organization to query the directory of another organization or vendor.

For example, one directory might define an objectclass named *Employee* and another directory might use the name *Worker* for a similar objectclass. The two directories might use different attribute names for identical data items such as office telephone number. One directory can send a query to the other asking it to enumerate its defined objectclasses and attributes. But without a standard naming system, there is no way to determine semantic equivalence of schema classes and attributes automatically.

In the absence of standard class and attribute names, directory vendors and user organizations have developed proprietary directory schemas that usually are incompatible. Even if a set of standard schema classes and attributes is developed, the existing base of installed incompatible directory schemas will hinder true directory interoperability for years to come.

Technology
Focus
Microsoft Active Directory

Microsoft Active Directory (AD) is the directory service and security system built into Windows 2000 Server. AD stores information about many network resources, including computers, I/O devices, and users (see Figure 13-18). AD incorporates security services that allow system administrators to limit resource access to specific users or user groups. AD supports large directories and distributed organizations with directory partitioning across Internet domains, directory replication across multiple servers, and automatic synchronization of replicated directories or directory partitions.

Figure 13-18 ▶

Active Directory
objects

Windows 2000 client software queries AD to locate and access network resources. For example, a user can use Windows 2000 Professional or AD client software installed under earlier Windows versions to search for users, computers, or shared directories matching specific criteria. Programs also can interact with AD by calling AD-specific service functions. A word processing program might query AD for all printers in a specific organizational unit, constructing and displaying the result to a user when he or she selects the Print function from the File menu.

Windows 2000 uses AD to store and access security information. Every user, group, and computer object is assigned a unique security identifier. Every AD resource or container object has an access control list that describes the access rights granted or denied to specific users, groups, and computers. Figure 13-19 shows the access control list for the Active Directory container object Pajaro. Members of the group Domain Admins have all permissions for Pajaro, and, by default, have all available permissions for all child objects.

Figure 13-19 ►

Access
control list

AD access control lists and permissions inheritance can be used to distribute system administration tasks across individuals and organizations. If AD objects are organized hierarchically using container objects, the directory administrator can delegate AD management functions to other users by assigning rights to container objects. For example, the directory administrator can assign rights such as Create All Child Objects and Modify Permissions to other users for lower-level container objects, so those users can administer one part of the AD.

AD is based on LDAP and the Internet Domain Naming Service (DNS). AD responds to standard LDAP information requests and uses LDAP concepts such as objectclasses and organizational units to store and hierarchically organize directory information. AD clients rely on a DNS server to locate an AD directory server. AD servers register their names and services by supplying a service record (SRV record) to a DNS server. AD clients query the DNS server for LDAP servers registered in their domain. The DNS server returns one or more IP addresses which the client uses to send TCP/IP messages to an AD server.

Like many other commercial products, AD doesn't implement the entire LDAP protocol. Unlike other directory service software, AD relies on DNS and requires DNS servers to accept dynamic updates. Although dynamic updates are defined as an Internet DNS standard, many DNS servers do not support them. Even if a DNS server does support dynamic updates, many DNS administrators are reluctant to enable them because they are a potential security and reliability threat. Microsoft's DNS server software does support dynamic updates, so most AD installations use Microsoft DNS server software.

AD does not support distributed or component-based software directly. Shared files, including executable files, and databases can be registered in AD, which could support locating and accessing programs and data sources. However, AD provides no standard way to "glue together" components and external data sources to create complex systems. Windows 2000 relies on a parallel directory and registration system to track COM+ components and provide component-related directory and interprocess communication services.

SUMMARY

► Modern information systems typically are distributed across many computer systems and geographic locations. Client-server, three-layer, and n-layer architecture define frameworks for decomposing applications into multiple layers that can execute on different computers. Multi-layer architectures require standard methods and services to communicate with one another. Systems software that implements communication standards and gives clients and servers the ability to interact is called middleware.

► A network protocol enables users and applications to interact with resources and applications on remote computer systems. The operating system service layer accepts resource access requests and routes them to appropriate remote or local service processes. Local requests to remote resources are routed through the protocol stack to distant servers and remote requests to access local resources are received through the protocol stack. Connections to remote resources may be static or dynamic. Dynamic connections are more flexible but they require a distributed registry of resource names and locations.

► Distributed processes must communicate with one another to exchange data and synchronize their activities. Peer-to-peer interprocess communication protocols include sockets, named pipes, remote procedure calls, and DCE. Sockets implement direct process-to-process communication via protocol stacks. With named pipes multiple clients can interact with a single server via a shared file system object. Remote procedure calls allow one process to execute another as a subroutine with parameter passing and format translation. DCE combines all of these approaches and adds security and minimal directory services.

► The Internet is a global network based on TCP/IP and many other protocols. The Web is the set of resources accessible over the Internet via standard protocols. Internet protocols define methods for sharing many types of resources including files, documents, and application programs. Multi-layer applications can be implemented over the Internet using a Web browser for the user interface and Internet and Web protocols to connect the layers.

- ► Component-based applications divide an application into many different cooperating processes or distributed objects. Each distributed object implements a public interface to the services it provides. Component-based applications require protocols and infrastructure for component registration, discovery, and communication. CORBA and COM+ are two standards that address the full range of component infrastructure and communication services.

- ► With directory services, users, resources, and components can find one another on the Internet. Modern directory services must span the range of different resource types and Internet protocols. LDAP is a widely deployed directory service standard that can track users, distributed resources, and objects. Although LDAP defines a standard method for defining and accessing directory structure, it doesn't define standard content templates, which limits interoperability among different LDAP directories.

Key Terms

Active Directory (AD)
business logic layer
client
client-server architecture
Common Object Request
   Broker Architecture
   (CORBA)
component
Component Object Model Plus
   (COM+)
data layer
directory services
distributed computing
Distributed Computing
   Environment (DCE)
distributed processing
dynamic connection
Extensible Markup Language
   (XML)
File Transfer Protocol (FTP)
Hypertext Markup Language
   (HTML)

Hypertext Transport Protocol
   (HTTP)
Internet
Internet Inter-ORB Protocol
   (IIOP)
Internet Message Access
   Protocol version 4 (IMAP4)
intranet
Kerberos
Lightweight Directory Access
   Protocol (LDAP)
location transparency
middleware
Multipurpose Internet Mail
   Extensions (MIME)
n-layer architecture
n-tiered architecture
named pipe
network transparency
Object Request Broker (ORB)
objectclass
peer-to-peer communication
   protocol

Post Office Protocol version 3
   (POP3)
protocol stack
remote procedure call (RPC)
resource registry
scripting language
server
service-oriented resource
   access
Simple Mail Transport
   Protocol (SMTP)
socket
static connection
Telnet
three-layer architecture
three-tier architecture
Uniform Resource Locator
   (URL)
view layer
Web
World Wide Web (WWW)

Vocabulary Exercises

1. _____ architecture divides an application into multiple processes, some of which send requests, some of which respond to requests, and others that do both.

2. System software that connects parts of a distributed application or can be used by users to locate and interact with remote resources is called _____.

3. A(n) _____ is a process that sends a request. A(n) _____ is a process that responds to requests.

4. A server process creates a(n) _____ that allows clients to send data or messages via a shared filename.

5. With the _____ protocol, a process on one machine can call a process on another machine as a subroutine with parameters passed in either or both directions.

6. Three-layer architecture divides an application into _____, _____, and _____ layers.

7. A(n) _____ to a remote resource must be initialized by the user or a system administrator.

8. A(n) _____ is a unique combination of an IP number and a port number.

9. _____ is the original Internet e-mail standard. More modern e-mail standards include _____ and _____.

10. The _____ is a global collection of networks that are interconnected using TCP/IP.

11. Web pages are encoded in _____ and delivered from a Web server to a Web browser via _____.

12. The _____ standard defines methods for embedding graphics and other nontextual data within e-mail messages and Web pages.

13. The _____ is a collection of resources that can be accessed over the Internet by standard protocols such as FTP and HTTP.

14. _____ is a family of component infrastructure and interoperability standards supported by Microsoft.

15. _____ defines a standard for describing and accessing directories of users and distributed resources.

16. A(n) _____ contains a protocol, Internet host, optional socket, and resource path.

17. _____ is a wide-ranging standard covering network directory services, file sharing services, remote procedure calls, remote thread execution, system security, and distributed resource management.

18. With the _____ protocol, a client can interact with a remote computer's command layer as if it were a directly connected character-based VDT.

19. A(n) _____ is a standardized and interchangeable software module that is executable, has a unique identifier, and has a well-known interface.

20. _____ is a family of component infrastructure and interoperability standards supported by a broad consortium of computer companies.

21. The set of software layers that provide low-level access to remote resources through a LAN or WAN is called a(n) _____.

Review Questions

1. Describe client-server, three-layer, and n-layer architecture. What are the differences between a client and a server? What is the function of each layer in a three-layer application? Why might more than three layers be used?

2. What is middleware?

3. What is a protocol stack? What are the components of a typical protocol in a client computer that can access many Web servers?

4. What are the differences between static and dynamic connections to remote resources? Which connection type requires a resource registry? Where should the resource registry be located?

5. A modern operating system acts as both client and server. How are software components organized to accomplish both functions at the same time?

6. List and describe three low-level peer-to-peer interprocess communication standards. What are the advantages and disadvantages of using these standards to implement distributed multi-layer applications?

7. Do the terms Internet and Web describe the same thing?

8. What are the components of a URL?

9. List and describe at least five standard Internet and Web protocols.

10. How can the Internet be used as a platform to implement distributed multi-layer applications? Which Internet and Web protocols are used and how are they used?

11. What is a component? Component-based design and development has been the norm in manufacturing durable goods for decades. Why has it only recently been adopted for designing and deploying information systems?

12. Describe the two major standard families for component infrastructure and communication. If you were forced to choose one standard to support a new large-scale information system, which would you choose? Why?

13. What are directory services? What types of information might be made available through directory services? Describe the LDAP directory services standard.

Research Problems

1. The Kerberos security model is a part of both the DCE and CORBA standard families. Investigate the capabilities and limitations of Kerberos. What degree of security does it provide? What infrastructure components must be present to implement it? Does it address all security issues in distributed applications and resources?

2. The CORBA and LDAP standards are rapidly evolving. Access the Web sites www.corba.org and www.ietf.org to determine what features have been added by recent standard revisions. What needs are those new features intended to satisfy? Are the newer features being implemented widely? Why or why not?

3. Distributed.net (www.distributed.net) is an organization that coordinates research on distributed computing applications. Individuals and organizations can join the organization and contribute idle computing power to ongoing research projects that require or use massively parallel approaches to solving complex computational problems. Investigate the projects currently underway and the methods and protocols that coordinate software on members' computers.

Chapter 14

System Administration

Chapter Goals

- ► Describe system administration responsibilities and tasks

- ► Explain the process of acquiring computer hardware and system software

- ► Describe tools and processes for evaluating application resource requirements and computer system performance

- ► Define a system security model and describe how it can be implemented

- ► Understand issues related to installing and protecting computer hardware

In previous chapters, we explored the inner workings of computer hardware and system software. In this chapter, we take a step back from bits, bytes, and instructions and examine hardware and software from the point of view of a system administrator. A system administrator's job includes many varied tasks—describing all of them in detail easily could fill an entire book. This chapter concentrates primarily on tasks that require in-depth knowledge of the topics covered in earlier chapters (see Figure 14-1).

Figure 14-1 ▶

Topics covered in this chapter

SYSTEM ADMINISTRATION

Within the context of information systems, **system administration** covers a range of activities and responsibilities. The primary responsibility of a system administrator is to ensure efficient and reliable delivery of information services (IS). A typical system administrator performs many specific tasks to meet this responsibility. These tasks can be grouped into the following broad categories:

▶ Acquiring new IS resources

- Maintaining existing IS resources
- Designing and implementing an IS security policy

The assignment of these responsibilities to specific individuals varies considerably in different organizations. In very small organizations, the user or organizational administrator might assume these responsibilities directly. In very large organizations, these responsibilities might be partitioned among a large number of people. Medium-sized organizations might have a single technical specialist who assumes all system administration responsibilities.

Each activity area entails a variety of specific system administration tasks. Some of these tasks are mundane and address a relatively short time frame, such as maintaining user accounts and performing file system backups. Other tasks require many different skills and address a longer time frame, such as IS planning and resource acquisition.

Strategic Planning

IS resource acquisition and deployment should occur only in the context of a well-defined strategic plan for the organization as a whole. For the purposes of this chapter, a **strategic plan** is defined as a set of long-range goals and a plan to attain those goals. The planning horizon is typically three years and beyond. Minimally, the goals must include:

- Services to be provided
- Resources needed to provide those services

The strategic plan addresses the following issues with respect to achieving stated goals:

- Strategies for developing services and a market for them
- Strategies for acquiring sufficient resources for operations and growth
- Organizational structure and control

In all cases, strategic plans must address the basic question of "How do we get there from here?" with respect to the strategic goals.

The information systems component of an organization is only one part of the entire organization, and the IS strategic plan is only part of an organization's strategic plan. All portions of its plan must be evaluated in concert with those of other organizational units. Information systems are normally a support service for other organizational units and functions such as customer services, accounting, and manufacturing. Thus, the strategic plan for information systems tends to follow, rather than lead, the strategic plans of other units in the organization.

Hardware and Software as Infrastructure

The resources devoted to most organizational activities can be classified into two categories—capital expenditures and operating expenditures. **Capital expenditures** are used to purchase **capital resources,** which are assets or resources that are expected to provide benefits for more than the current operating period, such as a fiscal year. **Operating expenditures** are expenditures that provide benefits for a single operating period or less.

Typical capital expenditures include buildings, land, equipment, and research and development costs. Computer and software purchases also are considered capital expenditures, even though the expected useful lifetime of computer hardware and software has decreased in recent years due to rapid technological change. Typical operating expenditures include salaries, rent, lease payments, supplies, and equipment maintenance.

Many capital resources provide benefits to a wide range of organizational units and functions. An office building, for example, provides benefit to all of the units and functions that are housed within it. Such resources are referred to as **infrastructure** and have the following characteristics:

- ▶ Service to a large and diverse set of users
- ▶ Costs that are difficult to allocate to individual users
- ▶ Recurring need for new capital expenditures
- ▶ Significant operating costs for maintenance

The computer hardware and system software used to provide information systems services are infrastructure. This is obvious in organizations that rely extensively on large computer systems used by many units in the organization. It is less obvious but no less true in organizations that have highly decentralized computer hardware and system software. The IS strategic planning issues are similar to those in many infrastructure-based service organizations.

Examples of infrastructure-based service organizations include those that provide communications, electrical power, and water. In general, the strategic issues that must be addressed by such organizations are:

- ▶ What services will be provided?
- ▶ How will service users be charged?
- ▶ What infrastructure is required to provide the services?
- ▶ How can the infrastructure be operated, maintained, and improved at minimal cost?

As an example, consider the communication services provided by local telephone companies. The primary strategic question to be addressed is what types of services should be provided: basic telephone only, expanded telephone services, computer communication services, mobile services, information storage and retrieval, and so on. Answers to the primary question lead to decisions regarding the nature of the required infrastructure and its associated capital and operating expenditures.

Standards

Providing infrastructure-based services to a wide variety of users requires adopting **service standards**. However, standardization tends to stifle innovation and to produce solutions that are suboptimal for some users. Once again, consider the local telephone system as an example. All users agree on and abide by a number of standards for telephone service. These include acceptable user devices, basic service availability, standards for interacting with the infrastructure (for example, line voltage and signal encoding), and other standards. The nature of infrastructure requires standardization in order to provide service at a reasonable cost.

Standardization often causes problems for some users, especially for those who demand services at or near the leading edge of technology. For example, many telephones are connected to the nearest switching center with low-capacity copper wiring, which limits the availability of high-speed Internet services. This in turn slows the trend toward employees working at home. The provider of an infrastructure-based service constantly must balance the benefits of standardization in reduced costs and simplified service against its costs in terms of stifled innovation and failure to meet the needs of some users.

It is particularly hard to deal with the standardization issue when considering computer hardware and system software. There are many choices in both categories. In addition, although there has been some progress in hardware and software compatibility, it still is the exception rather than the rule. The issue is complicated further by the diverse set of components required for information processing within even a modest-sized organization.

Competitive Advantage

Discussion of computer hardware and system software strictly as infrastructure ignores certain types of opportunities. Infrastructure management tends to concentrate on the provision of short-term services at minimal cost. Such an outlook tends to preclude major technical innovations, as well as radical redefinition of services to be provided.

The term **competitive advantage** describes a state of affairs in which one organization employs resources so as to give it a significant advantage over its competitors. This can take a number of forms, including:

► Providing services that competitors are unable to provide
► Providing services of unusually high quality

► Providing services at unusually low price

► Generating services at unusually low cost

Computer hardware and system software can be applied to achieve competitive advantage in any or all of these areas. Examples include automated bank tellers, scanning grocery checkouts, and computer integrated manufacturing.

Note that each example of computer hardware/system software was initially a competitive advantage, but now is commonplace. This is the nature of applying technology for competitive advantage—rapid technology changes and adoption by competitors severely restrict the useful life of most technology-based competitive advantages. Furthermore, there are substantial risks in pursuing competitive advantage through new technology. The costs of technology can be high for developers and early adopters, but then decrease rapidly after introduction, so that late adopters may incur substantially lower costs while realizing most of the benefits of the technology. Early adopters also face the inefficiency of starting at the beginning of a learning curve. These factors often combine to create high costs with limited benefit.

Business
Focus
A Standard Hardware Platform?

Cooper State University (CSU) is a large school with over 1000 faculty members and a full-time enrollment of 20,000 students. Academic programs span a wide range of fields including hard sciences, humanities, engineering, education, management, law, and medicine. CSU offers many graduate degree programs and is highly ranked in terms of externally funded research.

CSU has a campus-wide computing organization called Computer Information Services (CIS). CIS is responsible for supporting the computing and information processing needs of both academic and administrative users. It supports administrative users by operating a large mainframe computer, several local area networks, and many administrative applications such as payroll, accounts payable, class scheduling, class registration, and student academic records. It supports academic users with shared minicomputers, shared file servers, and a large number of microcomputers in dedicated classroom labs and general-purpose facilities. CIS also operates the campus network that links most offices, classrooms, and buildings.

CIS has been under considerable budgetary pressure for several years. Demand for computing services has grown rapidly but funding has not kept pace. Declining hardware costs have been a budgetary bright spot but they have been more than offset by increases in support costs for CSU's wide variety of hardware platforms, software packages, and administrative applications. Integrating this disparate collection of hardware and software with the growing campus network has been a difficult and costly task.

CIS recently has proposed standardizing the hardware platforms and operating systems used for microcomputers and minicomputers to minimize support costs. They want to contract with a single vendor for each class of machines and negotiate volume pricing on hardware, operating software, and maintenance. All CSU microcomputer and minicomputer purchases, including those made with non-CIS funds, would be part of the contract.

Reactions to the proposal have ranged from indifference to near revolt. Most administrative departments and many academic departments are supportive, provided that CIS can assure them of a continued ability to meet their computing and information processing needs. In general, the supportive departments rely on CIS for most or all of their computing and information processing needs. Some academic departments that fund some of their own computer purchases are concerned with loss of control over the acquisition process.

The computer science, engineering, information systems, and physics departments are vehemently opposed to the proposal. They argue that their computing needs are unique and require hardware and software at the cutting edge of technology. They claim that standardized platforms would lag the cutting edge of technology and interfere with their teaching and research missions. They point out that most of their computing needs are met with funds derived from external sources such as research grants and contracts. They vigorously oppose restrictions on how that money is spent.

Questions

1. Are the benefits that CIS anticipates from a standardized platform likely to be realized?
2. Are the concerns of the computer science, engineering, information systems, and physics departments valid? Should any other departments share these concerns?
3. Should the proposal be adopted as is? Should the dissenting departments be given the option to opt out of the contract?

THE ACQUISITION PROCESS

The process of acquiring computer hardware and system software is ongoing in most organizations. New hardware and software can be acquired to:

► Support new applications
► Increase support for existing applications
► Reduce the cost of supporting existing applications

The exact nature of the acquisition process depends on which of these cases, or a combination thereof, motivates the new acquisition. It also depends on a number of other factors, including:

► The mix of applications that the hardware/software will support
► Existing plans for upgrade or change in those applications

▶ Requirements for compatibility with existing hardware and software

▶ Existing technical capabilities

The applications to be supported figure prominently in both the motivation for new acquisitions, as well as the factors for choosing among them. It must be remembered that hardware and software exist to support present and future applications. Planning for acquisition is little more than guesswork if you don't have a thorough understanding of present and anticipated application needs.

The acquisition process consists of the following steps:

1. Determine the applications that the hardware/software will support.

2. Specify detailed requirements in terms of hardware and software capability and capacity.

3. Draft a request for proposals and circulate it to potential vendors.

4. Evaluate the responses to the request for proposals.

5. Contract with a vendor or vendors for purchase, installation, and/or maintenance.

The acquisition process begins when a new system is implemented or an existing system is upgraded. In either case, hardware and system software requirements are determined early in the design phase of the systems development life cycle. The analysis phase provides detailed estimates of system activity such as transaction volume and file or database content. Decisions about what processes will be automated, type of user interfaces, operating systems and other system software, and selection of development tools are made early in the design phase. These decisions, combined with the activity estimates, are translated into detailed hardware requirements.

Determining and Stating Requirements

Computer system performance is measured in terms of application tasks that can be performed within a given time frame. Performance can be measured in terms of throughput, response time, or some combination of the two. The first step in stating hardware/software requirements is to state the application tasks to be performed, along with performance requirements for those tasks.

Different techniques will apply to determining requirements, depending on the motivation for the proposed acquisition. For existing applications, hardware and system software requirements might be based on measurements of existing performance and resource consumption. Requirements for new applications are more difficult to derive. A number of different techniques that might be used to determine requirements are discussed later in this chapter.

Although application requirements are the primary basis for hardware and system software requirements, other factors also must be considered, including:

- Integration with existing hardware/software
- Availability of maintenance services
- Availability of training
- Physical parameters such as size, cooling requirements, and disk space required for system software
- Availability of upgrades

These factors are integral to the overall requirements statement. Some requirements, such as physical parameters, are essential while others are less important bases for differentiating among potential vendors.

Request for Proposals

A **request for proposal (RFP)** is a formal document sent to vendors. It states requirements and solicits proposals to meet those requirements. It often is a legal document as well, particularly in governmental purchasing. Vendors rely on information and procedures specified in the RFP. They invest resources in the response process with the expectation that stated procedures will be followed consistently and completely. An RFP generally is considered a contract offer and a vendor's response is an acceptance of that offer. Problems such as erroneous or incomplete information, failure to enforce deadlines, failure to apply procedures equally to all respondents, and failure to state all relevant requirements and procedures can lead to litigation.

The general outline of an RFP is as follows:

- Identification of requestor
- Format, content, and timing requirements for responses
- Requirements
- Evaluation criteria

The identification describes the organization that is requesting proposals. Identification should include the name of a person to whom questions can be addressed. Physical and e-mail addresses, telephone numbers, fax numbers, and so on should be included.

The RFP should state the procedural requirements for submitting a valid proposal. Where possible, an outline of a valid proposal should be given with a statement of the contents of each section. In addition, deadlines for questions, proposal delivery, and other important events should be stated clearly.

The majority of the RFP comprises the requirements statement. Requirements should be categorized by type and listed completely. Relevant categories include:

► Hardware/software capability

► Related services such as installation and maintenance

► Warranties and guaranties

► Financial considerations, for example, acquisition cost, maintenance cost, lease cost, and payment options

Requirements should be separated into those that are absolute or essential, and those that are optional or subject to negotiation. For example, minimum hardware capacity generally is stated as an absolute requirement, whereas some related services might be only desirable.

Evaluation criteria are stated as specifically as possible. A point system or weighing scheme often will be used for the optional or desirable requirements. Weight also might be given to factors that are not stated as part of the hardware or software requirements. Such factors might include the financial stability of the vendor and good or bad previous experiences with the vendor.

Evaluating Proposals

Proposal evaluation is a multi-step process. The usual steps are:

1. Determine minimal acceptability of each proposal.

2. Rank acceptable proposals.

3. Validate high-ranking proposals.

Each proposal must be evaluated to determine if it meets minimal criteria of acceptability. These criteria include meeting essential or absolute requirements, financial requirements, and deadlines. Proposals that fail to satisfy minimal criteria in any of these categories are eliminated from further consideration.

The remaining proposals must be ranked. Proposals are ranked by evaluating the extent to which they exceed minimal requirements. This includes the provision of excess capability or capacity, satisfaction of optional requirements, and other factors. Measurements of subjective criteria such as compatibility, technical competence, and vendor stability also are considered at this stage.

A subset of the highly ranked proposals is chosen for validation. The subset should be small due to the length and expense of validation procedures. To validate a proposal, the evaluator must determine the correctness of vendor claims and the ability of the vendor to meet commitments in the proposal. Note that the ranking process relies primarily on assertions by the vendor in the proposal. The validation stage is where the accuracy of those assertions is determined.

Various methods and sources of information are applicable to proposal validation. The most reliable of these is a benchmark of the proposed system with actual applications. A **benchmark** is a performance evaluation of application software or test programs using actual hardware and systems software under realistic processing conditions. In years past, benchmarking often was difficult to perform due to the expense of the hardware/software configurations and the length and cost of installation. Currently, this is less of a problem due to cheaper hardware, streamlined installation procedures, and fierce competition among vendors. For all but the largest systems, it now is common for a vendor to deliver and install hardware and system software for customer evaluation.

Alternatives to on-site benchmarking with actual application software include:

► Benchmarking at alternate sites with actual applications
► Benchmarking of test applications
► Validation through published evaluations

With systems that are difficult to install, it might be possible to test applications at an alternate site. This could be on the premises of the vendor or of another customer.

It might not be possible to test configurations with actual applications if those applications are large, difficult to install, or not yet developed. In such cases, it might be possible to use standard benchmarking software or to construct benchmarking software that simulates the execution of the actual application. Many standard benchmarks are available, primarily for smaller computer systems and scientific applications.

Care must be taken when using benchmarks, especially if only one benchmark is executed at a time, testing only one type of performance. Overall system performance under mixed application demand can vary substantially from individual benchmark results. For standard configurations, benchmark results sometimes can be obtained from trade publications, periodicals, or independent testing agencies.

DETERMINING REQUIREMENTS AND EVALUATING PERFORMANCE

Information systems are a combination of computer hardware, systems software, and application software. One of the most difficult system administration tasks is determining hardware requirements for a specific set of application software. This task is difficult because:

► Computer systems are complex combinations of interdependent components.
► The configuration of operating and other system software can significantly affect raw hardware performance.
► The hardware and system software resources required by applications cannot always be predicted precisely.

The starting point for determining hardware requirements is the application software that will run on the hardware platform. The nature and volume of application software inputs must be described to determine what application code will be executed and how often it will be executed. This determines the demands that application software will place on system software, such as number and type of service calls, and on computer hardware, such as number of CPU instructions.

If the application software already has been developed, its resource consumption in terms of hardware and system software can be measured. A number of software and hardware tools can be used to monitor resource utilization while application programs are executing. Executing application programs with live data inputs will generate precise measurements of resource requirements.

Determining resource requirements is much more complex when application software has not yet been developed. User requirements are not fully defined until the conclusion of the SDLC analysis phase, and detailed software implementation plans are not developed until late in the design phase. Estimates of hardware and system software requirements will be only as precise as the specifications of application software requirements. General statements of user and software requirements can generate only rough estimates of hardware requirements. More accurate hardware requirements cannot be determined until late in the design phase.

Unfortunately, application developers seldom have the luxury of waiting until the end of the design phase to acquire hardware. This is especially true when implementing large systems. In order to go through the acquisition process described earlier in this chapter, sufficient lead time is required. The acquisition process typically is started early in the design phase, after the scope of the automated system has been determined. Thus, hardware and system software requirements must be determined based on earlier analysis phase descriptions of processing requirements.

Informal methods of estimating requirements based on analysis phase outputs sometimes can be employed. Such methods typically employ comparisons to similar application software and work volumes. For example, a developer of an online transaction processing system might attempt to identify a similar transaction processing system inside or outside the organization. The system is examined to determine the adequacy of its existing hardware and system software. If current hardware resources match current application demand, then those hardware resources can be used as a baseline for estimating the hardware requirements of the yet-to-be-developed system. Requirements can be adjusted to account for differences in transaction volume and application processing characteristics.

This basic approach to requirements estimation is used widely and successfully. It is particularly successful for smaller hardware platforms running "garden variety" application software. The risk of inaccurate estimates is minimized for smaller hardware platforms because they can be modified relatively quickly and inexpensively.

Larger hardware platforms and large-scale one-of-kind application software require more formal methods of estimating requirements. Lead times for hardware acquisition and configuration are longer for mainframes and large minicomputers than for smaller hardware classes. Hardware and software costs are large, which increases the economic risks of inaccurate estimates. Estimating requirements in this environment typically employs formal methods based on mathematical models.

Modeling Concepts

A **model** is an abstract representation of physical reality. Models can be of many types, including graphic representations, physical representations, mathematical representations, and combinations thereof. An artist's rendition or drawing of the facade of a building is a graphical model. A blueprint of a building is a combination of a graphical and mathematical model. A set of simultaneous equations is a purely mathematical model. A scale model of a building or sculpture is a physical model.

All models limit the features of reality that are represented within the model. The process of choosing features to include in the model and of choosing an appropriate representation for those features is called **abstraction**. Different types of models abstract different features. For example, a drawing of a building facade abstracts visual features while ignoring many others such as construction materials. A blueprint models physical placement of construction materials in great detail while ignoring many issues of external appearance. A valid model accurately models important features and ignores unimportant ones.

Models are used to make a wide variety of decisions, and to avoid the expense of building and testing actual systems. To be useful, a model must be easier to build, understand, and manipulate than the corresponding reality. For example, a blueprint is a useful model of a building because it is far easier to create and modify than is an actual building. A model also must be valid for its intended purpose. The model must represent accurately all features that are important to the decisions that the model will support.

Models often are used as prediction tools, as shown in Figure 14-2. When used for prediction, a model may be manipulated by modifying one or more parameters. A mathematical model can be modified by changing the numerical values (parameters) assigned to one or more model features (variables). For example, an architect might change assumptions about the materials used to construct a certain part of a building. The new materials have a different mass, or weight, than the old. The mass of the new materials can be used as a new parameter value for various predictions such as the weight of a floor or required support strength. A model also can be used for prediction by changing model components, as opposed to parameter values. For example, an architect might add or delete a wall and assess or predict the effect of the change on the usability or structural integrity of the building.

Figure 14-2 ►

Model-based
prediction

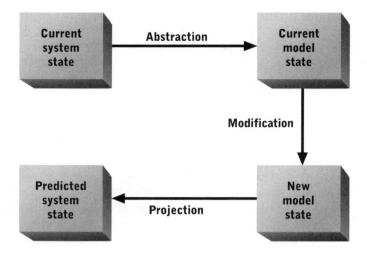

Predictions based on a model are made by projecting the results of model manipulation back onto the original system. In other words, it is assumed that modification of model parameters and components will produce results that mirror the effects of those same modifications to an actual system. Every projection from a model to reality carries a risk of incorrect prediction. The projection might be invalid due to any number of factors, including incomplete abstraction, improper representation of features, and incorrect model manipulation.

Computer System Models

Predictive models can be used to determine requirements and evaluate computer hardware and software performance. Models can be used to:

► Determine requirements for new hardware and system software to support a new application

► Evaluate the ability of an existing configuration to handle a new application or additional workload from an existing application

► Determine the change in application performance resulting from changes in hardware and/or system software

► Formulate long-range plans for expanding or acquiring computer capacity

A generic modeling strategy can be applied to each of these scenarios. The general steps to develop and evaluate a model are as follows:

1. List each distinct type of application processing task and its stated performance requirements.

2. Determine the hardware and system software resource requirements for each task.

3. Determine the frequency and distribution of each task.

4. Construct a composite model of application demand.

5. Construct a model of hardware-resource availability for each configuration of hardware and system software under consideration.

6. Test each configuration model against the application demand model.

Application Demand Models

An **application demand model** or **workload model** describes the relationship between user processing activities and demands for hardware and system software resources. Examples of user-level processing tasks include transaction updates, database queries, generation of reports, and batch file updates. Each of these processing tasks places different demands on hardware resources, and each requires a separate model of resource utilization. For example, a database query typically will require many accesses to secondary storage with little use of the CPU. A batch update might require many storage accesses for a large file and/or many CPU cycles due to complex calculations.

Resource demand by application tasks can be measured in many different ways, from very high-level measurements such as average number of CPU instructions executed, to detailed low-level measurements of individual machine actions and operating system service requests. Frequently used demand variables include:

▶ Average number of CPU instructions executed

▶ Bytes read from and/or written to secondary storage

▶ Bytes read from and/or written to specific I/O devices

▶ Bytes read from and/or written to servers or clients over a network

CPU instructions include only those instructions executed to perform computational and logical processing functions. In practice, this is a difficult number to estimate, because it is affected by the type of processing performed in the application, the efficiency of executable or interpreted code, and other factors. These factors are difficult to estimate for applications that have yet to be developed or tested. Estimates can be derived for existing applications through the use of program profiling, as described in a later section.

Bytes of secondary storage I/O represent demand for access to mass storage devices and channels, as well as the CPU (operating system service layer) overhead necessary to implement those accesses. Communication to and from I/O and network devices also includes channel capacity and related CPU overhead. For these reasons, I/O to devices or files frequently is measured in terms of specific operating system service calls. For existing systems, these calls can be profiled in much the same manner as application code, in order to derive accurate estimates of CPU overhead.

A composite model of application resource demand is a combination of workload models for individual processing tasks and estimates of the frequency or volume and distribution of those tasks over time. Processing volume can be measured in transactions per minute or hour, reports per day or week, queries per minute, and many other units of measurement. Distribution measures show the expected norms and variations in frequency. For example, sales transaction volume might average 200 per hour in a normal week. Typical volume, however, might vary by time of day, day of week, time of year, or other factors. Because of these sources of variation, estimates of processing volume rarely can be stated accurately as a single number. Instead, they require that the statistical distribution of processing demand be known.

Resource Availability Models

A **resource availability model** can describe either a hardware system or a combination of hardware and system software. The latter approach is more accurate, although it introduces additional complexities and possibilities for error. The raw capabilities of hardware represent only a potential. For most applications, access to that capability is obtained exclusively through the system software, especially the operating system. The efficiency with which system software manages and delivers hardware resources to applications is a primary factor in determining resource availability.

Resource availability for a configuration of hardware and system software must be expressed in the same terms as for application tasks. As discussed earlier, resource availability generally is measured in terms of CPU instructions and the ability to move data to and from secondary storage, I/O, and network devices. Time intervals and units of measure for resource demand must match those used in the application demand model.

CPU capability is measured in millions of instructions per second (MIPS) or millions of floating point operations per second (MFLOPS). Several problems can arise from the use of these measures, including:

► The execution of long instructions by an application

► Wait states due to memory access bottlenecks

► Incompatibility between the instruction sets of competing hardware platforms

To simplify the resource availability model, application demand for CPU instructions generally is standardized to a least common denominator, for example, 32-bit integer or floating point instructions. Configuration capabilities also are standardized to these generic instructions, and wait states are assumed to be zero or a fixed percentage of instructions executed.

For access to secondary storage and I/O devices, resource availability generally is measured in terms of sustainable data transfer rates. These rates depend on the characteristics of the devices, device controllers, and communication channels in use. Difficulties can be encountered when substantial differences exist between burst mode data transfer and sustainable data transfer rates. For example, with a caching disk controller, short bursts of extremely fast data transfer are limited only by the speed of cache access. However, for large transfers that exceed cache capacity, the sustainable data transfer rate would be limited by the physical characteristics of the storage device. An average rate sometimes is used if it can be derived via performance testing.

Model Manipulation and Analysis

Using mathematical models for application demand and resource availability allows well-established computational techniques to be applied to model evaluation. Computational approaches applied to computer performance modeling are of two types—static and dynamic analysis. Each can be implemented by specific computational techniques.

Static analysis refers to the use of computational techniques that assume that demand for resources and the ability to supply them are constant. Variation over time in application demand or in the ability of a configuration to supply resources is ignored. In contrast, **dynamic analysis** explicitly accounts for variation over time in the values of model parameters. As you might guess, computational techniques for dynamic analysis are much more complex than for static analysis.

Static Analysis When static analysis is used to evaluate models, all parameter values are assumed to be constant. Application tasks are assumed to have a constant frequency and resource demand. Configurations of hardware and system software are assumed to have a constant ability to supply computing resources. These assumptions usually are valid for resource delivery, but often are unrealistic for application demand.

Average resource demands for each application processing task are summed over an interval of time. For example, assume an application processes two types of user tasks—A and B. Assume further that resource demand for each transaction can be expressed by the following formulae:

$$A: 15000X_c + 300X_d + 200X_i$$
$$B: 10000X_c + 750X_d + 500X_i$$

where:

X_c = thousands of CPU cycles

X_d = KB of disk I/O

X_i = KB of display device I/O

The composite resource demand model is the sum of these two formulae, weighted by their expected volumes over a stated time interval. If peak volumes for tasks A and B are 100 and 250 per hour, respectively, the peak composite resource demand is:

$$A: 100(15000X_c + 300X_d + 200X_i) = 1500000X_c + 30000X_d + 20000X_i$$

$$B: 250(10000X_c + 650X_d + 500X_i) = 2500000X_c + 162500X_d + 125000X_i$$

$$\text{Composite:} = 4000000X_c + 192500X_d + 145000X_i$$

For each configuration under consideration, a resource availability model is constructed using the same variables, units of measure, and timing conventions. The capacity of each configuration is expressed as a set of maximum values for each resource class during a one-hour period.

Note that the maximum capacities of resource classes are interdependent due to tradeoffs among various types of processing. System overhead associated with mass storage and I/O accesses consumes CPU cycles and bus capacity. A model of resource availability must account for these complexities. For example, resource availability for a configuration might be stated as:

$$X_c \leq 5000000000$$

$$X_b \leq 25000000$$

$$X_d \leq 8000000$$

$$X_i \leq 1000000$$

$$X_d = X_b + X_s + 200X_c$$

$$X_i = X_b + X_v + 150X_c$$

where:

X_b = kilobytes of bus I/O

X_s = kilobytes of disk device I/O

X_v = kilobytes of display device I/O

The intermediate variable X_b represents the bus capacity consumed. Resource utilization for each unit of application I/O to disk or display device is expressed as a formula describing its use of CPU, bus, and device resources. The composite variables X_s and X_i represent the sum of all resources consumed for each byte of disk and display device I/O. These composite variables would be substituted for X_s and X_i in the application demand model.

The two models can be analyzed via a number of techniques such as a linear programming to test the feasibility of meeting application demand with a given configuration. More complex models can be used to determine detailed configuration requirements, for example, number of disk drives or I/O channels, or to determine a least-cost configuration among several competing configurations.

Dynamic Analysis In the static analysis example, all parameter values were single valued. In dynamic analysis, the distribution of values is modeled explicitly. This implies using more model information than with static analysis. In theory, this additional information will yield more accurate results. However, this additional information must be accurate to guarantee valid predictions. Although dynamic analysis promises more accurate predictions of system behavior, that accuracy is obtained at the expense of substantially more complex model inputs.

Dynamic analysis of a computer system generally is implemented via simulation, which is based on queuing theory and statistics. A **simulation model** is composed of processing tasks, processing elements, and a set of interconnections between them. Computer hardware and system software is modeled as a linked set of server processes and hardware devices, as shown in Figure 14-3. Each server is dedicated to a specific processing function such as executing program instructions, file I/O, and display I/O. Each server accepts input from a queue of pending requests, as shown in Figure 14-4, and generates a request to one or more hardware devices.

Figure 14-3 ▶

Network diagram of a simulation model

Figure 14-4 ▶

Individual
processing node
within a
simulation model

The arrival rate of processing tasks to initial system nodes (the dispatcher, in this example) is stated as a statistical distribution. This distribution is usually expressed as a distribution type (such as normal or Poisson), mean, and standard deviation. The distribution is used to determine the arrival rate for each processing task. Queuing theory can be used to determine the composition of and arrival times for combined queues. Each processing node accepts requests from the input queue and satisfies those requests. The amount of time needed to satisfy a request depends on the request. The time needed by a processor to satisfy each type of request is stated as part of the model inputs, often as a statistical distribution.

A number of software products are available that implement and analyze simulation models. Simulation software accepts a description of the modeled system as input. The operation of the system over time is simulated by advancing a computer-generated clock. Each time the clock is advanced, the program generates a set of task arrivals and hardware actions based on the statistical distributions of task arrival and hardware operation. The current state of the system (hardware states and queue sizes) is updated based on these calculations. Iterations continue until the simulation model reaches steady state operation. This state simulates the normal operation of the modeled system.

Simulation software can produce reports showing many kinds of system information. Examples include:

▶ Average queue size for each processor

▶ Average waiting time for each processing request

▶ Average processing time for each processing request

▶ Proportion of idle time for each processor

This information can be used to evaluate the performance of the system and/or to identify performance bottlenecks.

Measuring Resource Demand and Utilization

Both dynamic and static models require accurate data inputs. Accurate model analysis depends on the quality of application demand and resource availability estimates. Automated tools to measure resource demand and utilization include:

► Hardware monitors

► Software monitors

► Program profilers

These tools generate model input data to determine new system requirements. They also can be used to generate data for evaluating upgrades or modifications to existing applications or hardware/software configurations. The information generated by these monitors describes the behavior of specific devices, resources, or subsystems over some period of time.

A **monitor** is a program or hardware device that detects and reports processing or I/O activity. A **hardware monitor** is a device that is attached directly to the communication link between two hardware devices. It monitors the communication activity between the two devices and stores communication statistics or summaries. This data can be retrieved and printed in a report. Hardware monitors often are used to monitor the use of communication channels, disk drives, and network traffic.

A **software monitor** is a program that detects and reports processing activity or requests. Software monitors are typically included within operating system service routines and can be activated or deactivated by a system administrator. They can monitor high-level processing requests such as file open, read, write, and close, or low-level kernel routines such as flushing file or I/O buffers or virtual memory paging. When activated, a software monitor generates statistics of service utilization or processing activity. These statistics can be displayed in real time or stored in a file for later analysis.

Software monitors can alter the activity being measured. For example, a software monitor that measures system-wide CPU activity will generate inflated measurements because the software monitor itself requires CPU cycles to execute. Similar problems can exist for display I/O if the monitor displays a continuously updated activity graph, and disk I/O if the monitor logs activity measures to a file. Resources consumed by the software monitor must be estimated or measured and subtracted from the overall measurements to determine correct measurements for processes other than the monitor.

Monitors can operate either continuously or intermittently. A continuous monitor records all activity as it occurs. An intermittent, or sampling, monitor checks for activity on a periodic basis. The advantage of a sampling monitor is that fewer computer resources are expended executing the monitor process itself. Another advantage is that less data is accumulated in output files, which can become very large. Continuous utilization statistics can be estimated based on the sampled activity. Continuous monitors provide complete information on activity, but their operation can consume excessive system resources.

Monitors can be used to identify performance bottlenecks as a precursor to configuring hardware and/or system software for maximal performance. For example, monitoring I/O activity on all secondary storage channels can indicate that some disks are used continuously while others are not used as much. Based on this information, a system administrator might decide to move frequently accessed files from heavily used disks to less used disks, or to reallocate disks among controllers or I/O channels. An entire system can be tested at full load to determine its maximum sustainable resource delivery.

A **program profiler** is used to determine the resource or service utilization of an application program during execution. Typically, a set of monitor subroutines are added to the program's executable image during link editing. As the program executes, these subroutines record system service requests in a file. They also can record other statistics such as elapsed (wall clock) time to complete each service request, CPU time consumed by service calls, and CPU time consumed by program subroutines. This information can be used to derive a resource demand model for the application. It also can be used to identify segments of the program that might be inefficiently implemented.

Monitors and profilers are commonly used tools for performance evaluation and requirements determination. By providing accurate data on resource demands and utilization, they accurately model computer system configurations, which improves the accuracy of predictions based on those models.

Technology
Focus
Windows 2000 Performance Monitoring

The Windows 2000 operating system is used both for workstations and for servers. In workstation environments, it supports applications that demand high performance and reliability. In server environments, it provides distributed file system, database, and World Wide Web services. Workstation and server environments both might need to optimize hardware and software resources to intended applications.

Windows 2000 provides a performance monitoring utility called System Monitor (SM). With SM, a system administrator can monitor hardware and software resource use in real time. SM defines a number of system objects about which performance and utilization data can be captured. A partial display of these objects is shown in Figure 14-5. Monitored objects can be individual hardware devices such as the CPU, groups of hardware devices such as all physical disks, operating system services such as a network name service, or resource management data structures such as a service queue or a virtual memory paging file.

Figure 14-5 ▶

Windows 2000
System Monitor

Each object has a set of data items called counters. A counter monitors a specific performance or resource utilization aspect of an object. For example, there are 19 counters for the object PhysicalDisk, including current queue length, average seconds per disk read or write operation, number of disk read or write operations per second, and number of bytes read or written per second. Using different counters, a system administrator can monitor either aggregate or highly specific aspects of performance.

Resource utilization and other system activity can be displayed in various formats. Figure 14-6 shows a line graph displaying total CPU utilization (the thin line) and virtual memory page faults per second (the thick line). Statistics such as mean, minimum, and maximum are displayed for the currently selected counter. The display can be configured for any number of counters and each counter is assigned a specific color and type of line. A similar set of display options are available for other output formats such as bar graphs and text reports. Performance data also can be captured in a file for later analysis.

Figure 14-6 ▶

Real-time display of performance data

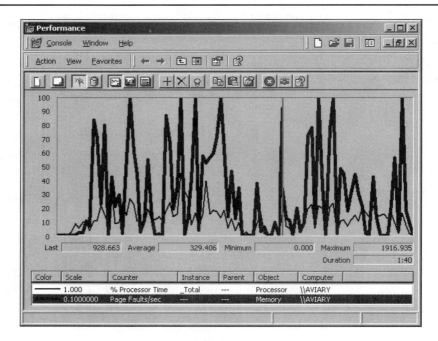

A system administrator needs accurate data to make system configuration decisions. For example, optimal decisions about changes to the size and/or location of virtual memory page files require accurate data concerning current virtual memory utilization. Optimal decisions about upgrading CPU capacity require data describing current CPU utilization and the utilization of related resources such as memory and bus I/O. SM provides a flexible and powerful means of acquiring such data.

Accurate data is not the only requirement for good configuration decisions, however. Interpreting that data requires a thorough understanding of the operating system and hardware components, which allows the system administrator to associate cause and effect and understand the trade-offs among various types of system performance. Without such an understanding, performance assessment is little more than guesswork, and configuration decisions can produce suboptimal or detrimental effects on overall system performance.

SECURITY

The information processing resources of many organizations represent a sizable investment. Some of the resources such as specific items of hardware and software are tangible and have well-defined dollar values. Other resources such as databases, user skill, and reliable operating procedures are less tangible but also are of considerable value. As applied to information systems, the term security describes all measures by which the value of these investments are protected. These measures include physical protection against loss or damage and economic protection against loss of value through unauthorized disclosure.

A well-integrated approach to system security includes measures to:

► Protect physical resources against accidental loss or damage

► Protect data and software resources against accidental loss or damage

► Protect all resources against malicious tampering

► Protect sensitive software and data resources against unauthorized access and accidental disclosure

Physical protection of resources against accidental loss or damage is discussed in a later section of this chapter. Protection of data and software against accidental loss or damage primarily is implemented via file system backup and recovery procedures as described in Chapter 12. Controls over malicious tampering and unauthorized access generally are implemented through a combination of physical and software access controls. Physical access controls are beyond the scope of this text. Software access controls are discussed in the following sections.

Access Controls

Multiuser and network operating systems require users to establish their identity prior to accessing resources. The identification process sometimes is called **user authentication**. A user enters a user name or other identifier and a password to prove that they are who they claim to be. The operating system verifies the user name and password against a security database. The security database can be maintained by the operating system on the local machine or located on a network server and accessed by network service calls.

An authenticated user name or identification is the basis for determining that user's ability to log on and to access resources managed by the operating system. Specific access controls can include restrictions on:

► Locations from which a user can log on to the system or access its resources

► Days/times when access is granted or denied

► The ability to read, write, create, and delete directories and files

► The ability to execute programs

► The ability to access hardware resources such as printers and communication devices

The system administrator establishes these access controls to implement the security policies of the organization.

Modern access control systems usually are based on a complex directory structure as described in Chapter 13. Figure 14-7 shows an example of a hierarchical directory structure for Novell Netware. Many types of objects are represented, including organizational roles such as Chairman, user groups such as Faculty, organizational units such as Information Systems, server computers such as Server1, file systems such as Applications, printers and printer input queues such as LaserJet 4Si and PrimaryQueue, and client computers such as ASM2104. A system administrator can define access controls and privileges for any object or group of objects. The variety of objects in the directory structure makes it possible to define complex and overlapping security policies.

Figure 14-7 ▶

Novell Directory Services (NDS) tree

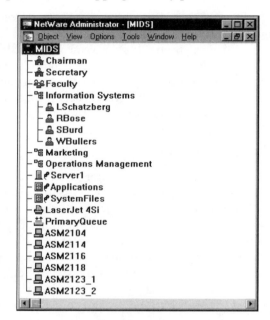

Password Controls and Security

User authentication is a critical component of any access control system. A failure in authentication procedures effectively disables most software-based security measures, so the authentication process must be highly reliable. It must ensure that only authorized users are able to log on and gain access to system resources and services. For optimal reliability, passwords must be securely stored and be difficult for unauthorized users to guess.

Modern operating systems and directory services employ a number of methods to enhance password security, including:

▶ Restrictions on the length and composition of valid passwords

---

System Administration

- ► Requirements that passwords periodically be changed
- ► Analysis of password content to identify passwords that are guessed easily
- ► Encryption of passwords in files and during transmission over a network
- ► Restrictions on locations from which a user can access the system

With most operating systems, the system administrator can create and enforce password policies on a per user, per group, or per system basis. Figure 14-8 shows the password controls that can be set in Windows 2000, including restrictions on password length, age, and uniqueness. A related set of policies, shown in Figure 14-9, can be set to deal with failed attempts to log on. In Windows 2000, accounts can be deactivated after a specific number of failed logons. The account can be reactivated manually by the system administrator or automatically after a specified time interval. Some of the password controls, such as password age restrictions, can be overridden for specific accounts.

Figure 14-8 ►

Windows 2000 password policies

Figure 14-9 ►

Windows 2000 account lockout policies

Locking out accounts after a specified number of failed logon attempts prevents unauthorized users from repeatedly attempting to guess correct passwords for valid user accounts. Software programs sometimes are used to attempt remote logon with words and/or names taken from a dictionary. Because many users choose common words or names as their passwords, it is possible for such programs to find a correct password eventually by trying every word in their dictionary. Automatic account deactivation coupled with delayed reactivation severely limits the number of guesses that can be made in any time period.

Some operating systems employ similar password guessing programs against their own password files on a periodic or continuous basis. When passwords are guessed successfully, the operating system forces the user to change their password at the next logon. Such programs also can be employed by the program used by users to change their passwords. An attempt to change a password to an easily guessed value is denied.

System security might be compromised if a copy of the password file or database somehow is distributed to or stolen by an unauthorized individual. The password file usually is protected by access controls but some degree of accessibility must be granted for normal logon processing. Most operating systems further protect the password file by encrypting part or all of its content. However, a stolen password file can be decrypted successfully if sufficient computing resources are employed to crack the encryption key.

PHYSICAL ENVIRONMENT

Installing computer hardware requires special attention to many aspects of the physical environment. Some of these are a matter of convenience and others are a matter of protection for the equipment itself. Particular issues to be addressed include:

► Electrical power
► Heat dissipation
► Moisture
► Cable routing
► Fire protection

Each of these is discussed in the next sections.

Electrical Power

Computer hardware is very sensitive to fluctuations in power levels. Computer circuitry is designed to operate at a constant low power level. Fluctuations can cause momentary loss of operation if power levels drop or can cause damage to electrical circuits if power levels rise. Fluctuations can be of several types:

► Momentary power surges
► Momentary power sags
► Long-term voltage sags
► Total loss of power

Power surges can be caused by a number of events. Lightning strikes in power generation or transmission facilities tend to cause the most dangerous types of power surges. Dangerous spikes can be caused by the failure of power transformers or other transmission equipment. These events lead to brief power surges of very high intensity. Because the surges are brief, they might not engage standard protection devices such as fuses or circuit breakers before significant damage has occurred. Standard fuses or breakers for a floor or entire building do not provide adequate protection for computer equipment.

Power sags normally occur when a device that requires a large amount of power is started. This can be seen in the home when devices such as air conditioners, refrigerators, and electric dryers are started, causing a momentary dimming of lights. Small power sags almost always are present when multiple devices share a single electrical circuit. Large power sags are a symptom of overloaded circuits. Using a single circuit for equipment with large variations in power requirements is particularly troublesome.

Longer-term power sags often are caused by the power provider itself. The common term for this event is a **brownout**. Brownouts occur when the demand for electricity exceeds the generation and transmission capabilities of the provider. This commonly occurs during peak demand periods such as hot summer days. The power provider will reduce the voltage level temporarily on a system-wide basis to spread the available power evenly.

Most computer equipment is designed to operate reliably over a range of voltage levels. In the United States, transformers within smaller computer systems are designed for 110 volt alternating current power input. Larger computer systems in the United States can use 220 or 440 volt power inputs. Computer systems in other countries use different voltage levels. Computer system power transformers usually tolerate input voltage variations of up to 10 percent. This characteristic provides some protection against surges and sags due to the startup or shutdown of other equipment, as well as brownouts.

Other provisions must be made to protect equipment against more severe power variations. These include mechanisms to deal with powerful surges and total power loss. Equipment can be protected against high power surges by the use of a **surge protector**. These devices detect incoming power surges and within nanoseconds divert them to ground. Surge protectors differ in the speed at which they react, the intensity of the surge that can be grounded, and whether the device can be reused after a surge. Surge protectors can be purchased as separate devices or can be a component of an integrated device such as a line conditioner.

By itself, total power loss rarely causes damage to computer hardware. However, a tripped circuit breaker or blackout often is accompanied by one or more power surges. The primary problem with power loss lies in the loss of data. Data held in RAM, including process data areas, secondary storage buffers, and communication buffers, is lost when power is interrupted.

Protection against power loss requires the use of an auxiliary power source. This can take several forms, including the use of secondary circuits, auxiliary generators, battery backup, or some combination of these. An **uninterruptible power supply** (UPS) is a device, usually battery based, that provides power to attached devices in the event of external power failure. These devices vary in their power delivery capacity, switching time, and duration of operation between power loss and power restoration. Surge protection normally is incorporated into a UPS. This is somewhat of a necessity because switching between external and internal power supplies might introduce surges.

Some UPSs, particularly those with short delivery times, are designed to work in concert with a computer's operating system. They are attached to a communication port of the computer system or directly to the system bus via an interface card. When a power failure is detected an interrupt is generated. This informs the operating system of the power failure and provides a mechanism for initiating protective actions prior to a total loss of power. Typically, the operating system will initiate a normal or emergency shutdown procedure when such an interrupt is detected.

Heat Dissipation

All electrical devices generate heat as a byproduct of normal operation. Excessive heat can cause intermittent or total failure of electrical circuits. Because of this, all computer equipment requires some means of **heat dissipation**. In equipment that generates little heat, vents in the equipment cabinet normally are sufficient to dissipate heat. Care must be taken to ensure that vents do not become blocked, so there is free movement of air through the cabinet.

Many hardware devices use fans to move air through the unit. Fans either force cool exterior air into the cabinet or draw hot interior air out. Either method requires at least two vents and a clear pathway for air movement. Vents normally are positioned at opposite corners of the cabinet to ensure that all components receive adequate cooling. Fan cooling also requires some filtering to remove dust and other contaminants.

When heat is dissipated from an equipment cabinet it collects in the room in which the cabinet is located. Some mechanism then must be provided to dissipate heat from the room as well as the cabinets. This is especially true when many hardware devices are situated in a relatively small space, such as in a server closet. Normal room or building air conditioning might be sufficient, or supplemental cooling capacity might be required. Supplemental cooling more effectively counteracts heat buildup and provides additional protection if the primary cooling system fails.

In extreme cases, auxiliary cooling can be provided within an individual equipment cabinet. This can take the form of a refrigerant-based heat exchanger, a liquid cooling system, or even a liquid nitrogen system. Such measures often are used in equipment where semiconductor devices such as supercomputers are operated at extremely high clock rates or access speeds. Liquid cooling has been used in some mainframe computer systems.

Moisture

Excessive moisture is an enemy of electrical circuitry due to the danger of short circuits, which result when water conducts electricity between two otherwise unconnected conductors. Short circuits can lead to circuit damage and also are a fire hazard. Water can damage computer equipment even when power is turned off. Impurities in the water remain on computer components as the water evaporates. These impurities can corrode exposed electrical contacts or other hardware components such as disk platters, printed circuit boards, and metal cabinets. They also can form a short circuit.

Well-designed cabinets are one defense against the dangers of moisture, but they protect only against overt spills and leaks. Another protective measure is to mount cabinets or devices within cabinets above floor level. This minimizes the danger from roof leaks, broken pipes, and similar problems that can lead to standing water.

Protection also must be provided against condensation due to excessive humidity. Low humidity also is a problem. Low humidity levels increase the buildup of static electricity, which increases the likelihood of circuit damage due to inadvertent static discharges. In general, the humidity level of a room containing computer equipment should be near 50 percent.

Humidity generally can be controlled automatically by optional components of heating, ventilation, and air conditioning systems. It is possible to use a household dehumidifier in smaller rooms, but be aware that when the unit powers on, it might cause a brief dip in power level if it is on the same circuit as computer equipment. Also, these units usually remove the humidity from the air by condensing it into water in a pan, which must be emptied regularly. An overflowing pan can be a safety or electrical hazard.

Cable Routing

Computer facilities must be designed to provide protection for data communication lines. Because the configuration of these lines changes frequently, ease of access also is very important. Computer facilities tend to deal with this problem in two ways—by using raised floors and dedicated cable conduits.

A raised floor generally will be used in a room that contains multiple hardware cabinets, for example, a single mainframe computer system or multiple minicomputers. The raised floor serves several purposes. The primary purpose is to provide an accessible location for cables connecting different devices. The floor is a set of load-bearing supports upon which a grid of panel supports is laid.

The surface of the floor consists of solid panels that can be installed or removed easily from the grid. Cables typically are routed in a straight line under walkway areas, such as between hardware cabinets. This makes cabling easily accessible by removing floor panels.

Secondary reasons for raised floors include protection from standing water and movement of chilled air. Several inches of water can accumulate without reaching the level of equipment cabinets. Moisture sensors typically are placed below the floor panels to detect the buildup of standing water. The space between the actual floor and the floor panels also can be used as a conduit for chilled air. When used in this manner, equipment cabinets are vented at the bottom and top. Chilled air is forced through the floor into the bottom of the cabinets and heat is dissipated through the top.

Dedicated cabling conduits normally are used to provide cable access between rooms or floors of a building. To prevent electromagnetic interference, these conduits should not be used to route both electrical power and electrical data communication lines—although fiber-optic lines are immune to interference from electrical power lines. In addition, access panels should be provided at regular intervals for the addition, removal, or rerouting of cables. Conduits should be shielded to limit external electromagnetic interference.

Fire Protection

Fire protection is an important consideration both for safety of personnel and the protection of expensive computer hardware. As with cooling, the normal fire protection mechanisms incorporated within buildings are inadequate for rooms that house computer hardware. In fact, such measures actually increase the danger to both personnel and equipment. This primarily is due to the extensive use of water, such as with automatic sprinklers.

Carbon dioxide, fire retardant foams and powders, and various gaseous compounds are alternate methods of fire protection. Carbon dioxide generally is unacceptable because it is a suffocation hazard and because it promotes condensation within computer equipment. Fire retardant foams and powders are unacceptable due to their moisture content, and powders generally have corrosive properties.

Many large computer facilities use halon 1301 gas. It does not promote condensation and also doesn't displace oxygen to the extent that carbon dioxide does. This gives personnel adequate time to evacuate a room or floor. Unfortunately, halon gas is a chlorofluorocarbon (CFC), a class of gases with ozone-depleting properties. Most industrialized countries have signed the Montreal Protocol, which has phased out CFC production and bans importation of newly produced CFCs. Various alternatives currently are under study, and include halocarbon compounds and inert gas mixtures. None has yet emerged as a clear successor to halon. (See Research Problem #2 at the end of this chapter for halon-related Web links.)

Fire detection also is a special problem within computer facilities. Electrical fires often do not generate heat or smoke as quickly as do conventional fires and normal detection equipment might be slow to react. Fast detection is an economic necessity. A fire within one item of computer equipment can spread quickly and/or cause damage to attached equipment through power surges.

Normal building fire detection equipment typically is supplemented within a computer room. Additional smoke detectors generally are placed near large concentrations of equipment. Smoke detectors also should be placed below raised floors to detect fires in cabling quickly.

Disaster Planning and Recovery

Because disasters such as fire, flood, and earthquakes cannot be avoided totally, plans must be made to recover from them. Disaster planning is particularly critical in online systems and systems where extended downtime will cause extreme economic impact. A number of measures normally are taken, including:

► Periodic data backup and storage of backups at alternate sites
► Backup and storage of critical software at alternate sites
► Installing of duplicate or supplementary equipment at alternate sites
► Arrangements for leasing existing equipment at alternate sites, such as with another company or a service bureau

The exact measures that are appropriate for any given installation depend heavily on local characteristics.

SUMMARY

► The primary responsibility of a system administrator is to ensure efficient and reliable delivery of IS services. Broad categories of system administration tasks include acquiring new IS resources, maintaining existing IS resources, and designing and implementing an IS security policy. There is considerable variation among organizations as to the assignment of specific tasks.

► IS resource acquisition and deployment should occur only in the context of a well-defined strategic plan for the organization as a whole. IS strategic plans tend to follow, rather than lead, the strategic plans of other units in the organization. IS resources can be considered organizational infrastructure. Many IS resources also are capital assets because they are used during multiple years. Strategic issues relevant to IS include services to be provided, charging for services, infrastructure composition, and infrastructure improvement and maintenance.

- Providing infrastructure-based services to a wide variety of users requires the adoption of service standards. However, standardization tends to stifle innovation and to produce solutions that are suboptimal for some users. Standardization often causes problems for users who need services at or near the leading edge of technology. Managing IS resources as infrastructure can result in ignoring opportunities to employ IS resources for competitive advantage.

- Planning for IS resource acquisition requires a thorough understanding of present and anticipated application needs. The acquisition process includes determining the applications that the hardware/software will support, specifying detailed hardware/software requirements, drafting and circulating a request for proposals (RFP), evaluating RFP responses, and negotiating a purchase and/or support contract.

- Determining hardware requirements is one of the most difficult system administration tasks. Hardware requirements depend on the hardware and system software resources required by application software. If the application software already has been developed, then its hardware and system software resource consumption can be measured. Determining resource requirements is much more complex when application software has not yet been developed. In this case, hardware and system software requirements must be determined based on the detailed analysis phase descriptions of processing requirements. Estimates can be based on comparisons to similar application software or on detailed modeling of application resource requirements.

- Computer hardware and software performance evaluation and requirements determination can be performed with mathematical models. Two separate models are required—an application demand model and a resource availability model. An application demand model describes the relationship between user-level processing activities and demands for hardware and system software resources. A resource availability model describes the ability of hardware and system software to supply computing resources to application programs.

- Computational approaches applicable to computer performance and application requirements modeling are of two types—static and dynamic. Static analysis assumes that all parameter values are constant. This assumption usually is valid for resource delivery but not for application demand. Dynamic analysis explicitly models the variability of model parameters. Dynamic analysis uses simulation and promises more accurate predictions of system behavior at the cost of substantially more complex model inputs and analysis.

► Automated tools such as hardware monitors, software monitors, and program profilers are available to measure resource demand and utilization. A hardware monitor is a device that is attached directly to the communication link between two hardware devices to report the communication activity between the devices. A software monitor is a program that detects and reports processing activity or requests. A program profiler describes the resource or service utilization of an application program during execution.

► The information processing resources of many organizations represent a sizable investment. A well-integrated approach to system security protects these resources against accidental loss or damage, malicious tampering, unauthorized access, and accidental disclosure. System security is enforced through the operating system by access controls and user authentication procedures. User authentication procedures are a critical component of any access control system. Passwords are the key component of a user authentication system. Operating systems employ a number of methods to enhance password security, including restrictions on password length and composition, requirements that passwords be changed periodically, periodic or continuous attempts to guess passwords, encryption of passwords in files and during transmission, and restrictions on locations from which a user can access the system.

► The installation of computer hardware requires special attention to many aspects of its physical environment. Particular issues to be addressed include electrical power, heat dissipation, moisture, cable routing, and fire protection.

Key Terms

abstraction
application demand model
benchmark
brownout
capital expenditure
capital resource
competitive advantage
dynamic analysis
hardware monitor
heat dissipation

infrastructure
model
monitor
operating expenditure
power sag
power surge
program profiler
request for proposal (RFP)
resource availability model
service standard

simulation model
software monitor
static analysis
strategic plan
surge protector
system administration
uninterruptible power supply
   (UPS)
user authentication
workload model

Vocabulary Exercises

1. An estimate of resource delivery by a hardware/software configuration can be obtained by a tested or published _____ .

2. _____ are expected to provide service over a period of years.

3. Integrated analysis of the operation of application software, system software, and computer hardware requires a(n) _____ model and a(n) _____ model.

4. _____ uses _____ theory to determine the arrival of service requests.

5. Features and capabilities of an application or computer system are _____ to create a model.

6. _____ analysis explicitly considers variations in model parameters over time, while _____ analysis ignores those variations.

7. Because IS resources can be considered _____ , issues of service standards, cost, and cost recovery are important components of the IS strategic plan.

8. A(n) _____ is a formal (legal) document that solicits proposals from hardware and/or software vendors.

9. _____ based on passwords is the most critical process in a security system.

10. A(n) _____ detects and reports hardware or software actions.

11. Opportunities to exploit IS resources for _____ might be missed if those resources are managed only as infrastructure.

12. A(n) _____ provides auxiliary power during blackouts and can notify the operating system when it is activated.

13. Computer hardware must be protected against _____ and _____ in electrical power.

14. _____ is the process of ensuring efficient and reliable delivery of IS services.

15. A(n) _____ is an abstract representation of reality.

16. Long-range acquisition of computer hardware and software should be made in the context of a(n) _____ for the entire organization.

17. The resource demands of an existing application can be measured with a(n) _____ .

Review Questions

1. What is infrastructure? In what ways do computer hardware and system software qualify as infrastructure?

2. What are the basic strategic planning questions that are addressed with respect to infrastructure?

3. What are the advantages and disadvantages of standardization in computer hardware and systems software?

4. What is a request for proposal (RFP)? How are responses to an RFP evaluated?

5. What problems are encountered when attempting to determine hardware and system software requirements for application software that has not been designed or developed fully?

6. How is computer performance evaluation related to applications?

7. What are some purposes of performance evaluation?

8. What are the essential elements of a simplified performance model, and how can these elements be organized?

9. What is the difference between static analysis and dynamic analysis? What are the comparative advantages and disadvantages of each?

10. What steps are used in model-based requirement determination and performance evaluation?

11. What parameters might be included in a measurement of application resource demand?

12. What is a monitor? List the various types of monitors and the information that they provide.

13. What is user authentication? Why is user authentication required to implement access controls?

14. List the types of access controls implemented by a typical operating system. How is access control information maintained by the operating system and system administrator?

15. Why are conventional methods of fire protection inadequate or dangerous for computer equipment?

16. What problems associated with electrical power must be considered in planning the physical environment of computer hardware?

Research Problems

1. Some manufacturers and vendors of large minicomputers and mainframes have developed documentation and software to assist engineers, account representatives, and purchasers to configure computer hardware and system software. Some of the software can assist users to match software demands against the capabilities of particular hardware configurations. Investigate the online offerings of several large computer companies, such as www.ibm.com, www.compaq.com, and www.unisys.com.

2. Halon use extends well beyond computer rooms. Other users include manufacturers of electrical equipment and the U.S. Navy. All these users have large investments in halon fire protection equipment. There is a significant economic incentive to developing halon alternatives that work with existing equipment. Investigate the current state of halon replacements with particular attention to cost, availability, fire protection ability, protection of computer equipment, and compatibility with existing fire protection equipment. Organizations that monitor halon replacements include The Halon Alternatives Research Corporation (www.harc.org), The Navy Shipboard Environmental Information Clearinghouse (http://www.navyseic.com/), and the U.S. Environmental Protection Agency (http://www.epa.gov/ozone).

3. Locate one or more benchmarking programs for personal computers on the Web. Download and install the benchmarks and test the performance of your own computer system. What do the results tell you about your computer? Do they provide you with any guidance for improving system performance? What types of application tasks do the benchmark programs simulate? Would similar benchmark programs be useful with minicomputers or mainframes?

Appendix

Measurement Units

Many measurement units describe the capabilities of computer system components such as storage devices, I/O devices, device controllers, and computer networks. This appendix describes common measurement units, their abbreviations, and common mistakes in use and interpretation.

TIME UNITS

Time intervals, such as cycle or access times, are expressed in fractions of a second, represented by the abbreviations and terms in Table A-1.

Table 1 ▶

Time measurements

Abbreviation	Term	Fraction
ms	millisecond	10^{-3} (one thousandth)
ms	microsecond	10^{-6} (one millionth)
ns	nanosecond	10^{-9} (one billionth)
ps	picosecond	10^{-12} (one trillionth)

Note that both millisecond and microsecond have the same abbreviation. Microseconds are rarely used when describing computer hardware, so you can generally assume that ms refers to milliseconds.

CAPACITY UNITS

Bits and bytes are the smallest data storage and transmission units. A bit is a binary digit containing the value zero or one. A byte contains eight bits. The terms bit and byte are often quantified with a prefix to indicate larger magnitudes. Such prefixes are all based on the value 2^{10} (1024_{10}) because it approximates traditional units based on powers of ten such as thousands, millions, and billions.

Table A-2 shows the abbreviations and prefixes that are employed in capacity measurements.

Table 2 ▶

Capacity measurements

Abbreviation	Prefix	Value
K or k	kilo	1024
M or m	mega	1024^2 (1,048,576)
G or g	giga	1024^3 (1,073,741,824)
T or t	tera	1024^4 (1,099,511,627,776)

Uppercase and lowercase letters are both used though uppercase is more common. The values represented by these prefixes are slightly larger than their decimal counterparts, which can cause confusion when an abbreviation based on 1024 is used to describe a quantity based on 1000.

For example, consider the capacity of a "standard" 3.5-inch PC disk which is frequently (and incorrectly) described with a capacity of 1.44 megabytes. The actual capacity is 1,474,560 bytes (2 sides × 80 tracks/side × 18 sectors/track × 512 bytes/sector), 1440 kilobytes (1,474,560 ÷ 1024), or 1.40625 megabytes (1,474,560 ÷ 1024^2). Hard disk capacities are often stated with similarly flawed units of measure.

Another source of confusion arises when the terms *bit* and *byte* are represented by the letter B or b. For example, the terms megabit and megabyte can both be abbreviated as MB. A common method of distinguishing between bits and bytes is to abbreviate the term *bit* using a lowercase letter and the term *byte* using an uppercase letter. Thus, the abbreviation GB would be interpreted as gigabyte or gigabytes and the term Gb as gigabit or gigabits. Unfortunately, this convention is not universally followed. Any abbreviation containing the letters b and B deserves careful scrutiny to determine whether bit or byte is intended.

DATA TRANSFER RATES

Data transfer rates are used to describe the capacity of communication channels and the input speed, output speed, or throughput of I/O devices and device controllers. A data transfer rate is always expressed as a quantity of data per time interval. For example, LAN data transmission capacity might be described as 100 megabits per second. Printer output speed might be described as four pages per minute.

Communication channel capacities are usually expressed in bits or bytes per second. Magnitude prefixes such as *kilo* and *mega* can be used (for example, 500 kilobits per second and 100 megabytes per second). Bit and byte capacity measures are abbreviated as described in the previous section, and the phrase per second, abbreviated as ps, is appended to the measurement unit. For example, kilobits per second is abbreviated as Kbps and megabytes per second is abbreviated as MBps. As with capacity measures, any abbreviation containing the letter b or B deserves careful scrutiny to determine whether bit or byte is intended.

Printer output rates are generally expressed in characters per second, abbreviated as cps, or pages per minute, abbreviated as ppm. A character is generally assumed to be eight bits, though 16 bits are required for Unicode characters. The number of bytes in a page varies widely depending on the page content and data encoding method. Video output rates may be expressed in frames per second, abbreviated as fps. The number of bytes per frame depends on frame size, image content, and data encoding method.

GLOSSARY

1GL — Acronym of first generation language.

2GL — Acronym of second generation language.

3GL — Acronym of third generation language.

4GL — Acronym of fourth generation language.

5GL — Acronym of fifth generation language.

absolute addressing — Term that describes memory address operands that refer to actual physical memory locations.

abstraction — (1) With respect to software, the conceptual difference between the physical reality of computer hardware and the abstract or logical view of that same hardware provided by system software and/or a programming language. (2) With respect to computer system modeling, the process of selecting and representing computing resources and application demands within a model.

access arm — Device within a disk drive containing one or more read/write heads mounted at one end. The other end is attached to a motor that allows the read/write head(s) to be positioned over a single track of the disk platter.

access time — Elapsed time between the receipt and completion of a read or write command by a storage device.

acknowledge (ACK) — ASCII control character sent from receiver to sender to indicate successful receipt of transmitted data.

ACK — Mnemonic for acknowledge.

ACM — Acronym of Association for Computing Machinery.

active matrix display — Liquid crystal or electroluminescent display that uses one or more transistors for every pixel.

ADC — Acronym of analog-to-digital conversion.

AD — Acronym of Active Directory.

ADD — (1) Boolean operation that causes the arithmetic sum of two numbers to be computed. (2) Central processing unit instruction that causes the arithmetic sum of two numbers to be computed and stored in a register or memory location.

additive colors — Red, green, and blue.

address — Physical location of a data item (or the first element of a set of data items) within a storage or output device.

address bus — Portion of a bus used to transmit the address of a storage location to be read from or written to.

address mapping — Synonym of address resolution.

address resolution — As performed by the control unit, converting relative or segmented address references into corresponding physical memory addresses.

addressable memory — Maximum amount of memory that can be addressed physically by the CPU.

Advanced Intelligent Tape (AIT) — Magnetic tape standard developed by Sony based on Digital Audio Tape technology.

AIT — Acronym of Advanced Intelligent Tape.

AITP — Acronym of Association for Information Technology Professionals.

algorithm — (1) Series of processing steps that describes the solution to a problem. (2) The sequence of instructions within a computer program that implements a complex processing operation.

allocation unit — Smallest unit of storage that can be allocated to an object such as a file or a program.

ALU — Acronym of arithmetic logic unit.

AM — Acronym of amplitude modulation.

American Standard Code for Information Interchange (ASCII) — Standard 7-bit coding scheme used to represent character data and a limited set of I/O device control functions.

amplifier — Hardware device used to increase the amplitude or power of a signal or waveform.

amplitude — Magnitude (highest value) of wave peaks of a waveform.

amplitude modulation (AM) — Data transmission method that encodes data values as variations in the amplitude of a carrier wave.

amplitude shift keying (ASK) — Synonym of amplitude modulation.

analog signal — (1) Signal in which any value of an infinite set can be encoded by modulating one or more carrier wave parameters in direct proportion to the encoded value. (2) Signal that varies continuously in one or more carrier wave parameters, such as frequency or amplitude.

analog-to-digital converter (ADC) — Processing device that generates a numerical representation of an analog waveform, *see* sampling.

AND — (1) Boolean operation that generates the value false unless both values are true. (2) Central processing unit instruction that generates the value false unless both input values are true, and stores the result in a register or memory location.

applet — Program written in the Java programming language that is executed by another program such as a Web browser.

application demand model — Model of the demand for computer resources by one or more application programs.

application development software — Software tools such as compilers and CASE tools used to develop application software.

application layer — Application programs (software), as compared to systems software and computer hardware.

application program — Program that addresses a single specific need of a user or a narrowly defined class of information processing tasks.

application programmer — Person who creates or maintains application software.

application software — Class of software consisting of all application programs.

arithmetic logic unit (ALU) — Circuitry within a central processing unit that performs computation, comparison, and boolean logic operations.

arithmetic SHIFT — SHIFT operation used to perform division or multiplication.

array — Set of data items stored in consecutive storage areas and identified by a common name.

ASCII — Acronym of American Standard Code for Information Interchange.

ASK — Acronym of amplitude shift keying.

assembler — Program that translates assembly language source code instructions into object code or central processing unit instructions.

assembly language — Programming language that uses mnemonics to represent central processing unit instructions and storage address operands.

Association for Computing Machinery (ACM) — Professional organization for computer scientists, programmers, and engineers.

Association for Information Technology Professionals (AITP) — Professional organization for information system managers and specialists.

asynchronous transfer mode (ATM) — Protocol for high-speed data transmission that uses hardware-based packet switching and routing and provides quality of service guarantees.

asynchronous transmission — Any transmission method in which sender and receiver do not use continually synchronized clocks.

ATM — Acronym of asynchronous transfer mode.

attenuation — Reduction in signal amplitude or power that occurs as the signal travels through a transmission medium.

audio response unit — Device that provides natural language instructions or responses to input from the keypad of a telephone or other device.

average access time — Statistical average (or mean) elapsed time required by a storage device to respond to a read or write command.

back-end CASE tool — Software tool that creates source code instructions from a system model.

band — Portion of the message-carrying capacity of a carrier wave or transmission medium, typically expressed as a range of frequencies, such as the band between 200 and 1,000 hertz.

bandwidth — Range of carrier wave frequencies that can be transmitted over a transmission medium.

bar code scanner — Device that optically scans surfaces for numbers encoded in sequences of dark bands or squares on a light background.

base — Multiplier that describes the difference between one position and the next in a numbering system.

---

baseband — (1) Transmission method that uses the entire bandwidth of a communication channel to carry one signal or data stream. (2) Local area network transmission method in which multiple digital signals are carried over a single communication channel through time-division multiplexing.

BCC — Acronym of block check character.

BCD — Acronym of Binary Coded Decimal.

benchmark — Program or program fragment and related data inputs that are executed to test the performance of a hardware component, a group of hardware or software components, or an entire computer system or network.

big endian — Term describing the architecture that stores a multibyte data item with the most significant byte in the storage location with the lowest-numbered address.

Binary Coded Decimal (BCD) — Antiquated 6-bit character coding method used by early IBM computer systems.

binary compatibility — Characteristic of two CPUs or computer systems that allows machine language programs to execute on both with equivalent behavior.

binary signal — Signal in which one of two values is encoded by modulating a carrier wave parameter.

bit — (1) Value represented in a one position of a binary number. (2) Number that can have a value of zero or one. (3) Abbreviation of a binary digit.

bit map — Image representation in which each pixel is represented by one or more bits.

bit string — Sequence of bits that describes a single data value.

bit time — Time interval during which the value of a single bit is present within a carrier wave.

block — (1) Series of logical records grouped on a storage device for efficient processing, storage, or transport. (2) Unit of data transfer between a storage device and another computer hardware device. (3) Portion of a program that is always executed as a unit.

block check character (BCC) — Extra 8-bit value containing error-detection information that is appended to the end of a set, or block, of transmitted data items.

block checking — Synonym of longitudinal redundancy check.

block size — Number of bits, bytes, or records within a single storage block or allocation unit.

blocked state — State of a process that is waiting for an event such as completion of another process, arrival of data input, or correction of an error condition.

blocking — Storage of multiple data items or data item fragments within a single storage block or allocation unit.

blocking factor — The number of bits, bytes, or logical records grouped within a single physical record, storage block, or allocation unit.

Bluetooth — Wireless networking standard used to form ad-hoc networks using radio frequency transmission over distances up to 10 meters.

boolean data type — Data item that may store only the value true or false.

boolean logic — Logic that a data item may have only the values of true and false.

BRANCH — (1) Processor operation that causes an instruction other than the next sequential instruction to be fetched and executed. (2) Processor instruction that implements a branch operation by overwriting the current value of the instruction pointer.

bridge — Device that copies or repeats packets from one network segment to another.

broadband — (1) Communication channel with a wide frequency range and high data-transmission capacity. (2) Network transmission method in which multiple digital signals are carried over a single channel through frequency division multiplexing.

broadcast mode — Transmission method that sends a message from a sender to all devices on a communication channel.

brownout — Reduction in voltage by an electrical power provider, usually because of temporary excess of power demand over available supply.

buffer — Primary storage used to temporarily store data in transit from a sender to a receiver.

buffer overflow — Receipt of more data than can be stored in a buffer.

bus — Communication channel shared by multiple devices within a computer system or network.

bus arbitration unit — Processor that mediates competing demands for control of or access to a bus.

bus clock — Clock circuit that generates timing pulses, which are transmitted to all devices attached to a bus.

bus cycle — (1) Period of time required to perform one data transfer operation on a bus. (2) Elapsed time between two consecutive pulses of a bus clock.

bus master — (1) Any device attached to a bus that can initiate a data-transfer operation or send a command to another device. (2) Device that controls a data-transfer operation during a specific bus clock cycle.

bus protocol — Communication protocol used by all devices attached to a bus.

bus slave — (1) Recipient of a command or data from a bus master. (2) Device that is incapable of acting as a bus master.

bus topology — With respect to data communication networks, a network implementation that uses a bus as the communication channel.

business logic layer — Software layer or component that implements the rules and procedures of business processing.

byte — String of eight bits.

cache — Area of high-speed memory that holds portions of data also held within another storage device. A cache is used to improve the average speed of read and write operations to or from its associated storage device.

cache controller — Processor or software that manages cache content.

cache hit — Access to data in a storage device also stored in a cache.

cache miss — Access to data in a storage device that is not also stored in cache.

cache swap — Operation that copies data from a cache to its associated storage device and then overwrites the cache copy with new data.

call — Instruction that executes a function, subroutine, or procedure.

capital expenditure — Funds expended to obtain a capital resource.

capital resource — A resource expected to provide benefits beyond the current operating period.

Carrier Sense Multiple Access/Collision Detection (CSMA/CD) — Media access control and transmission protocol that allows message collisions to occur but defines a recovery method.

carrier wave — Energy wave such as electricity or light that carries encoded messages within a communication channel.

CASE — Acronym of computer assisted software engineering.

Category 5 — Widely implemented twisted pair wiring standard that consists of four twisted pairs that transmit at speeds of up to 1 Gbps.

cathode ray tube (CRT) — Video display device that uses electron beams to excite phosphor(s) on the inner surface of a vacuum tube.

CD-DA — Acronym of compact disc–digital audio.

CD-R — Acronym of compact disc–recordable.

CD-ROM — Acronym of compact disc read-only memory.

CD-RW — Acronym of compact disc–read/write.

central processing unit (CPU) — Computer system hardware component that fetches and executes instructions.

character — (1) Primitive and indivisible component of a written language. (2) One byte of ASCII-encoded data or two bytes of Unicode-encoded data.

channel — (1) Synonym of input/output channel. (2) Shortened form of communication channel.

character framing — Serial data communication technique that surrounds encoded characters with specific bit patterns for purposes of transmission and error detection.

child process — Process created and controlled by another executing process (the parent).

chief information officer (CIO) — The manager charged with responsibility for planning, maintenance, and operation of all information-processing resources within an organization.

chromatic depth — Synonym of chromatic resolution.

chromatic resolution — (1) Number of bits used to describe the color of each pixel in a graphic display. (2) Number of different colors that may be assigned as the value of any single pixel.

CIO — Acronym of chief information officer.

circuit switching — Technique for allocating transmission lines or circuits to senders and receivers whereby a sender and receiver are granted exclusive use of a communication channel for a time period.

CISC — Acronym of Complex Instruction Set Computing.

class — Data structure that contains both traditional data elements and the software that manipulates the data elements.

client — Program or computer that requests and receives services from another program or computer through network communication.

client-server architecture — Organization of software and hardware components as clients and servers that interact through a communication network.

clock cycle — Frequency at which a system clock generates timing pulses.

clock rate — (1) Rate at which clock pulses (clock ticks) are generated by a clock, stated in hertz. (2) With respect to a central processing unit, the time required to fetch and execute the simplest instruction.

CMY — Acronym of cyan, magenta, yellow.

CMYK — Acronym of cyan, magenta, yellow, black.

coaxial cable — Transmission medium composed of a single strand of wire surrounded by an insulator, a braided return wire, and a tough outer coating.

code — (1) Act of creating program instructions. (2) A program or set of related source or machine code instructions.

code generator — Synonym of back-end CASE tool.

coder — Synonym of programmer.

coercivity — Ability of an element or compound to accept and hold a magnetic charge.

collating sequence — Order of symbols produced by sorting them according to a numeric interpretation of their coded values.

collision — Result of simultaneous transmission by two senders on a shared communication channel.

COM+ — Acronym of Component Object Model Plus.

command language — A language used to direct an operating system's actions.

command layer — System software component that accepts and processes user commands to manipulate software, stored data, and hardware resources.

Common Object Request Broker Architecture (CORBA) — Nonproprietary standard that allows software components to locate and communicate with one another.

communication channel — Combination of a transmission medium and communication protocol.

communication protocol — Set of communication rules and conventions encompassing data and command representation, bit encoding and transmission, transmission media access, clock synchronization, error detection and correction, and message routing.

compact disc–digital audio (CD-DA) — Standard format for storing and distributing music on compact disc.

compact disc–read only memory (CD-ROM) — Standardized 120 millimeter read only optical disc.

compact disc–read/write — Standard phase-change optical storage method and medium.

compact disc–recordable (CD-R) — Compact disc containing a laser-sensitive dye that can be written with a laser switched between high and low power.

compaction — Process of reallocating storage locations assigned to a file to eliminate empty storage locations.

competitive advantage — Use of resources to provide better or cheaper services than others can provide with those same resources.

compiler — Program that translates high-level programming language source code instructions into object code.

compiler library — Set of object-code modules intended to be linked with object-code modules produced by a compiler.

complete path — Named path from the root node of a storage volume or name space to a specific file or object through intervening layers of a hierarchical directory structure.

Complex Instruction Set Computing (CISC) — Processor architecture that uses complex instructions that embed many primitive processing operations, in contrast to reduced instruction set computing.

complex instructions — Instructions that combine primitive processing operations.

component — Executable software module or object with a well-known interface and a unique identifier.

Component Object Model Plus (COM+) — Microsoft proprietary standard that allows software components to locate and communicate with one another.

compression — Reduction in data size using a compression algorithm.

compression algorithm — Algorithm by which data inputs may be translated into equivalent outputs of smaller size.

compression ratio — Ratio of data size before and after applying a compression algorithm.

computer assisted software engineering (CASE) tool — Program or integrated set of programs used to analyze, design, and develop computer software.

computer network — Hardware and software that allows multiple users and computer systems to share information, software, and other resources.

computer operations manager — Manager responsible for the operation and maintenance of a large data-processing center or information system.

computer science — Study of the implementation, organization, and application of computer software and hardware resources.

concurrent execution — Execution of multiple processes or threads within the same immediate time frame through interleaved execution on a single processor.

condition — A comparison or other logical operation that produces a boolean result.

conditional BRANCH — (1) BRANCH operation that is performed only if a boolean variable is true. (2) Processor instruction that implements a conditional BRANCH operation.

conductivity — Ability of a transmission medium to allow an energy wave such as electricity or light to pass through.

conductor — Element or compound that exhibits conductivity.

connection-oriented protocol — Communication protocol that explicitly requires a connection to be established between sender and receiver prior to transmitting any data.

connectionless protocol — Communication protocol that does not require a connection to be established between sender and receiver prior to transmission of any data.

contiguous — Storage of a set of data elements in sequential physical storage locations.

control bus — Portion of a bus used to transmit command and status signals among devices.

control structure — High-level programming or job control language statement that describes the selection or repetition of other program statements, such as an If-Then or While-Do statement.

control unit — Component of a central processing unit that is responsible for moving data, accessing instructions, and controlling the arithmetic logic unit.

CORBA — Acronym of Common Object Request Broker Architecture.

core memory — Antiquated form of primary storage implemented as a lattice of wires with iron rings wrapped around each wire junction point.

CPU — Acronym of central processing unit.

CRC — Acronym of cyclical redundancy checking.

CRT — Acronym of cathode ray tube.

CSMA/CD — Acronym of Carrier Sense Multiple Access/Collision Detection.

current directory — Directory that serves as the origin for file and directory search operations.

cursor — Visual symbol on a video display, such as a colored box or underline, that indicates the current input or output position.

cycle — Single operation, function, or transfer within a series of similar operations, functions, or transfers, as performed by a processor, storage device, or communication channel.

cycle time — Duration of one cycle.

cyclical redundancy checking (CRC) — Technique for detecting errors in data transmission based on redundant data computed with a complex algorithm and appended to the original message.

DAC — Acronym of digital-to-analog converter.

DAT — Acronym of Digital Audio Tape.

data bus — Portion of a bus that transmits data among devices.

data declaration — Programming language statement that declares the existence, name, type, and other characteristics of a data item or structure.

data layer — Software layer or component that manages and interacts with a permanent data store.

data link layer — Layer of network system software responsible for media access and bit transmission.

data operation — Any processing operation that uses or alters data content.

data striping — Data storage method that improves read/write performance by dividing data into smaller segments and storing the segments on multiple disks.

data structure — A data item, such as an array or linked list, that contains multiple primitive data elements or other data structures.

data transfer rate — Rate at which data is transmitted through a medium or communication channel, as measured in data units per time interval.

database administrator — Manager responsible for ensuring data integrity, reliability, security, and availability.

datagram — Smallest unit of data transfer defined by the Transmission Control Protocol.

DCE — Acronym of Distributed Computing Environment.

DDS — Acronym of Digital Data Storage.

debugging tools — Type of application development software that simulates program execution and traces errors.

debugging version — Program that contains data and instructions to aid in locating and correcting errors.

decimal point — Period or comma in the decimal numbering system that separates the whole and fractional parts of a numeric value.

decoding — Process within the central processing unit of extracting the op code and operands, routing data and control signals, and signaling the ALU.

decompression algorithm — Algorithm that returns compressed data to its original state.

design model — Architectural blueprint for system implementation that is an output of the design phase.

device controller — Processor that controls the physical actions of one or more storage devices.

differential backup — Redundant data store containing copies of changes to data content that have occurred since the most recent incremental or full backup.

Digital Audio Tape (DAT) — Early magnetic tape technology on which Digital Data Storage standards are based.

Digital Data Storage (DDS) — Family of magnetic tape standards developed by Hewlett-Packard and Sony, and based on Digital Audio Tape.

Digital Linear Tape (DLT) — Magnetic tape standard developed by Quantum.

digital signal — (1) Signal in which one of a discrete, or countable, set of values can be encoded by modulating one or more carrier wave parameters. (2) Synonym of binary signal.

digital signal processor (DSP) — Processor that manipulates continuous streams of digital data, such as audio or motion video.

digital versatile disc (DVD) — Synonym of digital video disc.

digital video disc (DVD) — Standard optical disc format for distributing movies and other audiovisual content.

digital video disc random access memory (DVD-RAM) — Standard for rewriteable digital video discs with a capacity of 2.4 GB for single-sided discs and 4.8 GB for double-sided discs.

digital video disc read only memory (DVD-ROM) — Standard format for general-purpose read-only data storage on digital video disc.

digital-to-analog converter (DAC) — Processing device that generates an analog signal from a numerical representation of the signal.

digitizer — (1) Device that captures the position of a pointing device as input data. (2) Synonym of optical scanner.

DIMM — Acronym of double in-line memory module.

DIP — Acronym of dual in-line package.

direct access — Synonym of random access.

direct memory access (DMA) — Method of data transfer that allows direct data movement between main memory and a storage or I/O device, bypassing the central processing unit.

direct sequence spread spectrum (DSSS) — Radio transmission method established in the IEEE 802.11 standard that was enhanced in the 802.11b standard to provide raw data transmission at 5.5 and 11 Mbps.

directory — Data structure that contains information about files and other directories.

directory services — System software (middleware) that enables user and components to locate and access remote resources.

discrete signal — Synonym of digital signal.

disk defragmentation — Reorganization of data on a disk drive so that all portions of an individual file are stored in sequential sectors, tracks, and platters.

disk mirroring — Storage technique that enhances reliability by storing a redundant copy of all data on a second disk drive.

diskette — Synonym of floppy disk.

dispatching — (1) Act of giving a process or thread control of the central processing unit. (2) Act of moving a process or thread into the running state.

display attribute — Feature of printed or displayed output, such as background color and boldface type.

display list — Group of high-level commands that describe a printed or displayed image.

display objects — Primitive graphic object, such as a line or rectangle, as described or manipulated within a display list.

distortion — The undesirable alteration of one or more carrier wave characteristics resulting from interaction between signal energy and the transmission-medium or signal-propagation equipment.

distributed computing — Synonym of distributed processing.

Distributed Computing Environment (DCE) — Standard for distributed processing that covers network directory services, file sharing services, remote procedure calls, remote thread execution, system security, and distributed resource management.

distributed processing — Distribution and execution of processes or threads over multiple nodes of a computer network.

distribution version — Synonym of production version.

dithering — Simulation of continuous color tones using patterns of dots containing a finite set of colors.

DLT — Acronym of Digital Linear Tape.

DLL — Acronym of dynamically linked library.

DMA — Acronym of direct memory access.

DMA controller — Dedicated processor that handles the processing of data transfers during direct memory access operations.

dot matrix printer — Impact printer that generates printed characters using a matrix of pins and an inked ribbon.

dots per inch (dpi) — Measure of print or display resolution (pixel density) for a printer or video display device.

double in-line memory module (DIMM) — Small printed circuit board with memory chips and electrical contacts on both sides.

double precision — Representation of a numeric value with twice the usual number of bit positions for greater accuracy or numeric range.

doubly linked list — Set of stored data items in which each element contains pointers to both the previous and next list elements.

DRAM — Acronym of dynamic random access memory.

drive array — Set of disk drives that are managed as if they were a single storage device.

DSSS — Acronym of direct sequence spread spectrum.

DSP — Acronym of digital signal processor.

dual in-line packages (DIPs) — Early form of packaging for processor or memory circuits with two rows of electrical contact pins

DVD — Acronym of both digital video disc and digital versatile disc.

DVD-RAM — Acronym of digital video disc random access memory.

DVD-ROM — Acronym of digital video disc read-only memory.

dynamic analysis — Use of models or model analysis techniques that incorporate parameter variability and random events in representing system behavior.

dynamic linking — Linking process performed during program loading or execution.

dynamically linked library (DLL) — (1) A library of reusable software modules organized for dynamic linking. (2) A specific Microsoft Windows file format for storing reusable software modules.

dynamic connection — Connection between a client and a server or remote resource that is not established until the client accesses the server or resource.

dynamic random access memory (DRAM) — Type of electronic memory circuit that implements bit storage with transistors and capacitors.

early binding — Synonym of static linking.

EBCDIC — Acronym of Extended Binary Coded Decimal Interchange Code.

EEPROM — Acronym of electronically erasable programmable read only memory.

EDRAM — Acronym of enhanced dynamic random access memory.

effective data transfer rate — Error-free throughput of a communication channel under normal conditions with a specific communication protocol.

EL — Acronym of electroluminescent.

electroluminescent (EL) display — Display device that generates light by applying electrical power to solid-state elements coated with phosphors.

electromagnetic interference (EMI) — Alteration of electrical carrier wave characteristics caused by external electrical or magnetic phenomena, such as electric motors or sun spots.

electronically erasable programmable read-only memory (EEPROM) — Read-only memory device that can be erased and written by sending appropriate control signals.

EMI — Acronym of electromagnetic interference.

enhanced dynamic random access memory (EDRAM) — Type of electronic memory circuit primarily composed of DRAM but supplemented with a small SRAM cache.

Ethernet — Widely implemented local area network standard family.

even parity — Error-detection method that appends a parity bit to each transmitted or stored character. The value of the parity bit is one if the number of one valued bits in a character is even; it is zero otherwise.

excess notation — Coding method that represents integer values as bit strings such that the value zero (the midpoint of the numeric range) is represented by all zero bits except for the most significant bit (which contains a one).

exclusive OR (XOR) — (1) Boolean operation that generates the value true if only one of its inputs is true. (2) CPU instruction that generates the value true if only one of its inputs is true.

executable code — Program that consists entirely of CPU instructions that are ready to be loaded and executed.

executing — Process of carrying out an instruction within the central processing unit.

execution cycle — (1) Process of carrying out an instruction in the central processing unit and completing the operation it specifies. (2) Same process described in (1) as applied to all instructions in a program.

explicit priority — Process scheduling through explicitly stated priority levels and a corresponding scheduling algorithm.

Extended Binary Coded Decimal Interchange Code (EBCDIC) — Standard coding system for representing character data in an 8-bit format, most commonly used with IBM mainframe computers.

extensible markup language (XML) — Standard Web format that encodes document structure and content.

external function call — Placeholder for missing executable code.

FAT — Acronym of file allocation table.

father — Original version of a file, after updates have been applied to generate a new version (the son).

fault tolerance — Characteristic of a computer system or file management system that allows rapid recovery from the failure of a hardware component without data loss.

FCFS — Acronym of first come, first served.

FDM — Acronym of frequency division multiplexing.

ferroelectric random access memory — Type of nonvolatile random access memory that uses iron to store bit values.

fetch cycle — Portion of a central processing unit cycle in which an instruction is loaded into the instruction register and decoded.

FHSS — Acronym of frequency hopping spread spectrum.

field — Component data item of a record.

fiber-optic cable — Transmission medium for optical signal propagation, generally consisting of one or more plastic or glass fibers sheathed in a protective plastic coating.

fifth generation language (5GL) — Nonprocedural programming language used to develop software that mimics human intelligence.

file allocation table (FAT) — Data structure used in MS-DOS and early Microsoft Windows versions to record the allocation of storage device locations to files and directories.

file association — Relationship between file types and the programs or operating system utilities that manipulate them.

file close operation — Operation that severs the relationship between a file and a process, flushes file I/O buffers, and deletes operating system data structures associated with the file.

file control layer — System software layer that accepts file manipulation requests from application programs and translates them into corresponding low-level processing commands to the central processing unit and secondary storage device drivers.

file management system (FMS) — Entire collection of system software that performs file management, usually part of the operating system.

file migration — Management technique for secondary storage in which older versions of a file are moved automatically to less costly storage media or devices such as magnetic tape.

file open operation — Process of associating a file with an active process by allocating buffers and updating internal tables.

File Transfer Protocol (FTP) — Older Web protocol that specifies a client-server request and response language for copying files from one Internet host to another.

firmware — Software that has been permanently stored in read-only memory devices.

first come, first served (FCFS) — Scheduling method that services processes in order of their request arrival.

first generation language (1GL) — Synonym of machine language or central processing unit instruction set.

fixed length instruction — One member of an instruction set in which all instructions contain the same number of bits.

flag — Single bit within the program status word that represents one data item or condition.

flash memory — Form of electronically erasable read-only memory that requires relatively little time to update memory contents.

flash RAM — Synonym of flash memory.

flat memory model — Memory organization and access method in which memory locations are described by single unsigned integers that corresponds to linear position.

flat panel display — Display device that is thin, flat, and lightweight.

floating point notation — Method of encoding real numbers in a bit string consisting of two parts – a mantissa and exponent.

floppy disk — Small, removable magnetic disk storage medium encased in a protective cover.

FM — Acronym of frequency modulation.

FMS — Acronym of file management system.

font — Named set of display formats for printable characters and symbols with a similar appearance or style.

formula — Processing problem that can be solved by executing a fixed sequence of instructions.

fourth generation language (4GL) — Programming language that supports nonprocedural programming, database manipulation, and advanced I/O capabilities.

fragmented — File or storage device characteristic in which storage locations allocated to one or more files are scattered in noncontiguous locations.

frequency — Number of complete waveform transitions (change from positive to negative to positive energy peak) that occur in one second.

frequency division multiplexing (FDM) — Communication channel-sharing technique in which a single broadband channel is partitioned into multiple baseband subchannels (frequency bands), each of which can carry a separate data stream.

frequency hopping spread spectrum (FHSS) — Outdated RF transmission method established in the first IEEE 802.11 standard.

frequency modulation (FM) — Data transmission method that encodes data values as variations in the frequency of a carrier wave.

frequency shift keying (FSK) — Synonym of frequency modulation.

front-end CASE tool — Program that supports creating and modifying system models.

front end processor — Synonym of I/O channel, particularly as applied to minicomputers and mainframes not manufactured by IBM.

FSK — Acronym of frequency shift keying.

FTP — Acronym of File Transfer Protocol.

full backup — Data store containing copies of all data in a directory or storage device.

full-duplex — Type of communication channel composed of two transmission media that allows simultaneous communication in both directions.

fully qualified reference — Synonym of complete path with respect to a specific file or object.

function — Named segment of a high-level language program that is always executed as a unit.

gallium arsenide — Compound used to implement devices that have both electrical and optical properties.

gate — Processing device that implements a primitive Boolean operation or processing function by physically transforming one or more input signals.

gateway — Computer system or hardware device that is connected to two network segments and forwards packets from one segment to another.

general-purpose processor — Processor with capabilities that can be used to perform a wide variety of processing tasks.

GHz — Abbreviation of gigahertz.

GIF — Acronym of Graphics Interchange Format.

Gigabit Ethernet — High-speed Ethernet standard based on the 802.3z and 802.3ab standards.

gigahertz (GHz) — Unit of wave or clock frequency, one billion cycles per second.

grandfather — Version of a file prior to the father version.

graph directory structure — (1) Directory structure in which allowable relationships between directories can be represented as a graph. (2) Directory structure in which a file or subdirectory can be contained within two or more directories.

Graphics Interchange Format (GIF) — Standard format for storing compressed images.

Grosch's Law — Outdated mathematical relationship between computer size and cost per unit of instruction execution that states that cost per executed instruction decreases as computer system size increases.

half-duplex — Type of communication channel that uses only a single transmission medium where sender and receiver take turns transmitting data to one another.

hard disk — (1) Storage medium consisting of a rigid platter coated with a metallic oxide on which data are recorded as patterns of magnetic charge. (2) Synonym of hard disk drive.

hardware independence — Independence of a program or processing method from the physical details of computer system hardware.

hardware monitor — Program or device that records and reports processing or communication activity within or between hardware devices.

HALT — Instruction that suspends the execution of any further instructions by the central processing unit.

head-to-head switching time — Elapsed time required to switch shared read/write circuitry between two adjacent read/write heads.

heat dissipation — Act of conducting heat away from a device, thus reducing its temperature.

heat sink — Thermal mass placed in direct contact with a processor or other device to improve heat dissipation.

helical scanning — Tape recording method in which data is read and written by rotating the read/write head at an angle, moving from tape edge to tape edge.

hertz (Hz) — Unit of measure for signal frequency defined as one cycle per second.

hexadecimal notation — Numbering system with a base value of 16, which uses digit values ranging from 0 to 9 and from A to F (corresponding to decimal values 0 to 15).

hierarchical directory structure — Multilevel system of directories in which directories and files may be related to one another in a hierarchy or inverted tree.

high order bit — Synonym of most significant digit.

hit ratio — Ratio of cache hits to read accesses.

home directory — Primary directory associated with and owned by a single user.

Home Radio Frequency (HomeRF) — Home networking standard for devices connected within a 50 meter range.

HomeRF — Acronym of Home Radio Frequency.

HTML — Acronym of Hypertext Markup Language.

HTTP — Acronym of Hypertext Transport Protocol.

hub — Network interface device that serves as the connection point for multiple computer systems to a network.

Hypertext Markup Language (HTML) — Device-independent formatting language that is used to describe Web documents and in which links to other documents can be embedded.

Hypertext Transport Protocol (HTTP) — Protocol that specifies the language by which clients request HTML documents and how servers respond to those requests.

Hz — Abbreviation of hertz.

IA5 — Acronym of International Alphabet Number 5.

IC — Acronym of integrated circuit.

ICMP — Acronym of Internet Control Message Protocol.

IEEE — Acronym of Institute for Electrical and Electronics Engineers.

IEEE 802 standards — Standard family for voice, data, and computer networks.

IIOP — Acronym of Internet Inter-ORB Protocol.

image description language — Language that symbolically describes the content of a printed or displayed image.

IMAP — Acronym of Internet Message Access Protocol.

inclusive OR — Synonym of OR operation.

incremental backup — Data store containing copies of data sets (files) that have been altered since the most recent incremental or full backup.

index — Stored set of paired data items. Each pair contains a key value and a pointer to the location of the data item possessing (or corresponding to) that key value.

indirect addressing — Any addressing method where memory references made within a program do not necessarily correspond to physical memory storage locations.

infrared — Electromagnetic energy waves approximately within the frequency range 1 to 100 terahertz.

Infrared Data Association (IrDA) — Home networking standard based on infrared transmission.

infrastructure — Facilities or resources that provide services that are pervasively available and commonly used by a large numbers of users.

ink-jet printer — Printer that produces printed images by spraying small drops of liquid ink onto paper.

input output (I/O) unit — Device that enables communication between a user and software or among computer systems or their components.

Institute for Electrical and Electronics Engineers (IEEE) Computer Society — Subgroup of the IEEE that specializes in computer and data communication technology.

instruction — Bit string containing an operation code or one or more operands that cause a processor to perform a processing operation.

instruction cycle — Synonym of fetch cycle.

instruction explosion — Correspondence of a single high-level programming language statement to multiple central processing unit instructions.

instruction format — Length and order of the operation code and operands within a machine language instruction.

instruction pointer — Register that stores the address of the next instruction to be fetched from primary storage.

instruction register — Register that holds an instruction prior to decoding.

instruction set — Set of all machine language instructions that can be executed by a central processing unit.

integer — Whole number, or a value that does not have a fractional part.

integrated circuit (IC) — Semiconductor device, manufactured as a single unit, which incorporates multiple gates.

interleaved execution — Scheduling technique by which a processor alternates instruction execution among multiple active processes.

International Alphabet Number 5 (IA5) — International equivalent of ASCII.

International Standards Organization (ISO) — International body with functions similar to those of the American National Standards Institute.

Internet Control Message Protocol (ICMP) — Internet protocol for exchanging messages among gateways and routers.

Internet — Global collection of networks interconnected using TCP/IP.

Internet Inter-ORB Protocol (IIOP) — Component message passing protocol in CORBA.

Internet Message Access Protocol version 4 (IMAP4) — Protocol that extends Post Office Protocol 3 to enable permanent storage and management of e-mail messages on a server.

Internet Protocol (IP) — Internet protocol for packet switching and routing.

interpretation — Process for translating and executing source code instructions that interleaves instruction translation and execution.

interpreter — Program that performs interpretation.

interrupt — Signal to the CPU that some event requires its attention.

interrupt code — Numerically coded value of an interrupt, indicating the type of event that has occurred.

interrupt handler — Program or subroutine that is executed in response to an interrupt.

interrupt register — Register in the control unit that stores an interrupt code received over the bus or generated by the central processing unit.

intranet — Internal, private network that uses Internet protocols but is only accessible to a limited set of internal users.

I/O — Acronym of input/output.

I/O channel — (1) Device controller dedicated to a mainframe bus port that allows many devices to share access to (and the capacity of) the port. (2) Specific hardware component of an IBM mainframe computer system.

I/O port — Communication pathway from the CPU to a peripheral device.

I/O wait state — An idle processor cycle consumed while waiting for data transmission from a secondary storage or input/output (I/O) device.

IP — Acronym of Internet Protocol.

IPv6 — Internet Protocol standard version 6 that uses 128-bit addresses.

IrDA — Acronym of Infrared Data Association.

ISO — Acronym of International Standards Organization.

Java — Object-oriented programming language used to construct programs that can be executed on different hardware and system software platforms.

Java virtual machine (JVM) — Interpreter that translates and executes Java programs.

JCL — Acronym of job control language.

job control language (JCL) — Control language used to direct the batch execution of programs or groups of programs.

JPEG — Acronym of Joint Photographic Experts Group.

Joint Photographic Experts Group (JPEG) — (1) Organization that promulgates image representation standards. (2) Standard method of compressed image representation for single images.

journaling — Synonym of transaction logging.

JVM — Acronym of Java virtual machine.

JUMP — Synonym of BRANCH operation.

jumps — Synonym of branches.

Kerberos — Security model that defines interactions between clients, services, and a security service.

kernel — Portion of the operating system that manages resources and directly interacts with computer hardware.

L1 — Acronym of level one.

L2 — Acronym of level two.

L3 — Acronym of level three.

label — Mnemonic that represents the memory address of a program instruction.

LAN — Acronym of local area network.

laser printer — Printer that uses a laser to charge areas of a photoconductive drum and paper.

late binding — Synonym of dynamic linking.

Latin-1 — Standard character coding table containing the ASCII character set in the first 128 table entries and most of the additional characters used by Western European languages in the second 128 table entries.

LCD — Acronym of liquid crystal display.

LDAP — Acronym of Lightweight Directory Access Protocol.

least significant byte — Byte within a multiple byte data item that contains the digits of least, or smallest, magnitude.

least significant digit — Bit position within a bit string that represents the least (smallest) magnitude.

level one (L1) cache — Within-CPU cache when all three cache types are used.

level two (L2) cache — Primary storage cache implemented within the same chip as a processor.

level three (L3) cache — When three levels of primary storage cache are used, the cache implemented outside the microprocessor.

Lightweight Directory Access Protocol (LDAP) — Formal Internet standard formed from the X.500 standard and adopted by the Internet Engineering Task Force.

line turnaround — Signal sent between a sender and receiver that causes them to reverse roles (the sender becomes the receiver and vice versa).

linear address space — Logical view of a secondary storage or I/O device as a sequentially numbered set of storage locations.

linear recording — Recording of data onto a tape in which bits are placed along parallel tracks that run along the entire length of the tape.

Linear Tape Open (LTO) — Proprietary magnetic tape standard developed by Hewlett-Packard, IBM, and Seagate.

link — (1) A type of connection between a directory and another directory or file used to implement a graph directory structure. (2) A pointer that connects two data items within a data structure.

link editor — Program that combines multiple object code modules into an integrated set of executable code with a consistent scheme of memory addresses and references.

link map — Synonym of memory map.

linked list — Data structure in which each data item contains a pointer to the previous or next data item.

liquid crystal display (LCD) — Display device that uses liquid crystals that can be changed from transparent to opaque.

little endian — Term describing a memory architecture that stores a multibyte data item with the least significant byte in the storage location with the lowest-numbered address.

load — Act of copying a word from primary storage to a register.

local area network (LAN) — Network that spans a limited area such as a single building or office floor.

location transparency — Characteristic of software such that resource access is implemented identically whether the resource is part of the local computer system or located within or controlled by a remote computer system.

logic instruction — Instruction that implements a boolean operation or comparison operation.

logical access — Access to a storage location within a linear address space.

logical record — Record formatted and described in terms of a logical (rather than physical) file structure.

logical SHIFT — Shift operation used to extract a single bit from a bit string.

logical topology — Path that messages traverse as they travel among network nodes.

long integer — Double precision representation of an integer.

longitudinal redundancy check (LRC) — Parity checking method in which parity bits are determined based on equivalent bit positions in a group of bytes or characters.

lossless compression — Compression algorithm in which data content is unchanged when compressed and then decompressed.

lossy compression — Compression algorithm in which data content is altered when compressed and then decompressed.

low order bit — Synonym of least significant digit.

LRC — Acronym of longitudinal redundancy check.

LTO — Acronym of Linear Tape Open.

MAC — Acronym of media access control.

machine data type — Synonym of primitive data type.

machine independence — Synonym of hardware independence.

machine language — Language consisting solely of instructions from the instruction set of a specific central processing unit.

machine state — Current processing state of a program or the central processing unit, as represented by the values currently held in registers.

magnetic decay — Loss in strength of a stored magnetic charge over time.

magnetic leakage — Reduction in strength of a stored magnetic charge because of interference from one or more adjacent magnetic charges of opposite polarity.

magnetic tape — Ribbon of acetate or polymer coated with a metallic compound used to store data.

magneto-optical — Secondary storage device that reads and writes data bits using a combination of magnetic and optical methods.

main memory — (1) Synonym of primary storage. (2) Set of devices that implement primary storage excluding cache.

mainframe — High-capacity computer system designed to support hundreds of interactive users and processes simultaneously.

Mammoth — Magnetic tape standard developed by Exabyte based on the Digital Audio Tape standard.

mark sensor — Input device that recognizes printed marks at predetermined locations on an input document.

media access control (MAC) — Set of rules that regulate access to a transmission medium.

megahertz (MHz) — Measurement of wave or clock frequency, one million cycles per second.

memory — (1) Synonym of primary storage. (2) Random access memory devices used anywhere within a computer system, peripheral device, or device controller.

memory allocation — Allocation of primary storage resources to active processes.

memory map — Listing of module and data memory addresses produced by a link editor.

message — (1) With respect to communication networks, a command, request, or response sent from one network node to another. (2) Request sent from one object to another.

method — Program within a class that manipulates data.

MFLOPS — Acronym of millions of floating point operations per second.

MHz — Abbreviation of megahertz.

microchip — Semiconductor device that implements integrated electronic components in a single unit.

microcomputer — Computer system that can meet low-intensity processing needs of a single user.

microprocessor — Microchip on which all of the components of a central processing unit are implemented.

Microsoft Active Directory (AD) — Built-in directory service and security system included with Windows 2000.

middleware — (1) Software that binds together components in a client-server application. (2) System software that allows client and servers or distributed components to locate and communicate with one another.

MIDI — Acronym of Musical Instrument Digital Interface.

millions of floating point operations per second (MFLOPS) — Measure of processor or computer system speed in terms of the number of floating point computation operations executed per second.

millions of instructions per second (MIPS) — Measure of processor or computer system speed in terms of the number of central processing unit instructions executed per second.

minicomputer — Computer system that meets the processing needs of a small- to medium-sized group of interactive users.

MIME — Acronym of Multipurpose Internet Mail Extensions.

MIPS — Acronym of millions of instructions per second.

model — Abstract representation of a system or other physical reality.

modem — Abbreviation of modulator-demodulator.

modulating — Varying the amplitude, frequency, or phase of a carrier wave.

modulator-demodulator (modem) — (1) Acronym of MOdulator-DEModulator. (2) Device that translates analog signals into digital signals (and vice versa), allowing computer hardware to use voice-grade telephone lines for data communication.

monitor — (1) Video display device. (2) Hardware or software element that monitors and reports processing or communication activity.

monophonic — Capable of generating only one audible frequency (note) or timbre (voice) at a time.

Moore's Law — Statement that the transistor density and power of microprocessors doubles every 18 to 24 months with no per unit cost increase.

most significant byte — Byte within a multiple byte data item that contains the digits of most, or largest, magnitude.

most significant digit — Bit position within a bit string that represents the most (largest) magnitude.

Moving Pictures Experts Group (MPEG) — (1) Organization that promulgates motion video-image representation standards. (2) Specific method of compressed motion video-image representation.

mouse — Interactive pointing device used to control cursor position and to select objects on a video display.

MOVE — Instruction that can copy data bits among any combination of registers and primary storage locations.

MP3 — (1) Audio encoding and compression standard. (2) Layer 3 of the MPEG-1 standard.

MPEG — Acronym of Moving Pictures Experts Group.

multilevel coding — Data transmission method that encodes data values as variations in the phase of a carrier wave.

multimedia controller — Device controller for multiple audio or visual I/O devices.

multimode cable — Fiber-optic cable type that surrounds plastic or glass fibers with a reflective coating to contain light.

multinational character — Character, such as ñ or é, similar to an English language character but used by Western European languages other than English.

multiple master bus — Bus in which more than one attached device may control access to the bus.

Multipurpose Internet Mail Extensions (MIME) — Protocol that allows non-text content to be included in e-mail messages and other Internet data transmissions.

multitasking — Ability of an operating system to support multiple active processes.

multithreaded — Process divided into two or more threads, each of which may be scheduled and executed independently.

Musical Instrument Digital Interface (MIDI) — Standard method of encoding control information for musical instruments and synthesizers.

n-layer architecture — Client-server architecture that uses more than three layers.

n-tiered architecture — Synonym of n-layered architecture.

NAK — Mnemonic of negative acknowledge.

named pipe — Pipe with a name permanently placed in a file system directory and the ability to communicate among processes on different computers.

native application — Application compiled and linked for a particular central processing unit and operating system.

needs analysis — Synonym of systems analysis.

negative acknowledge (NAK) — ASCII control character used by a receiver to indicate the unsuccessful receipt of transmitted data.

network administrator — Manager charged with operating and maintaining a local or wide area network.

network computer — Personal computer or workstation with no locally stored operating software, application software, or configuration information.

network interface card (NIC) — Synonym of network interface unit (NIU), particularly with respect to a microcomputer, network computer, or workstation.

network interface unit (NIU) — Physical network interface device used by a single computer system.

network layer — Layer of network system software that routes packets to their destination.

network topology — Structure defined by the organization of network devices, routing of network cabling, and flow of messages through network nodes.

network transparency — Synonym of location transparency.

NIC — Acronym of network interface card.

NIU — Acronym of network interface unit.

noise — Combination of electromagnetic interference and distortion as added to a carrier wave during transmission.

noncontiguous — Any storage allocation scheme under which multiple storage locations allocated to a file or program need not be physically adjacent to one another.

nonprocedural language — Programming language that describes a processing requirement without describing a specific procedure for satisfying the requirement.

non-volatile — Term describing storage devices that retain their contents over long periods of time.

NOT — (1) Boolean operation that generates the value false if its input is true, and true if its input is false. (2) Central processing unit instruction that implements the boolean NOT operation and stores the result in a register or memory.

numeric range — Set of all data values that can be represented by a specific data-encoding method.

object — One instance, or variable, of a class.

object code — Output of an assembler or compiler that contains machine instructions and unresolved references to external library routines.

object-oriented programming (OOP) — Software construction method or paradigm that configures software as cooperating objects.

Object Request Broker (ORB) — CORBA Component that maintains a directory of components and routes messages among them.

objectclass — A lightweight directory access protocol (LDAP) concept that defines the attributes common to all members of a class.

OCR — Acronym of optical character recognition.

octal notation — Numbering system with a base value of 8 that uses digit values ranging from 0 to 7.

odd parity — Error-detection method that appends a parity bit to each transmitted or stored character. The value of the parity bit is one if the number of one-valued bits in a character is odd, and zero otherwise.

on-off keying (OOK) — Synonym of pulse code modulation.

offset register — Register that holds a memory address offset, which is added to each explicit memory reference.

OOP — Acronym of object-oriented programming.

OOK — Acronym of on-off keying.

op code — Synonym of operation code.

Open Systems Interconnection (OSI) model — Standardized seven layer architecture for computer network software and hardware.

operand — Component field of an instruction containing a data value or address used as input to or output from a specific processing operation.

operating expenditure — Funds expended during the current operating period to support normal operations.

operating system — Set of software programs that manage and control access to computer resources.

operation code — Coded value or bit string representing the function or operation to be performed by a processor.

optical character recognition (OCR) — Methods, programs, and devices by which printed characters are recognized as computer system input or data.

optical scanner — Device that can scan printed graphic inputs and convert them to computer system input or data.

optical sensor — Device that senses optical input, usually by a reflected laser.

OR — (1) Boolean operation that generates the value true if either or both of its inputs are true. (2) Central processing unit instruction that generates the value true if either or both of its inputs are true.

ORB — Acronym of Object Request Broker.

OSI — Acronym of Open Systems Interconnection.

overflow — Error condition that occurs when the output bit string of a processing operation is too large to fit in the designated register.

packet — Fundamental unit of data communication in a computer network.

packet switching — Form of time division multiplexing in which multiple message streams are broken into packets and individually transmitted and routed to their destination.

page — Small fixed-size portion of a program, swapped between primary and secondary storage under virtual memory management.

page fault — Condition that occurs under virtual memory management when a program references a memory location in a page not currently held in primary storage.

page frame — Portion of primary storage designated to hold a page under virtual memory management.

page hit — Reference to a page held in memory in virtual memory management.

page table — Table of pages and information about them maintained by an operating system that implements virtual memory management.

palette — Set or table of colors used by a display controller or device.

parallel access — Simultaneous access to multiple portions of a data item through multiple communication channels or transmission media.

parallel transmission — Simultaneous transmission of multiple signals over multiple transmission media or channels.

parent process — Process that initiates and controls the execution of another (child) process.

parity bit — Bit appended to a small data unit, such as an ASCII character, that stores redundant information used for error checking. See even parity and odd parity.

parity checking — Act of validating data by recomputing the value of a parity bit.

passive matrix display — Liquid crystal display that shares transistors among rows and columns of pixels.

PC — Acronym of personal computer.

PCB — Acronym of process control block.

PCM — Acronym of pulse code modulation.

peer-to-peer communication protocol — Protocol that allows processes to communicate synchronously across a network.

peripheral device — Device on a system bus other than the central processing unit and primary storage.

peripheral processing unit — Synonym of I/O channel, particularly as applied to minicomputers and mainframes not manufactured by IBM.

personal computer (PC) — Synonym of microcomputer.

phase — Characteristic of an analog wave that indicates its current cycle position (in degrees) with respect to the cycle origin.

phase shift keying (PSK) — Synonym of phase shift modulation.

phase shift modulation — Data transmission method that encodes data values as variations in the phase of a carrier wave.

phoneme — Individual vocal sound comprising a primitive component of human speech.

physical layer — Layer of network system software responsible for physical transmission of signals representing data.

physical memory — Physical primary storage capacity of a computer system, in contrast to virtual memory capacity and addressable memory.

physical record — Unit of physical data transfer to or from a storage device.

physical topology — Physical placement of cables and connections in a network topology.

pipe — Shared memory region used to route the data output of one process to another process as input.

pixel — (1) Abbreviation of the term "picture element." (2) Single unit of data in a graphic image. (3) Single point on a display surface.

platter — Disk with a disk drive, the surface or surfaces of which store data.

plotter — (1) Device that generates printed output through the movement of paper and one or more pens. (2) An ink jet printer that can print wide sheets or rolls of paper.

point — (1) One seventy-second of an inch. (2) Unit of measurement for font size.

pointer — (1) Data element that contains the address (location in a storage device) of another data element. (2) Device used to input positional data or control the location of a cursor.

polyphonic — Capable of simultenously generating many audible frequencies (notes) or timbres (voices).

pop — Process of removing an item from the top of a stack and copying its content to processor registers.

POP3 — Acronym of Post Office Protocol version 3.

port — Physical connection point on a bus, communication channel, or input/output device identified by a unique integer number.

Post Office Protocol version 3 (POP3) — Protocol that standardizes the interaction between e-mail clients and servers.

PostScript — Language for representing and transmitting complex images and display contents.

power sag — Momentary reduction in the voltage or amperage of electrical power.

power surge — Momentary increase in the voltage or amperage of electrical power.

preemptive scheduling — Scheduling method that allows a higher-priority process to interrupt and suspend a lower-priority process.

presentation layer — Layer of network system software responsible for input and output to network nodes.

primary storage — High-speed storage within a computer system, accessed directly by the central processing unit, used to hold currently active programs and data immediately needed by those programs.

primitive data type — Integer, real number, character, Boolean, and memory address data types supported by a central processing unit.

priority-based scheduling — Scheduling method that determines which ready thread or process should be dispatched to the central processing unit.

procedure — Group of instructions that are always executed as a unit (similar to a function or subroutine).

process — (1) Program or program fragment that is separately managed and scheduled by the operating system. (2) To transform input data by the application of processing operations. (3) Representation of data or information processing on a data flow diagram.

process control block (PCB) — Data structure that contains information about a currently active process.

process family — Parent process and all its descendants.

process list — Synonym of process queue.

process queue — List of process control blocks.

processor — Any device capable of performing data transformation operations.

production version — Program that contains no debugging components and has been optimized for minimal consumption of hardware resources.

program — Sequence of processing instructions.

program counter — Synonym of instruction pointer.

program editor — Software that assists programmers in creating valid source code.

program profiler — Software utility that monitors and reports the activities and resource utilization of another program during execution.

program status word (PSW) — Bit string held in a register within the control unit that stores status information in single bits.

program translator — Program that translates instructions in one programming language or instruction set into equivalent instructions in another programming language or instruction set. See assembler, compiler, and interpreter.

programmer — Person who creates or maintains programs.

programmer's workbench — Integrated set of programs that support program creation, editing, translation, and debugging.

programming language — Any language in which computer-processing functions or instructions can be expressed.

protocol stack — Set of software programs, each of which implements one protocol in a layered set of protocols.

PSK — Acronym of phase shift keying.

PSW — Acronym of program status word.

pulse code modulation (PCM) — Signal coding method under which bit values are represented as bursts of light or electrical voltage.

push — Process of copying register values to the top of a stack.

quality of service — Guaranteed minimum data transfer throughput between a sender and receiver.

quantum measurement problem — Behavior where the multiple states of a qubit are disturbed by any attempt to measure their value.

QIC — Acronym of Quarter Inch Committee.

Quarter Inch Committee (QIC) — Committee that promulgates standards for cartridge magnetic tapes.

qubit — Data stored in a single quantum particle.

radio frequency (RF) — Electromagnetic radiation propagated through space.

radix — Base of the numbering system, such as 2 for the binary numbering system and 10 for the decimal numbering system.

radix point — Period or comma that separates the whole and fractional parts of a numeric value.

RAID — Acronym of Redundant Array of Inexpensive Disks.

RAM — Acronym of random access memory.

random access — Ability of a storage device to access storage locations directly, or in any desired order.

random access memory (RAM) — (1) Generic description of semiconductor devices used to implement primary storage. (2) Device used to implement primary storage that provides direct access to stored data.

raw data transfer rate — Data transfer rate of a communication channel exclusive of capacity used for header information, error detection data, and retransmission of incorrectly received data.

read-only memory (ROM) — Primary storage device that can be read but not written to.

read/write head — Mechanism within a storage device that reads and writes data to/from the storage medium.

read/write mechanism — Synonym of read/write head.

ready state — State of an active process that is waiting only for access to the central processing unit.

real number — Number that can contain both whole and fractional components.

real resource — Hardware or software resource that physically exists.

real-time scheduling — Any scheduling method that guarantees the complete execution of a program or program cycle within a stated time interval.

record — (1) Data structure composed of data items relating to a single entity such as a person or transaction. (2) Unit of data transfer. (3) Primary component data structure of a file.

recording density — Closeness or spacing of bit positions on a storage medium, typically as measured in bits, bytes, or tracks per inch.

Reduced Instruction Set Computing (RISC) — (1) Use of a small set of simple central processing unit instructions that cannot be decomposed into other processor instructions. (2) Processor architecture that emphasizes short, primitive instructions that reference operands contained in registers, in contrast to complex instruction set computing.

Redundant Array of Inexpensive Disks (RAID) — Set of techniques for ensuring data integrity using redundant disk storage.

refresh cycle — Period during a dynamic random access memory refresh operation when the storage device is unable to respond to a read or write request.

refresh rate — Number of times per second a refresh operation occurs on a video display device.

register — High-speed storage location within a central processing unit that can hold a single word.

relative addressing — Synonym of indirect addressing.

relative path — Path that begins at the level of the current directory.

remote procedure call (RPC) — Protocol that enables a process on one computer to call a process on another computer.

repeater — (1) Device that relays signals from one network segment to another. (2) Synonym of bridge.

request for proposal (RFP) — Formal document stating hardware or software requirements and requesting proposals from vendors to meet those requirements.

requirements analysis — Synonym of systems analysis.

requirements document — Output produced by systems analysis.

resistance — Phenomenon in electrical transmission where electrical energy is converted to heat, reducing signal strength.

resolution — Number of pixels displayed per linear measurement unit.

resource availability model — A model of the computing resources that can be provided by a given combination of computer hardware and system software.

resource registry — Database of services and resources available within a local or wide area network.

return — Instruction executed at the end of a function, subroutine, or procedure to return control to the calling function, subroutine, or procedure.

return wire — Transmission line that completes an electrical circuit between the sending and the receiving device.

RFP — Acronym of request for proposal.

RGB — Acronym of Red, Green, and Blue.

ring topology — Network configuration in which each network node is connected to two other network nodes, with the entire set of connections and nodes forming a closed ring.

RIP — Acronym of Routing Information Protocol.

RISC — Acronym of Reduced Instruction Set Computing.

Rock's Law — Observation that states that the cost of fabrication facilities for the latest generation of semiconductor devices doubles every four years.

ROM — Acronym of read-only memory.

rotational delay — Waiting time for the desired sector of a disk to rotate beneath a read/write head.

router — Device that examines packet destination addresses and forwards them to another network segment, if necessary.

Routing Information Protocol (RIP) — Internet protocol for the exchange of routing information among network nodes.

RPC — Acronym of remote procedure call.

run queue — Synonym of thread list.

running state — State of a process that currently is executing within the central processing unit.

sampling — (1) Process of measuring and digitally encoding one or more parameters of an analog signal at regular time intervals. (2) Process of digitally encoding analog sound waves.

sandbox — Protected area of a program within which a Java applet is executed.

scan code — Coded output generated by a keyboard for interpretation by a processor or keyboard controller.

scanning laser — Laser that is automatically swept back and forth over a predefined viewing area, as used in optical input devices.

scheduler — Program within the operating system that controls process states and access to the CPU.

scheduling — Process of determining and implementing process priorities for access to hardware resources.

scripting language — (1) Simple programming language that can be embedded within HTML pages. (2) Stored set of operating system commands or a job control language program.

SCSI — Acronym of Small Computer System Interface.

SDLC — Acronym of systems development life cycle.

SDRAM — Acronym of synchronous dynamic random access memory.

second generation language (2GL) — Synonym of assembly language.

secondary storage — Set of computer system devices that provide large-capacity and long-term data storage.

sector — (1) Smallest accessible unit of a disk drive. (2) For many disk drives, 512 bytes.

segment register — Register that holds the base address of a memory segment, as used in a central processing unit that uses a segmented memory model.

segmented memory model — Memory allocation and partitioning based on equal sized segments and segmented memory addresses.

semiconductor — Material with resistance properties that can be tailored between those of a conductor and an insulator by adding chemical impurities.

sequential access time — Time required to access the second of a stored sequential pair of data items.

serial access — Access technique where data items are read and written in an order corresponding to their position within allocated storage locations.

serial transmission — Method of transmission in which individual bits are transmitted sequentially over a single communication channel.

server — Computer system or process that manages hardware or software resources and makes those resources available via a network.

service call — Request for an operating system service by an application program.

service layer — Portion of the operating system that accepts service calls and translates them into low-level requests to the kernel.

service-oriented resource access — Operating system architectural or behavioral feature in which all resources are managed and accessed through server processes.

service standard — Standard for interacting with a shared resource.

session layer — Layer of network system software responsible for establishing and managing communication sessions.

shell — User interface (or command layer) of an operating system.

SHIFT — Operation whereby bits of a bit string are moved left or right by a stated number of positions, empty positions are filled with zeros, and bit values that shift beyond the bounds of the bit string are discarded.

shortcut — Synonym of link, as implemented within Microsoft Windows for graph directory structures.

shortest time remaining — Scheduling method that assigns highest priority to processes or requests with shortest time remaining until completion.

sibling process — Another child process created, or spawned, by the same parent.

signal — (1) Specific data transmission event or group of events that represents a bit or group of bits. (2) Message sent from one active process to another.

signal-to-noise (S/N) ratio — Mathematical relationship between the power of a carrier signal and the power of the noise in the communication channel, measured in decibels.

signal wire — Component of an electrical communication channel (circuit) used to transmit a carrier wave.

SIMM — Acronym of single in-line memory module.

Simple Mail Transport Protocol (SMTP) — Earliest e-mail protocol that defines how text messages are routed through the Internet.

simplex — Type of communication channel that allows transmission in only one direction.

simulation model — System model used to examine the dynamic behavior of a system such as a computer system.

sine wave — Waveform that varies continually between positive and negative states.

single in-line memory module (SIMM) — Small printed circuit board with memory chips on one or both sides and electrical contacts on one edge.

singly linked list — Data structure in which each data item contains a pointer to the next data item.

single-mode cable — Highest quality optical cable constructed so that light waves travel down the center of the fiber at up to 2 Gbps.

skew — Timing differences in the arrival of signals transmitted simultaneously on parallel communication channels.

Small Computer System Interface (SCSI) — Family of standard buses designed primarily for connecting secondary storage devices.

SMTP — Acronym of Simple Mail Transport Protocol.

S/N — Acronym of signal-to-noise.

socket — Internet standard for service identification that combines an Internet address with a service port number, such as 129.24.8.4:53.

software monitor — Program that monitors and reports the utilization of a software resource.

son — File containing the results of updating a master file (the father) based on the contents of a transaction file.

sound board — Synonym of sound card.

sound card — Hardware device containing components for sound input and output.

source code — Instructions or statements in a high-level programming language.

spawn — Creation of a child process by a parent process.

speaker dependent — Speech recognition system that recognizes input from only one user and which must be trained to recognize accurately that user's specific speech patterns.

special-purpose processor — (1) Processor with a limited set of processing functions. (2) Processor capable of executing only a single program.

speech recognition — Process of recognizing human speech as computer system input.

speech synthesis — Process of generating human speech based on character or textual input.

SRAM — Acronym of static random access memory.

stack — (1) List of data items maintained in last-in, first-out order. (2) Set of registers or memory locations used to store the register values of temporarily suspended processes.

stack overflow — Condition that occurs when an attempt is made to add data to a stack that is already at its maximum capacity.

stack pointer — Register containing the primary storage address of the last (most recently added) stack element.

star topology — Physical network connection pattern in which all nodes are attached to a central node.

start bit — Zero bits that are added to the beginning of each character in asynchronous serial data transmission.

static analysis — Use of models or model analysis techniques that do not incorporate parameter variability and random events in the representation of system behavior.

static connection — Mapping between a local resource name and a remote resource that does not change once initialized and which must be explicitly initialized prior to use.

static linking — Linking process in which library and other subroutines cannot be changed once they are inserted into the executable code.

static random access memory (SRAM) — Type of random access memory that implements bit storage with a flip-flop circuit.

storage allocation table — Table that correlates storage device allocation units with specific files, directories, or unallocated space.

storage input/output control layer — System software layer that controls input and output to and from storage devices.

storage medium — Device or substance within a storage device that serves as the physical repository for data.

store — Operation in which register content is copied to a memory location.

store and forward — Interconnected system of end nodes and transfer points used to route data among end nodes.

strategic plan — Long-range plan stating the services to be provided by an organization and the means for obtaining and using needed resources.

string — Ordered set of related data elements, usually stored as a list or an array.

subroutine — Named segment of a high-level language program that is always executed as a unit.

subtractive colors — Cyan (absence of red), magenta (absence of green), and yellow (absence of blue).

supercomputer — Computer designed for very fast processing of real numbers, typically implemented with some form of parallel processing.

supervisor — Operating system program that serves as the master interrupt handler.

surge protector — Hardware device that eliminates electrical power surges.

sustained data transfer rate — Maximum data transfer rate that can be sustained by a device or a communication channel during large or lengthy data-transfer operations.

swap space — Area of secondary storage used to hold virtual memory pages that cannot fit into primary storage.

switch — (1) Device that exists in one of two states and can be instructed to alternate between them. (2) Network node that quickly routes packets among network segments with switching technology.

symbol table — List of program module and data names and their allocated primary storage locations.

symbolic debugger — Debugger that can report program execution progress and errors in terms of symbolic module and data names used within program source code.

synchronous dynamic random access memory (SDRAM) — Read-ahead random access memory that breaks read and write operations into a series of simple steps that can be completed in one bus clock cycle.

synchronous idle message — Message without data content sent continually between sender and receiver to keep their internal clocks synchronized.

synchronous transmission — Reliable high-speed method of data transmission in which data are sent in continuous data streams.

system administration — Managerial actions required to ensure efficient, reliable, and secure computer system operation.

system bus — Bus shared by most or all devices in a computer system.

system development tools — (1) Software tool designed to aid in creating groups of programs constituting an entire system. (2) Approximate synonym of computer assisted software engineering tool.

system overhead — Resource(s) consumed for resource allocation and translation functions.

system requirements model — Model that provides the detail necessary to develop a system that meets the needs the user requires.

system software — Software programs that perform hardware interface, resource management, or application support functions.

systems analysis — Process of ascertaining and documenting information-processing requirements.

systems analyst — Person who performs systems analysis.

systems architecture — Computer system components and the interaction among them.

systems design — Process of determining the exact configuration of all hardware and software components in an information system.

systems designer — Person who performs systems design.

systems development life cycle (SDLC) — Process to develop an information system using survey, analysis, design, implementation, and evaluation.

systems implementation — Process of software construction, installation, and testing on a given hardware configuration.

systems programmer — Person who creates or maintains systems software.

systems survey — Process of initially ascertaining user processing requirements and determining the feasibility of meeting those requirements with computer-based solutions.

tape drive — Device that contains motors that wind and unwind tapes and read/write heads to access their content.

TCB — Acronym of thread control block.

TCP — Acronym of Transmission Control Protocol.

TCP/IP — Acronym of Transmission Control Protocol/Internet Protocol.

TDM — Acronym of time division multiplexing.

Telnet — Protocol in which users on one Internet host can interact with another host's operating system command layer.

terminal — Synonym of video display terminal (VDT).

thread — Subcomponent of a process that can be independently scheduled and executed.

thread control block (TCB) — Data structure in which an operating system keeps track of thread-specific information.

thread cycle — Set of program instructions that can be executed repetitively.

thread list — Set of threads awaiting execution represented by a list of thread control blocks.

three-layer architecture — Client server architecture composed of three layers: the view layer, the business logic layer, and the data layer.

three-tier architecture — Synonym of three-layer architecture.

time division multiplexing (TDM) — Technique that subdivides the capacity of a communication channel into discrete time slices, each of which may be allocated to a separate sender and receiver.

third generation language (3GL) — High-level programming language that does not possess advanced capabilities for interactive input/output, database processing, or nonprocedural programming.

timer interrupt — Interrupt automatically generated by the central processing unit in response to the passage of a specific interval.

token — Control packet used to regulate access to a communication channel.

token passing — Media access protocol that uses tokens passed between network nodes.

track — Set of sectors on one side of a disk platter that form a concentric circle.

track-to-track seek time — Time required to move a disk read/write head between two adjacent tracks.

transistor — Solid state electrical switch that forms the basic component of most computer-processing circuitry.

transaction — (1) Event or system input that initiates the execution of one or more program segments. (2) Real-world event from which data input is derived, such as a sale. (3) A general term describing any change to stored data, such as the addition of a record to a file.

transaction logging — Method of file system update in which all changes to file and directory content also are written immediately to a separate log, or journal, which is used to recover lost data in the event of a system crash.

Transmission Control Protocol (TCP) — Internet protocol for translating messages into packets and guaranteeing their delivery.

transmission medium — Physical communication path through which a carrier wave is propagated.

transport layer — Layer of network system software responsible for converting network messages into packets or other data structures suitable for transmission.

tree directory structure — Synonym of hierarchical directory structure.

truncation — Act of deleting bits that will not fit within a storage location.

twisted pair wire — Transmission medium consisting of two electrical conductors (wires) twisted around one another.

two's complement — Notation system that represents positive integers as an ordinary bit string, and negative integers by adding one to the bit string that represents the absolute value.

Type I error — (1) Percentage of data items containing errors that will be incorrectly identified as error free. (2) Percentage of a sample or population that will be incorrectly identified as satisfying the null hypothesis.

Type II error — (1) Percentage of data items without errors that will be incorrectly identified as containing errors. (2) Percentage of a sample or population that will be incorrectly identified as not satisfying the null hypothesis.

unblocked — Storage of a single logical record within a physical record (block).

unconditional BRANCH — Synonym for BRANCH instruction.

undelete operation — Act of restoring a record or file by re-creating its index information (directory entry) and recovering its previously allocated storage locations.

underflow — (1) Condition that occurs when a value is too small to represent in floating point notation. (2) Overflow of a negative exponent in floating point notation.

Unicode — Standard 16-bit character coding method.

Uniform Resource Locator (URL) — Unique name that identifies a Web resource and is composed of a protocol, host, port, and resource.

---

uninterruptible power supply (UPS) — Device that provides electrical power when normal power inputs are interrupted.

unresolved reference — Synonym of external function call.

UPS — Acronym of uninterruptible power supply.

URL — Acronym of Uniform Resource Locator.

user authentication — Process of verifying user identity, usually by validating a user name and password.

utility program — Program that performs a commonly used function, such as printing a text file or copying a file, typically provided by the operating system either as separate tool or as part of the service layer.

variable — (1) Mnemonic that represents a data item memory address in assembly language. (2) Name that represents a data item memory address in a high-level programming language.

variable length instruction — One instruction from a set of central processing unit instruction formats that vary in length.

VC — Acronym of virtual circuit.

VDT — Acronym of video display terminal.

vector — (1) Line segment that has direction and length. (2) One-dimensional array.

vector list — List of line descriptions used for the input or output of graphic images.

vertical redundancy checking — Synonym of parity checking.

victim — Memory page chosen to be swapped to secondary storage under virtual memory management.

video display terminal (VDT) — Input/output device comprising a keyboard and a video display surface.

view layer — Division of the client-server architecture that accepts user input and formats and displays processing results.

virtual circuit (VC) — Transmission path established between sender and receiver before data is transmitted.

virtual local area network — Two shorter local area networkss joined by a bridge, also called a bridged local area network.

virtual memory management — Mode of operating system memory management in which secondary storage is used to extend the capacity of primary storage.

virtual resource — Resource visible to a user or program, but not necessarily available at all times or physically extant.

volatile — Storage devices that cannot retain their contents indefinitely.

volatility — (1) Rate of change in a set of data items. (2) Lack of ability to permanently store data content.

wait state — Idle processor cycle consumed while waiting for a response from another device.

WAN — Acronym of wide area network.

Web — Synonym of World Wide Web (**WWW**).

wide area network (WAN) — Network that spans large physical distances, such as multiple buildings, cities, regions, or continents.

wire — Connection that allows electrons to travel from one place to another.

wiring concentrator — Simplest form of a hub that concentrates wiring connections at a single point.

word — Unit of data processed by a single central processing unit (CPU) instruction.

working directory — Synonym of current directory.

workload model — Synonym of application demand model.

workstation — (1) Powerful microcomputer designed to support demanding numerical or graphical processing tasks. (2) A synonym of personal computer or microcomputer.

World Wide Web (WWW) — Collection of resources accessed over the Internet by standard protocols such as FTP and HTTP.

WORM — Acronym of write once read many.

write once read many (WORM) — Type of optical storage media that is manufactured blank and can be written to once.

WWW — Acronym of World Wide Web.

XML — Acronym of eXtensible Markup Language.

XOR — Acronym of eXclusive OR.

INDEX

branches, 29
BRANCH instruction, 119, 133, 134
bridges, 339, 340
broadband, 305
broadcast mode, 299
brownouts, 555
buffering, 466–467
buffer overflow, 218
buffers, 218–221
bus(es), 202
    address, 203
    control, 203
    data, 203
    multiple master, 205–206
    system. *See* system bus
bus clock, 203–204
bus clock rate, 203–204
bus cycles, 203
business logic layer, n-layer architecture, 491
bus master, 205
bus network topology, 329–330, 332
bus protocol, 204–206
bus slaves, 205
bytes, 68
    most and least significant, 169

C

cable routing, 557–558
cabling
    electrical, 295–296
    optical, 296–297
cache(s), 221–227, 224–227
    primary storage caches, 224–226
    secondary storage caches, 226–227
    write caching, 222–223
cache controllers, 223
Cached DRAM, 166
cache hits, 223
cache misses, 223
cache swaps, 223–224
call instructions, 380

capacity
    storage devices, 161–162
    transmission media, 290
capacity units, 566–567
capital expenditures, 530
capital resources, 530
Carrier Sense Multiple Access with Collision Detection (CSMA/CD), 336–338
carrier waves, 278–281
CASE (computer assisted software engineering) tools, 395–396
Category 5 cable, 295
cathode ray tubes (CRTs), 253–254
CD-DA (compact disc digital audio), 189
CD-R (compact disc-read), 189, 190
CD-RW (compact disc-read/write), 189, 191–192
central processing units (CPUs), 32, 33–34. *See also* CPU *entries;* processors; *specific processors*
    electrical properties, 140–142
    gates, 138–140
    Intel, 51–53
    memory access. *See* CPU memory access
    operation, 110–113, 113–114
    processor fabrication, 142–144
    registers. *See* registers
    switches, 138
    technology capabilities and limitations, 145–147
CFCs (hlorofluorocarbons), 558
character(s), multinational, 87
character data, 81–87
character framing methods, 309
chief information officers (CIOs), 12
child processes, 414
chlorofluorocarbons (CFCs), 558
chromatic depth (resolution), 246
CIOs (chief information officers), 12
circuit length, CPUs, 142
circuit switching, 302, 303
CISC (Complex Instruction Set Computing), 127–129, 131–132

Ethernet, 350, 351–352

even parity, 312

excess notation, 74

executable code, 376

executable programs, 456–458
   debugging versions, 391
   production (distribution) versions, 392

executing instructions, 27

execution cycle, 111, 112–113

explicit priority scheduling, 422–423

Extended Binary Coded Decimal Interchange Code (EBCDIC) coding method, 82, 83

Extensible Markup Language (XML), 505

external function calls, 382

F

father (account master), 475

FATs (file allocation tables), 464–465

fault tolerance, 479, 480–483
   NTFS, 474

FCFS (first come first served) scheduling, 422

FDM (frequency division multiplexing), 305

ferroelectric RAM, 166

fetch cycle, 111–112, 113

FHSS (frequency hopping spread spectrum), 314

fiber-optic cable, 296–297

fiber optics, 290–291

fields, 454

fifth generation programming languages (5GLs), 367, 371–372

file(s), 98–99
   closing, 470
   contents, 456
   deleting and undeleting, 470–471
   executable. *See* executable programs
   NTFS, 473
   opening, 469
   recovering, 478–479
   types, 456–458

file allocation tables (FATs), 464–465

file association, 456–458

file attributes, NTFS, 473–474

file backup, 476–478

file close operation, 470

file compression. *See* compression

file control layer, 454

file management systems (FMSs), 451–484
   access controls, 471–472
   directories, 458–462
   fault tolerance, 479, 480–483
   file backup, 476–478
   file content and type, 456–458
   file manipulation, 469–471
   file migration, 475–476
   file recovery, 478–479
   logical and physical storage views, 454–456
   storage allocation. *See* storage allocation
   transaction logging, 478

file migration, 475–476

filenames, 459

file owner field, FMSs, 459

File Transfer Protocol (FTP), 347, 505

fire protection, 558–559

firmware, 167

first come first served (FCFS) scheduling, 422

first generation programming languages (1GLs), 367

fixed length instructions, 126

flags, 133, 134

flash memory, 167

flash RAM, 167

flat memory model, 89

flat panel displays, 254–255

floating point notation, 77–79

floppy disks, 181, 185–186

FM (frequency modulation), 282

FMSs. *See* file management systems (FMSs)

fonts, 244–245

formulas, 27–28

fourth generation programming languages (4GLs), 367, 369–371

fragmentation
  disks, 184
  memory, 432–433

frequency division multiplexing (FDM), 305

frequency hopping spread spectrum (FHSS), 314

frequency modulation (FM), 282

frequency of waves, 279
  transmission media, 290–291

frequency shift keying (FSK), 282

front-end CASE tools, 395

front end processors, 209

FSK (frequency shift keying), 282

FTP (File Transfer Protocol), 347, 505

full adder circuits, 139–140

full backups, 477

full-duplex mode, 300–301

fully qualified references, 460

function(s), 380

function calls, 380–382

G

gallium arsenide, 148

Gartner Group Web site, 15

gates, CPU, 138–140

gateways, 347–348

general-purpose processors, 27

general-purpose registers, 132–133

GHz (gigahertz), 129

GIF (graphics Interchange Format), 247

gigahertz (GHz), 129

grandfather (account master), 475

graph directory structure, 461–462

Graphics Interchange Format (GIF), 247

graphic video display devices, 253–255

grayscale displays, 245–246

Grosch, H. A., 38

Grosch's Law, 38–39

H

half adder circuits, 139–140

half-duplex mode, 299–300

HALT instruction, 120

hard disks, 181

hardware, 31–36
  functions, 31
  infrastructure, 530–531
  memory management, 437
  networks, 338–342
  standard platform, 532–533
  support for ASCII, 84, 86
  system components, 32
  Windows Me versus Windows 2000 Professional, 427

hardware interface, system software, 46–47

hardware monitors, 547

hardware personnel, 10

headers, NTFS file attributes, 473

head-to-head switching time, 182

heat
  CPUs, 141–142
  dissipation, 556–557

heat sinks, 141–142

helical scanning, 176–177

hertz (Hz), 129

hexadecimal notation, 69–70

high order bit, 68

hit ratio, 223

Hoff, Ted, 143

home directories, 460

home networks, 342–344

Home Radio Frequency (HomeRF), 342

HTML (Hypertext Markup Language), 505

HTTP (Hypertext Transfer Protocol), 347, 505

hubs, 339–340

humidity, 557

Hypertext Markup Language (HTML), 505

Hypertext Transfer Protocol (HTTP), 347, 505

Hz (hertz), 129

I

N

NAK (negative acknowledge) character, 299
named pipes, 500–501
native applications, 389–390
needs analysis, 5
negative acknowledge (NAK) character, 299
network(s), 53–56, 325–354
    addressing and routing, 333–336
    communication, 55–56
    external resources, 54–55
    hardware, 338–342
    home networking, 342–344
    MAC protocols, 336–338
    OSI model. *See* Open Systems
      Interconnection (OSI) model
    physical, 55–56
    software, 55
    standards, 349–351
    topology. *See* network topology
    upgrading capacity, 315–316, 352–353
network, Windows Me versus Windows 2000
  Professional, 427
network administrators, 11
network computers, 37
network interface cards (NICs), 339
network interface units (NIUs), 339
network layer, OSI, 346
network resource access, 493–498
    accessing remote resources, 495–498
    protocol stacks, 493–494
    remote resources, 495–498
network standards, 349–351
network topology, 326–332
    bus, 329–330, 332
    logical, 328
    physical, 328
    ring, 330–331, 332
    star, 329, 332
network transparency, 496
NICs (network interface cards), 339
NIUs (network interface units), 339

n-layer (n-tiered) architectures, 491–493
noise, 293–295
noncontiguous memory allocation, 433–434
nonprocedural languages, 371
NOT gate, 138, 139
NOT instruction, 115
NT File System (NTFS), 472–473
numbers, real, 76–81

O

object(s), 100, 373
    LDAP, 516–517
objectclass, 516
object code, 376
object-oriented programming (OOP)
    components, 512
    languages, 367, 373–374, 388–390
Object Request Broker (ORB), 514
OCR (optical character recognition) devices,
  261–262
octal notation, 70
odd parity, 312
offset registers, 171
on-off keying (OOK), 285
OOP. *See* object-oriented programming (OOP)
op code, 113
opening files, 469
open_pipe call, 444
Open Systems Interconnection (OSI) model,
  344–347
    application layer, 344–345
    data link layer, 346
    network layer, 346
    physical layer, 347
    presentation layer, 345
    session layer, 345–346
    transport layer, 346
operands, 113
operating expenditures, 530
operating systems, 47–48, 403–446. *See also*
  specific operating systems

CPU allocation. *See* CPU allocation
Intel CPU released, 51–53
interprocess communication, 441–444
layers, 406–408
management functions, 405–406
memory allocation. *See* memory allocation
process management, 413–415
resource allocation, 409–413
optical cabling, 296–297
optical character recognition (OCR) devices, 261–262
optical input devices, 24–25, 259–262
optical output devices, 24–25
optical processing, 148
optical scanners, 261
optical sensors, 259
optical storage devices, 187–192
    CD-ROMs, 187, 189–190
    CD-Rs, 189, 190
    DVD-ROMs, 189, 192
    magneto-optical drives, 189, 190–191
    phase-change optical discs, 191–192
    WORM discs, 190
ORB (Object Request Broker), 514
OR instruction, 116–117
OSI model. *See* Open Systems Interconnection (OSI) model
output devices, 243–259
    displays. *See* video displays
    image description languages, 248–251
    image storage requirements, 247
    matrix-oriented image composition, 243–247
    plotters, 259
    printers, 255–259
overflow
    real numbers, 79
    two's complement notation, 75–76

P

packets, 303
packet switching, 303–305
page faults, 435
page frames, 435
page hits, 435
page tables, 435
    Pentium processors, 438–439
palettes, colors, 246
parallel access storage devices, 160–161
parallel transmission, 301–302
parent processes, 414
parity bit, 312
parity checking, 312
passive matrix displays, 254–255
passwords, 552–554
PCB (process control block), 413–414
PCM (pulse code modulation), 285
PCs (personal computers), 36–37, 40
peer-to-peer communication protocols, 499
Pentium processors, 53, 136–138
    clock rate, 138
    data types, 136–137
    instruction set, 137–138
    memory addressing, 91
    memory allocation, 437–440
    word size, 138
performance, interrupts, 216–217
periodicals, 13–14
periods of waves, 279
peripheral devices, 202. *See also specific types of devices*
peripheral processing units, 209
personal computers (PCs), 36–37, 40
personnel, 8–12
    application development, 9
    hardware, 10
    systems management, 10–12
    systems software, 10
phase, waves, 283

phase of waves, 279
phase shift keying, 283
phase shift modulation, 283
phonemes, 263
photons, 24
physical environment, 554–559
    cable routing, 557–558
    disaster planning and recovery, 559
    electrical power, 554–556
    fire protection, 558–559
    heat dissipation, 556–557
    moisture, 557
physical layer, OSI, 347
physical memory, 169
physical network topology, 328
physical records, 465
pipes
    interprocess communication, 443–444
    named, 500–501
pits, in CD-ROMs, 190
pixels, 243
    color, 246
    content, 245–247
platters, 180
plotters, 259
point(s), 243
pointers, 93
pointing devices, 241–242
polyphonic input, 263
POP3 (Post Office Protocol version 3), 506
pops, stack processing, 216
port(s), 349
portability, storage devices, 161, 162
Post Office Protocol version 3 (POP3), 506
PostScript, 249–251
POWER CPU architecture, 41–42
power sags, 555
power surges, 555
precision of real numbers, 79–80
preemptive multitasking, 427
preemptive scheduling, 419–421
presentation layer, OSI, 345

primary storage, 32, 34–35
    comparison with other memory types, 35
primary storage devices, 163–168
    memory packaging, 167–168
    RAM, 164–166
    ROM, 167
    storing electrical signals, 163–164
primitive data types, 91
printers, 255–259
    dot matrix, 256–257
    ink-jet, 257–258
    laser, 258–259
priority-based scheduling methods, 422–423
procedures, 380
process(es), threads. *See* thread(s)
process control block (PCB), 413–414
process families, 414
process management, 413–415
    process control data structures, 413–414
    threads, 414–415
processors, 27–29. *See also* central processing
  units (CPUs); CPU *entries; specific processors*
    cost of fabrication facilities, 145
    manufacture, 143–144
    transistor density, 145
process queue (list), 413–414
production versions, 392
professional societies, 17–18
program counter, 133
program editors, 49
programmers, 366
    application, 9
    systems, 10
programmer's workbenches, 392–395
programming instructions, 366
programming languages, 366–375
    fifth generation, 367, 371–372
    first generation, 367
    fourth generation, 367, 369–371
    object-oriented, 367, 373–374, 388–390
    second generation, 367–368
    standards, 374–375

third generation, 367, 368–369
translators and, 49
program profilers, 548
program status word (PSW), 133, 134
program translators, 49
proposal evaluation, 536–537
protocol stacks, 493–494
pSeries 680 computers, 41, 42
PSW (program status word), 133, 134
pulse code modulation (PCM), 285
pushes, stack processing, 216

Q

QIC (Quarter Inch Committee), 179
quality of service level, 306
quantum computing, 25–26
quantum measurement problem, 25
quantum processing, 148–149
Quarter Inch Committee (QIC), 179
qubits, 25

R

radio frequency (RF), 288
radix, 65
radix points, 65
RAID (Redundant Array of Inexpensive
Disks), 480–483
random access memory (RAM), 34–35,
164–166
random access storage devices, 160
range
real numbers, 79
two's complement notation, 75
raw data transfer rate, 290
read-only memory (ROM), 167
flash, 167
read/write heads, 172
read/write mechanism, 156
ready state, threads, 416
real numbers, 76–81
floating point notation, 77–79
precision and truncation, 79–80

processing complexity, 80–81
range, overflow, and underflow, 79
real resources, 412
real-time scheduling, 423–424
record(s), 98, 454
physical, 465
recording density, magnetic storage, 174, 175
recovering files, 478–479
Reduced Instruction Set Computing (RISC),
127, 128–129, 131–132
Redundant Array of Inexpensive Disks
(RAID), 480–483
refresh cycle, 165
refresh rate, 254
registers, 33–34, 132–134
comparison with other memory types, 35
general-purpose, 132–133
offset, 171
segment, 171
special-purpose, 133–134
relative addressing, 171
remote procedure calls (RPCs), 501–502
remote resources, accessing, 495–498
repeaters, 297, 339, 340
requests for proposal (RFPs), 535–536
requirements analysis, 5
requirements documents, 5
resistance, CPUs, 141
resolution
chromatic, 246
of displays, 243
resource allocation, 409–413
goals, 410–411
multiple process, 410
processes, 411–412
real and virtual resources, 412–413
single process, 409–410
system software, 46–47
resource availability models, 542–543
resource demand and utilization, measuring,
547–548
resource management, 7–8